Who *Really* Invented
THE STEAMBOAT?

Robert Fulton's first American steamboat, commonly known as the *Clermont* but actually called the *North River Steam Boat*, is shown with many passengers on board making her first voyage to Albany, New York. This popular account including the passengers and shape of the bow, although inaccurate, still persists. (From Frank P. Bachman, *Great Inventors and Their Inventions* [New York: American Book, 1918], p. 42.)

Who *Really* Invented
THE STEAMBOAT?
Fulton's *Clermont* Coup

A History of the Steamboat Contributions of
William Henry, James Rumsey, John Fitch,
Oliver Evans, Nathan Read, Samuel Morey,
Robert Fulton, John Stevens, and Others

Jack L. Shagena, P.E.

Humanity
Books

an imprint of Prometheus Books
59 John Glenn Drive, Amherst, New York 14228-2197

Published 2004 by Humanity Books, an imprint of Prometheus Books

Inquiries should be addressed to
Humanity Books
59 John Glenn Drive
Amherst, New York 14228–2197
VOICE: 716–691–0133, ext. 207
FAX: 716–564–2711

08 07 06 05 04   5 4 3 2 1

Library of Congress Cataloging-in-Publication Data

Shagena, Jack L.
    Who really invented the steamboat? : Fulton's Clermont coup / Jack L. Shagena.
        p. cm.
    Includes bibliographical references and index.
    ISBN 1-59102-206-1 (hardcover : alk. paper)
    1. Rumsey, James, 1743?–1792. 2. Fulton, Robert, 1765–1815.
3. Steamboats—History—18th century. 4. Steamboats—History—19th century.
I. Title.

VM140.R8S53 2004
623.8'12044—dc22

                                                                    2004005748

Printed in the United States of America on acid-free paper

To my father,

who inspired in me
a love of things mechanical

Contents

Part 2. The Candidates Considered

Part 3. Conclusion and Afterthoughts

Preface

There is nothing more difficult to take in hand, more perilous to conduct, or more uncertain in its success, than to take the lead in the introduction of a new order of things.

—Niccolò Machiavelli

The title of this book poses a question that hopefully will be answered to the reader's satisfaction in this short history. Some historians have asserted that the steamboat was not an invention at all, but rather a technical evolution furthered by the contributions of several individuals, with the end result being Robert Fulton's commercially successful steamboat in August 1807.[1]

Others, sometimes with a genealogical or geographic connection, have argued that James Rumsey, John Fitch, Oliver Evans, Nathan Read, John Stevens, William Symington (the textbook favorite), Robert Fulton, or someone else was the rightful inventor.[2] Many writers have plowed many of the same historical records gleaning evidence to support their case but frequently come to different conclusions. Part of the problem lies in the concept of *invention* itself, as it is often shrouded in myth and is not well understood by researchers. Therefore a chapter in this work has been dedicated to this matter.

In reading these conflicting steamboat accounts, it has become apparent that almost none was written from a historical technologist's vantage point.[3] Most writers lacking a technical background have approached the subject as historians or biographers but without the requisite understanding of science, technology, and invention necessary to arrive at a conclusion that is grounded in a technical conscientiousness.

To bring a different perspective and hopefully avoid this pitfall, the underlying *systems engineering* development process that was intrinsically involved in the steamboat development, though definitely not understood at the time, is used as an overlay to objectively gauge and evaluate the efforts of William Henry, Rumsey, Fitch, Evans, Read, Samuel Morey, Fulton, Stevens, and others. With this vantage point the contributions of each individual and his relative importance to steam navigation is brought into better focus, allowing a few conclusions to be drawn. These are presented in the last chapter and in the epilogue.

Obviously the reader is free to apply his or her own knowledge to modify or rework the filter overlay and perhaps draw different conclusions. In any case a better understanding will likely result, and perhaps sometime in the future most researchers who delve into this subject will converge on a small number of deserving candidates, or even perhaps a single candidate worthy of recognition as *the* inventor.[4]

It is perhaps important to note that the scientist and the historian have a common interest, albeit in different fields. Each is intent on ferreting out the basic truth: the historian through a search for and analysis of original or primary documents; the scientist through observations of and experimentation with the laws of nature. When successful, information is added to our knowledge base, in the one case in literature, in the other in science.

On a parallel but different level the engineer and the serious nonfiction writer also share a common goal: each is intent on creating something new—the writer, a compelling story supported by primary and secondary sources; the engineer, a new and useful device, firmly established in basic science and technology with data gleaned from textbooks or handbooks. As the writer does not trace every date and fact back to an original document, the engineer does not re-derive basic equations or repeat experiments, relying instead on published information.

When the engineer turns to historical writing, as is the case here, the author has to make a deliberate decision to emulate the historian, who relies almost entirely on primary sources, or the serious nonfiction writer, who is comfortable with both primary and secondary sources. It is simply not in the makeup of most engineers to be scientists or historians as well; hence the only comfortable avenue is that of the serious nonfiction writer. The result is a fresh examination of existing literary knowledge from an engineering vantage point that yields new insight into the contributions of steamboat pioneers.

Another practical consideration has to do with the scope of the work. Had the number of individuals investigated been one, two, or perhaps even three, it would have been far more practical to depend almost exclusively on primary sources. However, when eight primary as well as several secondary individuals are considered, some presorting of the information was necessary to maintain a reasonable scope of work for a retired engineer with many interests.

The test is richly illustrated for two principle reasons: first, a picture or drawing can quickly convey more information to the reader than text (or, as thought of in an engineering context, it has a far wider bandwidth); second, a few of the illustrations have been included to relate to the reader the whimsical nature of how steamboats have been presented to the public over the years, even though in some cases, the device presented was technically flawed. In nearly all cases the illustrations are described with one or more sentences, so the reader can peruse the caption and obtain a greater insight into the information shown.

This effort started out as a paper for the Historical Society of

Cecil County, Maryland, focusing on James Rumsey, a native of this area recognized by some as the inventor of the steamboat. Perhaps this fact reveals a potential geographic bias on the part of the writer. As the Rumsey research evolved, however, it was realized that the subject was far more complicated than first envisioned and grew from a paper into a pamphlet and finally this book. It could have easily expanded into a much longer book, but the author took pains to achieve conciseness consistent with his engineering training while hopefully achieving clarity.

Notes

1. Roger Burlingame, *March of the Iron Men: A Social History of Union through Invention* (New York: Grosset & Dunlap, 1938), p. 193; Edwin T. Layton Jr., "James Rumsey: Pioneer Technologist," *West Virginia History* 48 (1989): 16; and James Croil, *Steam Navigation and Its Relation to the Commerce of Canada and the United States* (1898; reprint, Toronto: Coles, 1973), p. 29.

2. For Fitch, see Thomas Boyd, *Poor John Fitch* (New York: G. P. Putnam's Sons, 1935). The author's wife was Ruth Fitch Boyd, and he seems to have been predisposed to portray the subject as the inventor of the steamboat, no doubt because Fitch himself claimed the title. For Read, see David Read, *Nathan Read: His Invention of the Multi-tubular Boiler . . .* (New York: Hurd and Houghton, 1870), p. 4. In this account "by his friend and nephew," the biographer suggests Fulton's success was largely attributable to combinations of machinery first suggested by Nathan Read, implying therefore that he was the true inventor. For Symington, see H. Philip Spratt, *The Birth of the Steamboat* (London: Charles Griffin, 1958), pp. 8, 61. Spratt, of the Science Museum in London, says he sought in his book to avoid national bias but cites that in 1801, Englishman William Symington built the first practical steamboat, the *Charlotte Dundas*. For Fulton, see James Flexner, *Steamboats Come True: American Inventors in Action* (1944; reprint, Boston: Little, Brown, 1978), p. 378. In his comprehensive, interesting, and some would say definitive book, James Flexner concludes on the last page, "According to this definition, popular history is correct: Robert Fulton was the inventor of the steamboat." His definition, however, is closer to development than the original concept of invention.

3. See Layton, "James Rumsey: Pioneer Technologist," pp. 7–32; and Brook Hindle, "James Rumsey and the Rise of Steamboating in the United States," *West Virginia History* 48 (1989): 33–42. Both Layton and Hindle approached the contributions of James Rumsey from a technological viewpoint; however, they do not pursue the engineering aspect, which is so important to a system development such as the steamboat.

4. It has been suggested by some researchers that attempting to identify the inventor of the steamboat is not worthy of pursuit. To a community, county, or state, however, it can be important for a sense of identity. Bardstown, Kentucky, celebrates Fitch; Shepherdstown, West Virginia, Rumsey; and some parts of Pennsylvania and New York, Fulton. But this history is about much more than who invented the steamboat, for it brings into a single volume much of the extant knowledge about American steamboat pioneers and presents it with a new perspective.

Acknowledgments

A s with any significant research effort there are numerous individuals to thank. Jeanne Mozier, past president of the Museum of the Berkeley Springs, West Virginia, provided much insight into Rumsey as well as supplying a comprehensive bibliography of Rumsey's work. Nicholas Blanton, historian of the Rumseian Society in Shepherdstown, West Virginia, pointed me in the right direction on many occasions, and along with Daniel Tokar provided a good understanding of a replica of the steamboat Rumsey demonstrated on the Potomac River.

Jeff Korman and the staff at the Enoch Pratt Library in Baltimore were most helpful in locating and retrieving requested references, as were the librarians at Johns Hopkins University's Milton Eisenhower Library, audiovisual section. Dr. R. John Brockman, professor of English at the University of Delaware, brought to my attention two little-known steamboat pioneers, Green and Backus, who worked in Ohio.

Anne Carvel House, librarian for the Steamship Historical Society of American Collections in Baltimore, identified several sources that contributed to the effort. Eugene Ferguson and his wife Jo were most gracious in sharing their sources of information on Oliver Evans and early technology in America. A very special thanks goes to Kate Newell, Susan Hengel, and the staff of the Hagley Library, who on many occasions retrieved and cheerfully delivered old books and pamphlets to the author. The American Philosophical Society was the source of numerous old and obscure documents. Sarah Putnam, librarian of the Orford, New Hampshire, Social Library, identified and provided several publications on Samuel Morey.

Phillip J. Woodall, an internationally respected electronics systems engineer, a former classmate, and for many years an esteemed colleague, provided a technical review. He carefully read the draft and made many worthwhile suggestions on the technical information presented and the conclusions drawn. The author is most grateful to have received his valuable feedback. E. J. Raimondi, a retired Air Force Chief Engineer and friend, provided very valuable comments on the chapter on systems engineering. Good friend and former colleague William T. Sisson created all the portraits for the steamboat pioneers, as well as many of the other fine illustrations that are found throughout the book. His talented efforts are most appreciated.

Benjamin Keller provided the copyedit for the manuscript, and through his astute observations and able pen, the content and format have been much improved.

Finally, thanks to my loving wife Signe, whose patience and encouragement provided a nurturing environment that allowed me to pursue my steamboat passion.

Jack L. Shagena Jr.
Bel Air, Maryland
2004

Part 1.

Perspective and Background

Figure 1.1. The *Santa Maria*. Prior to the development of the steamboat, sailing vessels such as Columbus's flagship, depicted here, were totally dependent on the wind and the tides. (From William McDowell, *The Shape of Ships* [New York: Roy, n.d., ca. 1948], p. 59.)

1.

The Significance of the Steamboat

I will venture to affirm that History does not afford an instance of such rapid improvements in commerce and civilization as that which will be effected by steam vessels.

—Henry Bell (ca. 1810)

The basic needs of civilized society, among other things, include food, clothing, shelter, sanitation, information, and transportation, with the latter two being of fundamental importance to progress. Any technological development that promotes one of these needs is quickly advanced by those seeking profit or gaining advantage from its implementation, whereas it is opposed or rejected by those being negatively affected or displaced.

Some inventions, such as the cotton gin and sewing machine, were immediately and resoundingly hailed as labor-saving devices, freeing individuals from tedious work. Other developments, such as the spinning mills, threatened the English cottage industry and were initially attacked and destroyed by angry groups called the Luddites.

Even steam navigation seemed threatening to some, as Robert Fulton's steamboat was vandalized during construction. In his expense book is recorded the following entry: "$4.00 to the men for guarding the boat two nights and a day after the vessel ran against her."[1] When the *North River Steam Boat,* later called *Clermont,* was put into service, biographer Cynthia Owen Philip reports that "On her return from Albany in the second week of September, the sloop *Fox* rammed her wheels, then went about and hit her again, driving her on to a sand bar which broke her axle wheels."[2] Such were the perils faced by steamboat pioneers at that time.

For the dissemination of *information* the most important single development was the alphabet. This was followed much later by the printing press (see fig. 1.2), which provided a quantum leap in making information available to a wide audience. Information dissemination was and still is crucial to the advancement of civilization.

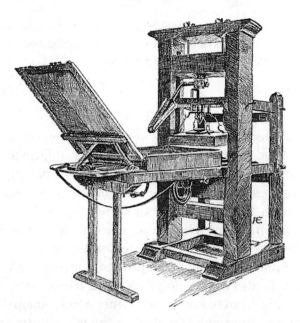

Figure 1.2. Benjamin Franklin's printing press. Franklin started his career as a printer and writer and went on to make many contributions to science, technology, and development, including lending support to James Rumsey for his inventions and his steamboat work. (From Lillian I. Rhoades, *Story of Philadelphia* [New York: American Book, 1900], p. 96.)

Improvements in the *transportation* field, such as the wheel, wagon, and carriage, along with ships and sails, made it incrementally easier to transport goods and people. The speed, however, was essentially limited to that of animal power, or in the case of water navigation, to the uncertainties of weather and wind along with the periodicity of the tides.

In our four-dimensional world (three spatial components plus time), any technical advance that enhances society's ability to work more efficiently or faster is routinely developed and adopted. The surveyor measuring a distance (one dimension) discovers a better way to pull a tape or chain accurately; the draftsman (two dimensions) crafts a special triangle, with angles unique to the drawing being prepared; or a blacksmith working in three dimensions fashions a better tool to handle and bend a hot piece of metal. In every case the new method is embraced, because it makes work easier, saves time, or improves quality. Although commonplace, these developments, improvements, innovations, or sometimes inventions move forward the lot of civilization at a steady pace. Each is important, for as everyone knowledgeable about business practices understands, time is money, and a savings in cost usually translates to improved sales at lower prices, or the same sales at an improved profit margin.

The fourth dimension of the world in which we live is time. Any invention or technical advance that *significantly* alters the amount of time needed to accomplish a task will unquestionably, and often drastically, disturb relationships within a society. Such was the case of the printing press that replaced the scribes, who heretofore had laboriously copied documents by hand. The cotton gin made it possible to remove the seeds quickly and efficiently from the fibrous cotton plant, thereby reducing the cost and freeing men and women from the laborious task. The labor in this case was put to work planting and harvesting more cotton. With the lower cost, demand went up and subsequently created much southern cotton plantation wealth.

Following the printing press in the faster dissemination of information were post riders, telegraph, telephone, radio, television, personal computer, and today the Internet. Each of these technological improvements or inventions has dramatically reduced the time it takes to exchange information. This has resulted in profound improvements in the base of knowledge available to society.

Transportation infrastructure is essential in any society for the effective interchange of ideas, goods, and the mobility of people. Movement of today's mail and packages depends on airplanes, ships, trucks with trailers, and finally the destination delivery vehicles. All of these means of transport rely totally on a transportation infrastructure—something assumed today, but this was not the case in colonial and early federal America (see fig. 1.3).

When people easily move from one place to another (as they frequently did in the Conestoga wagon, depicted in fig. 1.4), more extensive geographically diverse personal relationships become established. Prejudices are abated, business agreements are made, and sometimes conflict is averted. These face-to-face interactions are logically followed by commerce, with goods transferred between different places, resulting in increased demand and benefiting the traders as well as society.

Fulton certainly understood the importance of easy movement of people and goods. In chapter 2 of his small canal treatise of 1796, titled "Of the Importance of Canal Navigation, and the Benefits Arising to Society by Easy Communications," he encouraged governmental support for transportation infrastructure: "Like the government of China, the legislature of every country should be particu-

Figure 1.3. The stagecoach, pulled by four to six horses, provided colonial America with cross-country transportation at speeds up to eight miles per hour. (From D. H. Montgomery, *Student's American History* [Boston: Ginn, 1916], p. 167.)

Figure 1.4. Goods and often people moved in the sturdy Conestoga wagon pulled by four, six, and sometimes even eight horses over difficult terrain. (From Allen C. Thomas, *History of the United States* [Boston: D. C. Heath, 1901], p. 259.)

larly attentive to the reduction of the expense and delays of carriage, and to the formation of easy communications between different and distant provinces; as agriculture and commerce will improve, and happiness spread, in proportion as the facility of conveyance increases."[3] Fulton's plan was one of many small canals crisscrossing the countryside to achieve easy water communication, but as such was never adopted. Later, however, his steamboat legacy would achieve essentially the same effect for postcolonial America. In a similar fashion, John Stevens believed that "[t]he wealth and prosperity of a nation may be said to depend, almost entirely, upon the facility and cheapness with which transportation is effected internally."[4] This infrastructure visionary dedicated much of his life to achieving a network of roads, water routes, and rails.

Likewise, James Rumsey understood the potential economic benefits of steam navigation that would reduce the cost of goods conveyed and making it possible to go upstream on inland rivers:

And my machine [the steamboat], with all its misfortunes upon its head, is abundantly sufficient to prove my position; which was, "that a boat might be so constructed, as to be propelled through the water at the rate of ten miles in an hour, by the force of steam; and the machinery employed for that purpose, might be so simple and cheap, as to reduce the price of freight at least one half in common navigation; likewise, that it might be forced, by the same machinery, with considerable velocity, against stream of long and rapid rivers."[5]

A significant benefit of good transportation is in the mobility of a workforce that is able to travel longer distances, thereby geographically dispersing critical skills. This fosters innovation and improvements in efficiency, with the positive result of making more products available at a higher level of quality and at lower costs.

Although it has always been possible to move over land by foot, horseback, or wagon and over water by canoe, boat, or ship, the regularity and dependability of movement was very much dictated by the environmental elements of weather. Also, land movement was limited in power (hence weight of cargo of people and goods) to that of several pairs of horses, oxen, or other animals, whose performance in turn was affected by the factors of heat, cold, wind, rain, snow, and ice.

The steam engine propelling a boat upset this equilibrium between man and nature, which had existed since the advent of civilization. For the first time man was able, albeit in an elemental and primitive way, to conquer the wind, tides, and river currents; hence, in some measure he was able to become not only the master of what he could survey but beyond. Whether we call this new transportation a steamboat, steam vessel, steam tug, steamship, or just steamer, they all are fundamentally the same. They are derived from the desire for reliable transportation by waterborne vessels freed from the vagaries of wind and the limitation of muscle power to enable traveling long distances at good speed with few limitations imposed by forces outside the vessel itself.[6]

And this was only the beginning, as improvements in mechanical motion would continue with faster and larger steamboats later joined by railroads, automobiles, and trucks, with airplanes eventually reducing the four corners of the earth to hours—not days, weeks, or months—of separation. All advances brought new dangers causing some to resist change, but over time the new technologies were embraced and have become part of society's fabric.

It was the *steamboat*, however, that pioneered the initial shrinkage in the size of the world, precipitating at the time technological development difficulties that have yet to be fully understood by historians and philosophers. In the preface to *Steamboats Come True*, James Flexner ponders the period between the availability of the parts and the achievement of a practical steamboat: "Finally, all of the components of the steamboat existed: it was only necessary to combine them. One would foresee, after such seemingly endless expectation, a quick, triumphant surge. However, for twenty-five years of anguished effort, the combination was not effectively made. Why? That is the question that haunts the body of this volume."[7]

"Why did it take so long?" is asked, but the more operative question poised by a technologist might be, "How did it happen so quickly?" The answer is simple.

The steamboat was the *first* transportation system ever developed allowing a vessel propelled by found fuel to be self-sustained and independent of the routine elements of nature. It became possible for the very first time to achieve a very complicated interaction of chemical fuel, an efficient boiler, reduced-weight steam engine, power transmission, water propulsion, steering, and speed control of a watercraft that heretofore was considered by many impossible and against nature. In America it was more that twenty years from a technically feasible concept to commercial realization, a period of time that seems too long to some, but seems quite reasonable to technologists and engineers. It was the *first system* ever developed, and it was accomplished under very primitive technological conditions.

In writing about Fitch's steamboat development Greville Bathe provides a technical as well as social explanation of the delay:

> Fitch and Voight were ingenious men and capable mechanics, but they, and all the rest, depended too much on abstract philosophy and too little on the exact science of physics. And so much time and money were wasted by these early promoters in unnecessary experiments and empirical reasoning. Superficially, it may seem strange that, having gone so far with steamboats, little or no further progress was made to establish them in the public service, but things other than mechanical difficulties intervened and the impulse was lost for more than a decade, the public mind not yet

being fully aroused to the possibilities of engineering science, which had to wait its turn after the dawn of the new century.[8]

Another, much earlier (1845) Fitch biographer, Charles Whittlesey, observed about the steamboat that "the development was gradual and sublime,"[9] also noting American political and educational factors: "The inventive genius of the United States appears to have been simultaneously directed to the subject of steam in its various applications. This infant nation, in a state of political chaos, without order or strength, without science or institutions, became the theatre of intense speculation upon the greatest problem in the range of national science."[10] The steamboat was a major achievement, both technically and socially. It brought people closer together, uniting them in economic bonds that prevented civilizations from splintering, particularly in America. Its contribution in tying the eastern and western portions of the fledgling republic together into a nation was significant.

The paramount importance of transportation in postcolonial America was highlighted by Archer B. Hulbert when writing in the preface to his 1920 book, *The Paths of Inland Commerce*:

> If the great American novel is ever written, I hazard the guess that its plot will be woven around the theme of American transportation, for that has been the vital factor in the development of the United States. Every problem in the building of the Republic has been, in the last analysis, a problem in transportation. The author of such a novel will fine a rich fund of material in the perpetual rivalries of pack-horsemen and wagoner, of riverman and canal boatman, of steamboat promoter and railway capitalist. He will find at every point the old jostling and challenging the new: pack-horseman demolishing wagons in the early days of the Allegheny traffic; wagoners deriding Clinton's Ditch [the Erie Canal]; angry boatmen anxious to ram the paddle wheels of Fulton's *Clermont*, which threatened their monopoly. Such opposition has always been an incident of progress; and even in this new country, receptive as it was to new ideas, the Washingtons, the Fitches, the Fultons, the Coopers, and the Whitneys, who saw visions and dreamed dreams, all had to face scepticism and hostility from those whom they would serve.[11]

Forty-two years after Robert Fulton successfully demonstrated the first commercially successful steamboat on the Hudson River in 1807, Thomas Babington Macauley penned some prophetic words that were later to be inscribed on one side of the golden door of the Transportation Building at the World's Columbian Exposition, held in Chicago in 1893. The words on the door simply read,

Of all inventions,
the alphabet and printing press alone excepted,
those inventions which abridge distance
have done the most for civilization.[12]

Canadian author James Croil also recognized the steamboat's significant contribution in his 1898 book titled *Steam Navigation and Its Relation to the Commerce of Canada and the United States*: "When the history of the nineteenth century comes to be written, not the least interesting chapter of it will be that which treats the origin, the development, and the triumphs of Steam Navigation—that mighty combination of inventive genius and mechanical force that has bridged the oceans and brought the ends of the earth together."[13] The steamboat was entirely responsible for this initial abridgment of distance and brought together the ends of the earth to provide a significant advance in civilization. For American transportation, Brooke Hindle and Steven Lubar observe that the new technology served "a compelling need. . . . The steamboat converted the Mississippi into a national highway system."[14] It allowed the populated East Coast to be linked with the middle of the country.

Why in America?

The story of how the steamboat came about is fascinating, and it should start with a discussion of the unique set of circumstances in America that fostered its embryonic and occasionally abortive beginnings.

In the latter part of the eighteenth century, it certainly would have been logical for a more scientifically and technically advanced England, or one of the other European countries, to develop and introduce steamboat service. This task, however, was left to Americans, who were driven by a real or perhaps perceived need for westward expansion coupled with a burst of creativeness following the war for independence with the British. England already had a system of canals in place where barges could easily be moved with animal power, and their navy, which was the most powerful in the world, had little interest in steam-powered vessels. There were individuals who built steamboats, but they were not adopted, as the time in England and elsewhere was not yet right.

In America, however, the climate was drastically different. During the latter part of the century, pioneering individuals had started moving farther westward across the Allegheny Mountains into the Ohio Valley (see map, fig. 1.5). What had begun as a trickle during the Revolutionary War became a flood afterward, with some of them led by famous wilderness hunter and trailblazer Daniel Boone. They built homes, cleared the land, planted crops, and raised families.

The fertile lands in the Ohio Valley produced more food than was needed for consumption; the excess was loaded onto flatboats and floated a circuitous path down the Ohio River heading southwest into the Mississippi (see fig. 1.6). Thereafter, drifting south, they continued to ports such as Natchez or New Orleans, where the goods were sold. As it was very difficult—almost impossible—to pole the flatboats back upstream, they were disassembled into boards that were sold.[15]

To return home these pioneers often rode or walked north on an old Native American trail called the Natchez Trace, which was mapped by the French in 1733. Over four hundred miles long it went, from Natchez to Nashville, through difficult terrain often frequented by bands of thieves who would steal travelers' money and belongings. Along the Trace were rustic inns, called stands, which offered meager food and lodging, but the journey back home was nevertheless arduous and fraught with danger.[16] Eventually, however, the steamboat would come to their rescue.

The 1790 U.S. Census reported a population of slightly less than

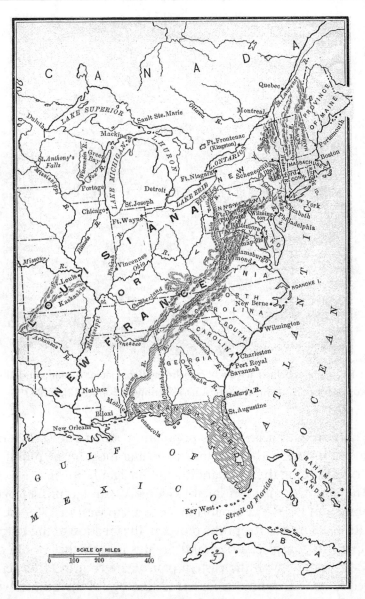

Figure 1.5. Map of the eastern United States. Being able to go up the Mississippi and Ohio rivers with the steamboat was very important for uniting the new republic into a cohesive country. As the mean annual speed of the Mississippi River's current below St. Louis was 2.3 miles per hour, it was necessary for a steamboat to have a speed of 4–5 miles per hour, to be practical. (From D. H. Montgomery, *Leading Facts of American History* [Boston: Ginn, 1891], between pp. 126 and 127.)

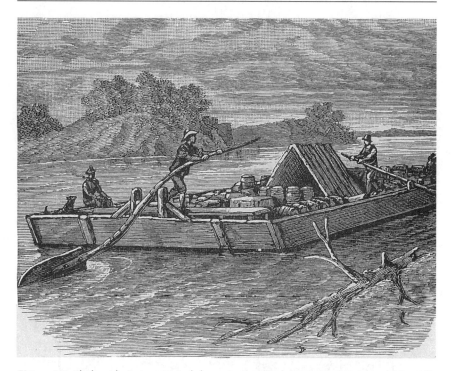

Figure 1.6. Flatboat being navigated downstream. (From James A. James and Albert H. Sanford, *American History* [New York: Charles Scribner's Sons, 1912], p. 275.)

four million with a geometric population center twenty-five miles east of Baltimore.[17] This was indicative that most of the population was located along the coast and inland navigable streams.

The country's new president, George Washington, had been a proponent of westward expansion since a trip west in 1759 (see fig. 1.7). In 1785 he was the prime mover in the creation of the Potomac Company to construct a river and canal waterway that would open the west to trade with the eastern part of the country. This venture proved to be too ambitious for the technology of the time. It would be another sixty-four years before the 185.5-mile-long Chesapeake and Ohio Canal made it as far west as Cumberland, Maryland, and by that time the railroad was already there.[18]

Nevertheless, land grants in what was called the Northwest Territory and Kentucky provided encouragement for pioneers to go west. The

Figure 1.7. From an early age, George Washington promoted westward expansion of the United States, and he played a pivotal role in James Rumsey's steamboat development. (From Montgomery, *Leading Facts of American History*, between pp. 134 and 135.)

expansion policy continued by Thomas Jefferson was successful, and lack of water transportation notwithstanding, the population center edged westward, moving essentially along a parallel about fifty miles every ten years for the next one hundred years (see map, fig. 1.8).[19]

With navigable streams in the east and the beckoning Mississippi River on the western boundary, but with the canal building era still almost fifty years away, America not only was ready for the steamboat but also needed it to unite the young country. As historian Roger Burlingame said, "America was the only country in the world which could have taken this device at the precise stage which it had reached and made it answer a national necessity."[20] British writer S. B. Woolhouse first takes a swipe at the technical capabilities of individuals in the United States, saying, "The projects of the Americans are seldom founded on the sober reasoning of science," but goes on to note, "Considering the importance to America of navigating her immense rivers, it is not surprising that the application of the power of steam to propelling vessels, should by persevering efforts have been first carried into successful practice in that continent."[21] Such was the environment into which American ingenuity was thrust, and the result was the world's first transportation system, the steam-powered boat capable of going upstream.

In a letter to George Washington on March 10, 1785, James Rumsey cautiously described his steamboat, secretly under development, writing with his poor spelling and lack of punctuation:

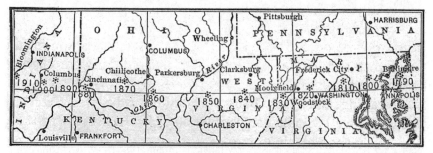

Figure 1.8. The snowflake stars show the progressive westward movement of the United States population from 1790 to 1910, essentially along a parallel. Starting near Annapolis in 1790, America's population center reached south of Indianapolis by 1910. The population is uniformly distributed around the stars. (From D. H. Montgomery, *Leading Facts of American History* [Boston: Ginn and Company, 1917], p. 180.)

I have taken the greatest pains to afect another kind of Boats upon the princeples I was mentioning to you in Richmond I have the pleasure to Inform you that I have Brought it to the greatest perfection It is true it will Cost Sum more that the other way [pole or stream boat] But when Done is more manageable and Can be worked by a few hands, the power is amence and I am Quite Convinced that Boats of passage may be made to go against the Current of the Mesisipia or ohio River, or in the gulf Stream from the Leward to the Windward Isslands from sixty to one hundred miles per Day I know it will appear Strange and Improbeble and was I to say thus much to most people in the neighbourhood they would Laugh at me and think me mad But I can ashore you Sir, that I have Ever Been Very Cautious have I aserted any thing that I was not Very Certain I could perform Besides it is no phenomena, when known, But Strickly agreeable To philosophy.[22]

The following May the Continental Congress promised to James Rumsey thirty thousand acres of Ohio land if he could construct a boat to travel fifty miles per day on the Ohio River.[23] With this incentive Rumsey certainly understood the importance of the steamboat to the government and prompted the initiation of his work and the hiring of brother-in-law Joseph Barnes to construct the vessel.

John Fitch, having explored the western part of America in his earlier life, was also cognizant of the importance of navigating upstream with a steamboat. In his petition to Congress, late in the year of 1785, for financial support of his development, he points out. "Sir:—The Subscriber begs leave to lay at the feet of Congress, an attempt he has made to facilitate the internal Navigation of the United States, adapted especially to the Waters of the Mississippi."[24]

It is clear from this message that Fitch perceived the importance of the steamboat to upstream navigation, for it was a significant selling point in his petition, but Congress failed to provide any encouragement as it had done for Rumsey. This distressed Fitch, making him even more determined to succeed and to "prove them to be but ignorant Boys."[25]

Fitch was not to be discouraged, as he and many other Americans represented a Yankee "can-do" attitude hell-bent on creating something new, sometimes with the anticipation of fame or perhaps riches, but always with the idea of making the country a better place in which to live. This was the frontier spirit, exuberant and optimistic that prompted an eighteenth-century Swedish visitor to remark, "This is America, which dares to undertake the impossible."[26]

Notes

1. Quoted in John Morgan, *Robert Fulton* (New York: Mason/Charter, 1977), p. 137.

2. Cynthia Owen Philip, *Robert Fulton: A Biography* (New York: Franklin Watts, 1985), p. 205. When Robert Fulton registered his first American steamboat in 1807, he simply called her *Steam Boat*, being the only one in existence. In his advertising, to identify the route, it was referred to as the *North River Steam Boat*; this more descriptive name is used here. After the hull was rebuilt and enlarged during the winter of 1807–1808, she was re-registered as the *North River of Clermont*, meaning that *Clermont*, the name of the estate belonging to his partner Robert R. Livingston on the North or Hudson River, was the home port. The name *Clermont* was embraced by the public; hence the confusion. See Randall J. LeBoeuf Jr., *Some Notes on the Life of Robert Fulton* (New York: South Seaport Museum, 1971), pp. 3, 13; Alice Crary Sutcliffe, *Robert Fulton and the* Clermont (New York: Century, 1909), between pp. 268 and 269; and H. W. Dickinson, *Robert Fulton, Engineer and Artist: His Life and Works* (London: John Lane, 1913), Appendix G, pp. 326–27.

3. Robert Fulton, *Treatise on the Improvement of Canal Navigation; Exhibiting the Numerous Advantages to Be Derived from Small Canals* (London: I. and J. Taylor, 1796), p. 11.

4. Quoted in Archibald Douglas Turnbull, *John Stevens: An American Record* (New York: Century, 1928), p. 465.

5. James Rumsey, *A Short Treatise on the Application of Steam, Whereby Is Clearly Shewn from Actual Experiments, That Steam May Be Applied to Propel Boats or Vessels of any Burthen against Rapid Currents with Great Velocity* (Philadelphia: Joseph James, 1788), second paragraph.

6. The author is indebted to P. J. Woodall for this observation.

7. James Flexner, *Steamboats Come True: American Inventors in Action* (1944; reprint, Boston: Little, Brown, 1978), p. xiv.

8. Greville Bathe, *An Engineer's Miscellany* (Philadelphia: Patterson & White, 1938), p. 54.

9. Charles Whittlesey, "John Fitch," *Library of American Biography: Lives of Ezra Stiles, John Fitch, and Anne Hutchinson,* conducted by Jared Sparks (Boston: Charles C. Little and James Brown, 1854), pp. 85–86.

10. Ibid., p. 115.

11. Archer B. Hulbert, *The Paths of Inland Commerce: A Chronicle of Trail, Road, and Waterway* (New Haven, CT: Yale University Press, 1920), pp. vii–viii.

12. Quoted in *Respectfully Quoted,* ed. Suzy Platt (Washington, DC: Library of Congress, 1989), p. 288.

13. James Croil, *Steam Navigation and Its Relation to the Commerce of Canada and the United States* (1898; reprint, Toronto: Coles, 1973), p. ix.

14. Brooke Hindle and Steven Lubar, *Engines of Change: The American Industrial Revolution, 1790–1860* (Washington, DC: Smithsonian Institution Press, 1986), p. 116.

15. *International Cyclopedia,* ed. Richard Gleason Green, 15 vols. (New York: Dodd, Mead, 1887), 10:43. The mean annual current of the Mississippi River below where the Missouri River joins near St. Louis is noted to be 2.3 miles per hour. During some times of the year with large rainfalls, the current would be even faster.

16. National Park Service, *Natchez Trace Parkway, Official Map and Guide* (Washington, DC: Government Printing Office, 1997).

17. D. H. Montgomery, *Leading Facts of American History* (Boston: Ginn, 1891), p. 194n.

18. National Park Service, *Chesapeake and Ohio Canal* (Washington, DC: Government Printing Office, 2000).

19. Montgomery, *Leading Facts of American History,* p. 194n.

20. Roger Burlingame, *March of the Iron Men: A Social History of Union through Invention* (New York: Grosset & Dunlap, 1938), p. 194.

21. Quoted in David Read, *Nathan Read: His Invention of the Multitubular Boiler . . .* (New York: Hurd and Houghton, 1870), p. 33.

22. Rumsey, "Letters of James Rumsey," ed. James A. Padgett, *Maryland Historical Magazine* 32 (1937): 19.

23. Jeanne Mozier, "James Rumsey: The Man and His Life" (manuscript, Museum of the Berkeley Springs, WV, 1986). The author is indebted to Ms. Mozier, former president of the Museum of the Berkeley Springs, for a copy.

24. Quoted in Thomas Boyd, *Poor John Fitch* (New York: G. P. Putnam's Sons, 1935), p. 137.

25. Fitch, *Autobiography of John Fitch,* ed. Frank D. Prager (Philadelphia: American Philosophical Society, 1976), p. 153.

26. Quoted in Russell Bourne, *Invention in America* (Golden, CO: Fulcrum, 1996), front dust cover.

Figure 2.1. Leonardo da Vinci (1452–1519)—painter, mathematician, sculptor, musician, scientist, architect, and perhaps the greatest inventor the world has ever known, spent most of his life working as a civil engineer. (Rendered by William T. Sisson after Leonardo's self-portrait when the latter was about sixty.)

2.

The Invention, the Inventor, and Society

A tool is but the extension of a man's hand, and a machine is but a complex tool. And he that invents a machine augments the power of a man and the well being of mankind.

—Henry Ward Beecher

I n order to determine who (if anyone) invented the steamboat, we must look at the very concept of invention. Did Leonardo da Vinci "invent" a flying machine or helicopter around the end of the fifteenth century because he described such devices in words and graphics? Clearly the answer is no!

We must be careful to recognize the difference between the dreamer or prophet, on the one hand, who recognizes that something could possibly be done, and, on the other, the individual or individuals who achieve a practical implementation of the idea. As James Flexner has pointed out, there is a vast difference between the armchair writer who has never walked up a hill, pointing out it should be possible to climb Mount Everest, and the climber who actually does it.[1] The world has many dreamers, some of them far removed in a science-fiction world, but a few of them dare to turn their dreams into reality.

Invention, or at least invention of a relatively complicated machine like the steamboat, must have some reasonable base in both science and technology, for these are the foundation pillars upon which new and complex ideas can be realized. The concept must be achievable with existing materials and processes or straight-forward extensions of technology and cannot depend on something yet to be developed or invented.

Science is founded in discovery. It explains the underlying principles of nature and as such is not invention. For example, the laws of gravity and force would have been the same if someone other than Newton had discovered and quantified them. Science, however, allows an inventor to assess the likely results of an experiment or development beforehand; with its introduction it was no longer necessary to rely exclusively on the trial-and-error method.

Technology, the accumulated knowledge of how to use found or raw materials to produce useful or artistic objects, is as old as civilization itself. This knowledge encompasses a wide diversity of information from relatively simple things such as carving a bowl from wood or stone, molding and firing bricks or a piece of pottery, or spinning and weaving, to more sophisticated projects such as designing and building a steamboat, locomotive, or airplane. This base of knowledge accrues at a slow but steady pace as artisans and engineers, the practitioners of technology, learn by doing, make improvements, and pass this information along to others.

This process of continuous improvement in technology, such as a blacksmith learning how to weld a stronger joint or a spinner producing a tighter yarn, is called *innovation*. Nearly every artisan or engineer who works with any technology on a prolonged basis progressively makes incremental, although often very simple, improvements in the way they do things. These improvements or innovations allow a product to be produced at a higher quality, a lower cost, or at a faster pace. In a sense each innovation is a very low-level *invention*, but since the improvements are usually "fairly obvious" to anyone practicing the art, they most often go unrecognized by society. However, peers in any particular area of work may revere the contributing individual.[2]

One way to make a distinction between levels of improvement is to recognize that innovation *improves* the way something is done, and invention *changes* the way.[3] On a boat or ship a new sail configuration might be considered an innovation, whereas installing an engine and means of propulsion for the first time would be regarded as an invention.

An example of innovation pointed out by historian Brooke Hindle and others is worthy of mention. When the westward-heading pioneers set out to establish a new American frontier they were almost solely dependent on their own resources. They built their dwellings, cleared the land, and planted and harvested their crops. The critical tool for survival, besides the rifle, was the ax; but British and other European axes were not well adapted to the task. Depending on the country of origin, the axes were somewhat unwieldy and inefficient, resulting in a number of blacksmiths making improvements, and through innovation a far better ax with a poll (a head weight behind the cutting edge) was evolved (see fig. 2.2). This new design with a better head and more centered handle could, in the hands of a capable woodsman, fell trees three times as fast.[4]

Although not understood from a technical or scientific point of view at the time, the innovative blacksmith working with the early axes knew they didn't have a "sweet spot." Anyone who has hit a ball

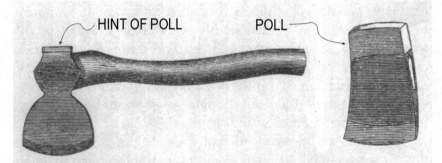

Figure 2.2. Placing a poll, a weight behind the cutting edge, on an ax created a better-balanced device with a "sweet spot." (From C. P. B. Shelley, *Workshop Appliances* [London: Longman's, Green, 1894], p. 34.)

with a golf club, tennis racket, or baseball bat has an understanding of the sweet spot: that place on the instrument where contact with the struck object produces little disturbance to the hands and arms of the swinger. The wrists are not rotated to the left or right, and the arms are not subjected to an acceleration or deceleration of the motion of the swing. In other words there is no sting to the human body—hence a comfortable impact. There have been many improvements, particularly in golf clubs and tennis rackets, to increase the area or size of the sweet spot, which in science is known as the center of percussion. These sporting instruments have met with good success in the marketplace with both professional and amateur players.

The ax, as an important and critical tool, is no different. Blacksmiths hammering at the forge and woodsmen swinging at the trees, through a process of evaluation and feedback, evolved a far superior cutting tool eventually known as the Anglo-American ax. Not only was the head improved, but eventually the handle was also curved with a butt end, producing a more comfortable, easier to use instrument. With a better tool, more useful work could be done in a given time, leading to a more secure lifestyle. This allowed more time for leisure, with the attendant benefit of having time to think of additional improvements to enhance life.

One might logically ask if these innovations leading to a better ax were not in fact inventions. On a minuscule level, the answer is definitely yes, but practically speaking, each was such a relatively insignificant step that it would be difficult to afford the inventor of each any special recognition or reward. Furthermore, the improvements often evolved simultaneously at different locations by different individuals, and although the final product was indeed superior to its predecessor, assigning credit for these piecemeal improvements would be almost impossible. The changes improved the way the ax worked as a felling and chopping tool, but it did not change the fundamental method of cutting.

Invention, and a patent thereof granted by a government, requires a substantial step, often a sidestep, in improvement to obtain legal protection. Innovation routinely follows once a deficiency has been recognized, and the improvements will usually take place regardless of the engineer or artisan involved, albeit at different rates because of differing levels of skill. The innovation, being more or less obvious to anyone working in the field, becomes a difficult concept to assign, as any competent individual would arrive at or near the same improvement. In other words, knowledgeable and current practitioners of the art would likely see the possibility of the innovation when presented with the same scenario. Although innovation versus invention seems a very difficult standard to apply, in actual practice it works well, as inventors, usually engineers, are assiduously honest with each other; therefore, when they are unable to convince their peers of the novelty of a new idea, they will not seek patent protection (see diagram in fig. 2.3).[5]

Most of what qualifies as science, the world's accumulated knowledge of the laws of nature, is relatively young when compared to ancient technology. About four to five thousand years ago, the first astronomers observed the movement of what they believed to be the earth-centered heavens and initiated scientific study. From their observations they were able to construct fairly accurate calendars, although the results were primarily empirical, as they did not have a basic understanding of the laws of nature.

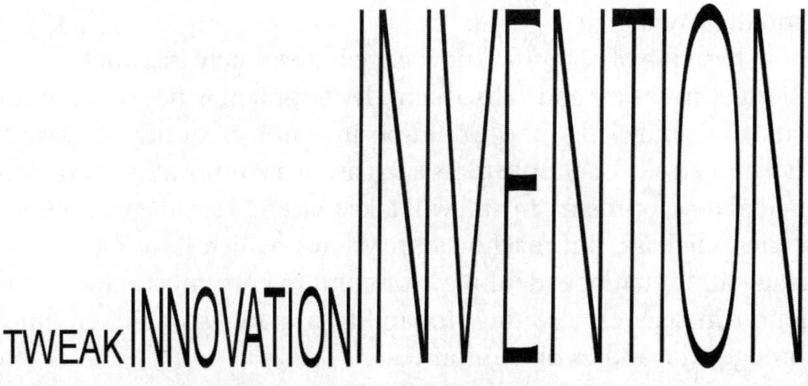

Figure 2.3. Diagram representing the relative importance of technological change: *tweaking* an existing machine or process improves performance with a given set of inputs; *innovating* a change further improves performance; and *inventing* means an altogether new way to achieve higher performance. A good example is an automobile engine: adjusting the carburetor is tweaking; replacing a two-barrel carburetor with a four-barrel model is innovating; and replacing the carburetor with electronic fuel injection constitutes invention. (Illustration by Jack L. Shagena.)

It was not, however, until after the studies of Kepler and Galileo that Isaac Newton codified the gravitational laws in his famous 1687 *Principia*, establishing for the first time the dynamics of the solar system. His study proceeded from examining movement in nature to investigation of the forces of nature to finally understanding and applying this knowledge to solve other problems. He determined that the "celestial" laws, which govern the motion of the skies, were the same as the terrestrial laws that govern an apple falling from a tree.

Newton's scientific breakthrough, along with the work of Robert Boyle, Denis Papin, Joseph Black, and others in thermodynamics became part of the Scientific Revolution, which established the foundation for the Industrial Revolution to follow in the latter part of the eighteenth century.[6]

Scientists, who seek to understand the world around them, employ an objective technique of hypothesis, followed by experimentation, observation, and analysis, which lead to validation or rejection of the hypothesis. Science is a process of discovery, and unlike invention, which takes on the technical idiosyncrasies or preferences of the inventor, the laws of nature are exactly the same, irrespective of the personality and background of the individual who discovers them.[7]

A baseline of scientific knowledge is absolutely essential to complicated invention and subsequent development, since without fundamental knowledge the would-be inventor can only advance by trial and error. For simple tools such as an ax, innovation and incremental development can and will take place at a steady pace without a scientific base. But machines or systems, which have many interacting and interdependent parts or components, must have a foundation in science, else the probability of success of the ensemble through trial and error is minuscule.[8]

Development is the process of taking an idea through an engineering process so as to transform a paper concept into something tangible and marketable, beginning with an examination of the market need, shown in figure 2.4 as a man staring at a crystal ball. This first step is crucial, because if the need is only perceived, not real, the product's market success, and thus its future, is doomed.

When the *need* is understood and the available technology and

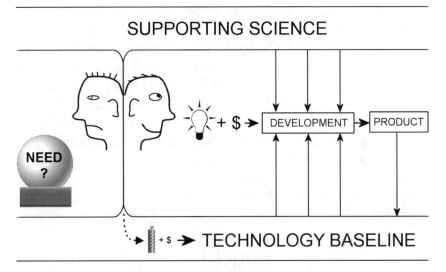

Figure 2.4. Drawing showing the relationships among need, science, technology, invention, innovation, finances, and development of a new product for the marketplace. (Illustration by Jack L. Shagena.)

supporting science have been carefully factored in, the cognitive process begins, and ideally a new and viable invention evolves. Once the idea has germinated and has been scrutinized, the mind does an about-face and, instead of looking back, now faces the future of transforming the idea through the development process into a viable product. In some cases much simpler ideas, represented in figure 2.4 by the candle at the bottom, fall out of the cognitive process. In such cases, it is likely that with the addition of a modicum of money, the small dollar symbol, the idea can be converted into an innovation, which can be added to the technology baseline.

Before the development process can start it is first necessary to obtain the requisite finances to pay for the labor and materials. Thereafter the process is bounded only by technology and science as it progresses through several phases of development further discussed in chapter 3. As technology is continuously being enhanced, it is interesting to observe that once the development is completed the product then becomes part of the available base as represented by the down arrow.

After successful completion of product development the device is tested in the marketplace as shown in the product validation model in figure 2.5.

One of the mistaken beliefs often held by first-time inventors is that the new idea or invention in and of itself is a valuable commodity. This logic extends to the belief that a company whose business could benefit from the invention would be willing to buy the idea from the inventor. Except in rare cases this is not true, for the company is fully aware that new ideas must be augmented by time and money and that they nearly always require the talent of a group of individuals, not just the single inventor. So a novel idea, while fascinating and intriguing, often remains as such because of lack of capital to transform it into a product suitable for market introduction.

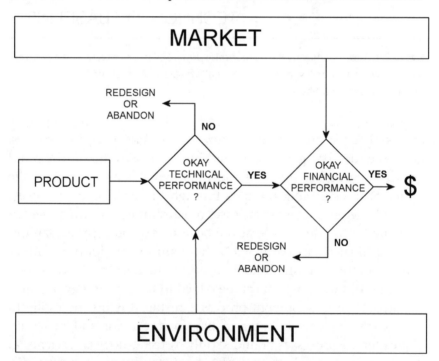

Figure 2.5. Product validation model. A new product is first evaluated in the environment in which it must operate to be acceptable to the potential users. If it is satisfactory, the users in the market are invited to embrace the product. A patent for the product provides the inventor with a short-term monopoly that rewards his or her efforts with the expectation of reaping a profit to work on another idea or further develop the current one. (Diagram by Jack L. Shagena.)

About Invention

Invention is the concentrated application of intellectual horsepower turning and churning the mechanisms of technology, generally enhanced but also occasionally constrained by scientific knowledge, to create an arrangement of the old into something decidedly new. Science can therefore be a double-edged sword. If an individual with a working knowledge of the underlying science incorrectly perceives that Mother Nature will not let a particular approach work, then he will be erroneously diverted from what may have been a workable solution where the less knowledgeable individual may have plowed ahead and been successful. (Except for relatively simple invention, however, this is not likely to occur.)

Invention is art, and there is a significant parallel between the conceptual workings of an engineer and those of an artist. Engineers, who are disciplined and scientifically trained, often resent being compared to artists, who often operate seemingly unencumbered by rules, but there is a striking similarity to the way each create. Both conceive a new idea in the mind's eye, each mull over combinations and permutations of achieving a completed product, but thereafter the parallel diverges.[9] Once the idea is formulated the artist *just does it*, adjusting to problems as she goes, whereas the engineer breaks the product into a number or manageable pieces or component parts, produces and tests each, then assembles or integrates the pieces into a working whole.

Another view of invention is creation. The inventor projects his conscience into producing a novel arrangement of the old into the new; hence, creating something that did not previously exist. To this extent invention is the servant of imagination, held hostage by a lethargic mind. It is only from the mind that new ideas arise, and it is science that gives these ideas a logical order and technology that gives them substance (see fig. 2.6).

In light of the need for both a technology base *and* scientific knowledge for complicated invention, it surprising how well the steamboat pioneers actually accomplished their objectives. When Rumsey lacked scientific data to support his activities he performed experiments to gather the necessary information. Fitch was much more dependent on trial and error, but he did have some rudimentary technical knowledge as well as the dedicated support of the ingenious Henry Voight. Likewise, Joseph Barnes, a capable shipwright and mechanic who later was Rumsey's attorney, equally supported him. Oliver Evans had an excellent base of science, honing and expanding his knowledge through writing and publishing a treatise on mills and steam engines. Samuel Morey was privy to a fair amount of technical knowledge and possessed the mind of a scientist, investigating and reporting to scientific journals his observations. Robert Fulton assimilated a goodly amount of science during his twenty years in England and France, but his real strength was not in original thinking but in entrepreneurial development and integration of the best ideas of others that he found available. John Stevens made an attempt through extensive reading to gather information about the latest technological developments and often had thoughts on improvements. He was primarily a transportation visionary, whereas his son Robert Livingston became an engineer and significant inventor.

Figure 2.6. Invention must rest firmly on the two pillars of science *and* technology, else it is simply prophesy and not a practical embodiment of a new or worthwhile idea. (Illustration by Jack L. Shagena.)

Are most engineers inventors? Sometimes yes and sometimes no! In many cases engineers just extrapolate the extant technology base, such as using an off-the-shelf bridge design to span a longer distance. At times, however, because of technology or cost constraints they are "forced" to abandon the conventional and "invent" a new design. Being generally conservative by nature, and with the absence of constraints, the engineer nearly always opts for the tried and true means (i.e., one with the lower risk) to accomplish a design. In high-tech industry, such as aerospace engineering, some engineers have a different bent, as it is the new—and, by definition, untried—idea that portends a technological advance and hence a competitive edge. Although the risks of failure are higher, these entrepreneurs and engineers are driven by market factors to attempt the very difficult and with success they reap the attendant financial rewards.

One of the reasons for a generally conservative approach is grounded in the nature of problem solving. Engineers are skilled at breaking down a complicated problem into manageable pieces, then assembling these pieces into a working whole through steps: A before B before C etc. This is a linear, vertical-thinking process, which, if rigidly followed, proves successful but seldom yields new insights, a requisite of invention. For invention, thinking *across* disciplines is often required, and for the engineer, this is simply not the normal way of doing things.

One practical and effective tool for sideways or lateral thinking, as it is sometimes called, is to employ two decks of cards with words printed on one side of each card. On the cards in one deck are printed the names of the technology disciplines, such as "electrical," "hydraulic," "pneumatic," "magnetic," "mechanical," etc.; on the cards of the other deck would be names of technologies one might be interested in, such as "lights," "switches," "motors," "grinders," "gauges," etc. The decks are shuffled separately of each other, and one card is drawn from each deck and paired with the other, creating, for example, "hydraulic" and "grinder." The individual is then challenged to invent a grinder based on the principles of hydraulics and is hence forced to think in a sideways fashion.[10]

A Brief History of Patent Protection

To better understand what constitutes an invention it is instructive to examine how the United States government provides legal protection for an invented device, especially as it pertains to the time when the steamboat was being developed. In America, as early as 1641 the General Court granted a process patent to Samuel Winslow for the manufacture of salt in the Massachusetts Bay Colony. A few years later, in 1645, the same General Court awarded the first mechanical patent to Joseph Jenckes for improved sawmills and scythes.[11] While monopolies in general were judged to be undesirable, it was recognized that providing protection for a short period of time was beneficial to society, as it stimulated innovation and invention. This basic principle would later become the basis for the United States patent system.

Petitioning to the colonies or later to the states for time-limited monopolies (patents) continued for over 130 years, even after America adopted the Articles of Confederation in 1778, which made it possible to obtain protection from Congress. In 1784 James Rumsey obtained from Virginia a ten-year patent for his pole boat, which moved upstream by the force of water rotating a paddle wheel that set poles against the stream bottom.

Upon adoption of the Constitution in 1787, Congress was given the power: "[t]o promote the progress of science and useful arts by securing for limited times to authors and inventors the exclusive right to their respective writings and discoveries."[12] This section was inserted into the Constitution at the insistence of James Madison (see fig. 2.7) and Charles Pickney.

Rumsey was in England supervising the construction of his second steamboat, the *Columbian Maid*, when he received news from Philadelphia in late April 1789 about the organization of the new federal government under the Constitution. From Dover on June 6, he wrote Thomas Jefferson in Paris encouraging that the committee appointed to address the patent legislation protect principles, such as his tube boiler, rather than a specific implementation or configuration of a device. Otherwise a slight change to the way a pipe is bent would allow another individual to also patent the boiler, discour-

Figure 2.7. James Madison played a prominent part in the formulation of the U.S. Constitution and along with Charles Pickney had the patent clause inserted into the document. Madison later became the country's fourth president. (From *Messages and Papers of the Presidents, 1789–1897*, ed. James D. Richardson, 10 vols. [Washington, DC: Government Printing Office, 1897], 1:464.)

aging the original inventor. With poor spelling he suggests that "The french method of haveing new invention Examammined by a Committee of philosophical Charactors, before grants Can be obtained, is certainly a good one."[13]

In his first annual address to Congress on January 8, 1790, President George Washington urged the speedy enactment of a patent law: "I cannot forbear intimating to you the expediency of giving effectual encouragement, as well to the introduction of new and useful inventions from abroad as to the exertion of skill and genius at home."[14]

It has been asserted by John Stevens's biographer, Archibald Turnbull, that Colonel Stevens was the prime mover behind the congressional passage of the first patent act; however, according to Dorothy Gregg, "The Stevens Collection revealed no support for such a claim."[15] Stevens had become interested in the steamboat after seeing Fitch's boat on the Delaware River and reading about the 1788 Rumsey-Fitch controversy as to who was the rightful inventor. He had his own ideas for steam navigation and was motivated to secure a more favorable position against current and future rivals.

A patent bill, introduced into Congress the previous year, was passed on April 10, 1790. Defined as patentable in the legislation was any "useful art, manufacture, machine, or device, or any improvement thereon, not before known or used."[16] An applicant was required to submit written specifications, drawings, and in certain cases a model of the invention to be patented. The law required that the invention be important and have utility. A fee of about $4.00 was charged.

Rumsey was not happy about the wording of the enacted patent legislation; writing Levi Hollingsworth on June 30, 1790, he lamented,

The Law of Congress respecting patent rights almost amounts to an Exclusion of my ever returning again to my own Country. If Judges should decide literally agreable to the Expression of that Law no Patent in America can be worth holding—according to that Law. The principles of no invention can be secured a man [illegible] rack his brain & spend his substance to discover & put into practice a Principle of great utility & for no better reason than because it can be varied into a different Shape it must be taken from him by such Plagiaries as Montgomery—If that proves to be the case I must seek a reward for whitening my hairs in a Strange land & shall bid adieu to (tho' my own) so inhospitable a[s] those to Science on which act. I shall at present suspend the sending forward my new Inventions & Improvements least they should be stolen from me by petty Projectors.[17]

Rumsey is expressing bitterness that Congress did not consider the concerns expressed in his letter to Jefferson, but as we shall note a bit later Jefferson himself did take heed.

For administration the law provided that a board, consisting of the secretary of state, secretary of war, and attorney general, should review the applications and issue patents for those deemed appropriate for a period not to exceed fourteen years.[18] The secretary of state, then Thomas Jefferson (see fig. 2.8), was the keeper of the records reviewing every patent application, and in practice headed the board later referred to as the board of patent commissioners. There were three patents issued in 1790: Samuel Hopkins for making pot and pearl ash; Joseph Simpson for the manufacture of candles; and Oliver Evans (see chapter 8) for manufacturing flour and meal.[19]

During the following year the patent applications for the steamboat inventors were addressed, but Congress was considering a modification to the patent law that would require an inventor to give up any rights granted by a state upon receiving a federal patent. Through their secretary the board advised each of the steamboat applicants of the delay. When the legislation failed to pass, the board addressed the steamboat petitions but was unable to resolve the relative priorities and merits of the designs of Rumsey, Fitch, Read, and Stevens. They ended up awarding patents on August 26, 1791, to all four, with conflicting claims.[20]

In general the new patent system did not work very well, partly because of the inadequacy of the law and partly because the members of the board had other pressing duties. Jefferson, an inventor himself, was diligent at least in not allowing later patents that conflicted with those already issued. The led to frustration on the part of many inventors who claimed that "the board was not in sympathy with those whom the law was intended to benefit, and they were by education and interest hostile to the industrial classes."[21] On February 21, 1793, the law was changed, eliminating the examination requirement and demanding little more than a registration of patents. The secretary of state was the granting authority, and the courts were left to resolve the disputes that arose.

In 1836, a third patent law was passed, providing for the system in place today—examining each application for newness, novelty,

Figure 2.8. Thomas Jefferson became secretary of state on September 26, 1789. He was the most influential member of the United States' first patent board, which consisted of three members. (From *Messages and Papers of the Presidents, 1789–1897*, 1:318.)

and conflicts with existing patents. As a result, a corps of patent examiners was established to provide this vital review function. It is this current approach that will later be examined to better understand what qualifies as an invention.

Why Does Invention Happen?

Another question that must be addressed at this point is why invention even happens. In examining the *why*, it is instructive to scrutinize one of the long-held but debatable premises that "necessity is the mother of invention." Necessity can be thought of in the sociobiological sense of addressing only the needs of air, water, food, shelter, and reproduction, or broadened perhaps by some to encompass, for example, the need for tools to produce food to satisfy hunger. Historians often argue, however, that the premise is false, because some inventions catch on without genuine need—for example, the automobile, which was invented at a time when there did not exist a shortage of horses and buggies.[22] They turn the phrase on its head, stating that *invention is the mother of necessity*, and argue that once something is invented those who seek to profit from the invention create a societal need through marketing. It seems fairly obvious that the laser (light amplification by stimulated emission of radiation), invented by Charles H. Townes and Arthur Leonard Schawlow and patented on March 22, 1960, falls into this category. For years it was an invention in search of an application; however, today its use is ubiquitous, found in survey equipment, wood and metal engraving machinery, ceiling tile installation tools, medical surgery instruments, and the compact disc player. The list goes on.

Beyond the basic needs of survival and procreation, the society in which we live also has other needs, influenced by ethnic diversity, language, geography, religion, and intellect. Using geography as an example, a need in one community may not exist in another: one village, located on gently sloping or flat terrain, may need animal power for grinding grain, whereas a similar community located near a hydraulic fall line may need water wheel technology to take advantage of the energy in falling water.

Another geographic, or stated more precisely, climatic factor, can be contrasted in colonial America between New England and the states of the south. With a longer growing season in the southern part of the Untied States, agricultural pursuits dominate most of the year, and the relatively short winter months were used to get ready for the next season. In New England, however, which has much longer winters, there was time available to pursue other activities, giving rise during the early federal period to the "Yankee tinkerers." These creative individuals designed and constructed useful mechanical devices that were sold throughout the country.

In a similar way, other social factors, as well as one's individual creative juices, affect the concept of need, and it is not the intent here to examine or quantify factors of religion, ethnicity, or even geography and climate. However, since social diversity creates different demands, those in a particular community who are creative will endeavor to innovate and sometimes invent new devices targeted at what they perceive as real needs. Sometimes the output is applicable only to a local need, for example, a cooper's gouge specifically adapted to the grain of the wood from a treee peculiar to the area; in other cases, such as the screw auger invented by William Henry for boring holes through lumber, a device finds universal application.

Sometimes a creative individual is driven to "invent" a novel device, but in the hands of the craftsman, who can benefit from its use, the object does not fit or integrate into the way he does things. An example of this is the universal screwdriver with multiple bits to fit any type and size of screw. These devices are popular with the handyman, who is never sure what device he will next encounter in need of repair, but the craftsman, working with a limited set of different screws, much prefers to use the specific tool designed for the particular job. When such novel devices do not find any or very limited application they are relegated to the bottom of the toolbox; in some cases they may even become museum artifacts, with curators asking visitors to try to identify their use. In this case the inventor foresees a need for which the invention was unsatisfactory, or perhaps for which the need was perceived but did not actually exist.

Most technologists are employed today by companies in the business of producing specific products, always with an eye to making them cheaper, better, or faster. The management of these corporations will challenge the workforce to come up with a better way of production, and it is the rare company that does not have a suggestion system with rewards for worthwhile ideas. Engineers are routinely charged with developing a new product with more features than the last one, in a smaller package, at a lower cost, with improved ease of assembly, and that will be easy to use and "never" fail either because of abuse by the customer or because of design flaws. In these cases, the need for the company to stay in business by outdistancing its competition propels necessity into the roll as mother of at least innovation and perhaps invention.

But some invention comes about as the result of mental gymnastics of an individual who has a built-in craving to create a new three-dimensional device, just as some are driven to solve a two-dimensional crossword puzzle. These individuals perceive a need that may or may not exist; they may in some cases be visionaries, or they may simply be people with an active mind full of stimulating thoughts. Back in the 1960s the author conceived of pause control for the automobile windshield wiper. The device had nothing to do with his job, but he was annoyed by continuous and sometimes noisy operation of wiper blades when there was little rain; an intermittent wiper control *seemed* like a good idea. Using a handful of electronic components, a prototype was constructed and installed on his automobile, and the design was published in a trade magazine. Not only did it allow for one to many seconds between wiping actions, but a second, usually preset, control allowed the wiping to be operative for one, two, or three cycles, just enough to make sure the windshield was cleared—a problem somewhat a function of the condition of the rubber in the wiping blades.

As the device was unrelated to the aerospace electronic work that employed the author, no invention disclosure was filed with the company and no patent protection pursued, but the design was published in an electronic trade magazine. Unbeknownst to the author, however, another inventor also created what was substantially the same device; about this same time, pause control first appeared on some automobiles. Many years later, the author received a telephone call from an attorney of one of the automobile manufacturers inquiring if the author's pause control work had significantly predated the article, as the company was involved in litigation with someone who claimed the invention and was probing for a better defense.

As it was some time before pause control was in widespread use, the historian studying this scenario would likely argue that invention was the mother of necessity, but the inventor could counter that he was a visionary who correctly foresaw the need that the marketplace later embraced.

In answering the question of *why* does invention happen, all of these scenarios must be considered. Necessity is undoubtedly the driving force in many cases, but not all of them. However, this statement does not satisfactorily address visionary inventors who are sometimes right, but can also be on the wrong track. Therefore a more operative *why* seems to be answered by the statement: *need, either perceived or real, is the mother of invention.* This definition widens the starting point of novelty considerably but still will not keep marketers or historians from turning the idea on its head to claim that invention is the mother of real or perceived need—the essence of the infomercial.

Patents Today

To obtain a U.S. patent, the inventor must be able to demonstrate in an application to the satisfaction of a patent examiner that

1. *The idea is new, novel, and first.* The submitted idea must be new and cannot have been published more than one year prior the application date. In addition, it must not be obvious to persons skilled in the art of the invention. It is customary to search the patent office files before incurring the expense of filing an application to make sure the idea has not already been patented. Public files, however, are those of awarded patents and not applications in process. Therefore it is possible to make a search and find nothing applicable, only to be advised later the same idea has already been filed by another individual.

2. *The idea can be described in practical terms.* To convey the invention or new idea to the patent examiner there must be a written description that generally takes the form of a diagram, where each of the components or parts are labeled and accompanied by an explanation of the function of each and how they work together. The parts that make up the invention must exist, or must at least be able to be made to exist, to the satisfaction of the examiner, else the idea is a dream or prophecy and not something for which a patent can be obtained. (It is this critical test that the invention must pass to become a patent.

Although working models may be submitted they are not generally required with the application unless the invention appears to violate the laws of science, such as in the case of perpetual motion machines. This frees the inventor from the sometimes prohibitive cost of reducing the idea to practice but does require that the inventor be grounded in the extant technology. Generally the inventor has a thorough understanding of the technology base, as it is the gestation and juxtaposition of this information that often produces the new idea.)

3. *Claims (rights of use) do not conflict with previously granted patents.* These are the legal rights to which the inventor can utilize the idea free from infringement by others for a specified period of time, usually seventeen years. It is this feature of a patent that protects the inventor and rewards her intellectual effort. Although the government provides legal protection against infringers, it is incumbent upon the inventor to enforce the protection.

Although a patent establishes the legal right for an inventor to perfect her invention without interference from others, it by no means ensures success. As the early steamboat inventors were to learn, there is a large leap of faith, time, and money between the idea and its practical realization. Sometimes a step in the development process is so difficult that it can only be made by a few individuals, or even one. The step itself might be the subject of a patent but would require revealing it to the world; hence, in these cases the inventor may keep the process as a trade secret, hoping no one else discovers the underlying principles. Such has been the case for the recipe for Coca-Cola.

Society and Invention

The foregoing is the federal government's approach to invention, but operating at the same time are societal beliefs regarding invention. Society has always been fascinated by stories such as James Watt watching his mother's teakettle boil and later inventing the steam engine (see fig. 2.9), or the overly simplistic story of Goodyear acci-

Figure 2.9. Young James Watt is depicted with the popular notion he invented the steam engine while watching his mother's teakettle boil. He did not invent the steam engine but did make a significant improvement in its operation by adapting a separate condenser. (From Frank P. Bachman, *Great Inventors and Their Inventions* [New York: American Book, 1918], p. 9.)

dentally spilling a compound on his stove and inventing rubber vulcanization. As far as can be reasonably determined, these are just stories with no foundation in fact, yet they capture the imagination of the public. The importance of society's role in shaping inventors should not be underestimated; unless an invention serves a public good, it is of little value and will be ignored or discarded. If it serves society well, the public will have a major say in who is to be given credit.

There has always existed some reluctance on the part of society to embrace new ideas. Some of this skepticism is healthy, as there have always been "snake-oil salesmen," who tout claims that fail to materialize. However, at times this skepticism went far beyond what is reasonable, as when Oliver Evans in the 1790s demonstrated his automated flour mill. The Brandywine millers exclaimed, "It will not do! it cannot do!! and it is impossible it should do!!!"[23] The early steamboat pioneers were similarly rebuffed; Fitch was considered "possessed," Rumsey was called "crazy," and Fulton was, of course, pursuing his "folly."

New ideas that have been integrated into the fabric of society in a reasonable period of time to the benefit of the inventor have been called *successful-invention*. The term is hyphenated to distinguish it from invention that did not serve the public need, either because of timing, utility, or practicality. Adapted from Russell Bourne, successful-invention has four characteristics that distinguish it: (1) it is a general combination of the skills of many people, producing a superior product; (2) it accomplishes a specific purpose for which it is promoted; (3) it is supported by capital throughout the development and marketing process; and (4) its introduction enjoys fortunate timing.[24] As an example, Fulton was able to borrow from, and improve on, the work of many individuals who preceded him, and using this knowledge along with the capital and political connections of his partner Robert R. Livingston, he introduced the steamboat in time to preserve a twenty-year monopoly (that was extended several times) on the waters of the state of New York. Everything fell into place for his *North River Steam Boat*, later called *Clermont*, especially the timing of its introduction, and passengers paid to travel on the boat between the cities of New York and Albany. He was therefore successful, and success brings fame and in his case recognition by the public that he was the inventor. He was certainly at the right place at the right time with the right product, but the question of whether or not he is the inventor is still debated.

How does an invention come about? The "flash of genius" or "moment of inspiration" is the popular belief, and while these phrases might somewhat accurately describe the *exact* moment of formulation of the invention, they fall far short of explaining the process.

An inventor is somehow *driven* to create something new and different. She studies all of the available information about a problem or idea until reaching a thorough grounding in the base of the technology and science surrounding the issue; she then starts to mull over a myriad of possible solutions. The majority of the solutions are dismissed in a few seconds or minutes as not being practical or feasible, but unrelentingly, the combination and permutations of ideas and thoughts continue to churn. This preoccupation pervades her time during work and play, while driving or going to sleep; and perhaps while sleeping, eating, and grooming, until a viable idea finally falls into place. When inventing a part of his automated flour mill called the "hopper boy," Oliver Evans faced this and wrote, "I persevered with a zeal and indefatigability peculiar only to inventors."[25]

Upon further reflection, the *apparently* good idea may also prove to be unworkable, but undeterred, the inventor continues the intellectual gymnastics. Sometimes the problem is unyielding, but at other times the dominoes suddenly topple, and the inventor knows instinctively she has stumbled, in Evans's words, "with infinite pleasure" onto something new and workable.[26]

This period of time, sometimes called immersed gestation, intellectual preoccupation, or intense introspection, has been examined by Elmer Sperry, inventor of the aircraft gyropilot. In a 1930 essay, "The Spirit of Invention in an Industrial Society," which appeared in the book *Towards Civilization*, Sperry called this gestation time "brooding meditation." Robert C. Post, a historian in the National Museum of History and Science, has aptly pointed out, "'Brooding meditation'— a haunting phrase, but not overly melodramatic, for this trait comes as close as any to being a universal among inventors. Indeed, it comes much closer to helping us frame useful questions as to what the inventive process is all about than the phrase 'a flash of genius.'"[27]

Then who is an inventor? She is someone who is thoroughly grounded in the science and the technology of the subject she is pursuing and who creates something new and novel from things that exist or can readily be made to exist. In bringing his *North River Steam Boat* or *Clermont* to working condition, Robert Fulton certainly understood the importance of immersing himself into the project. With perceptive insight he wrote, "Therefore the mechanic should sit down among levers, screws, wedges, wheels, etc. like a poet among the letters of the alphabet, considering them as the exhibition of his thoughts; in which a new arrangement transmits a new idea to the world."[28]

If the individual obtains the first patent for a new device, this document can serve as proof that she is entitled to the designation of "inventor," but society can and will nevertheless sometimes assign an inventor's name to a product based on its manufacture or distribution, which may be totally unrelated to who deserves the actual credit. For example, Elias Howe is actually the inventor of the sewing machine and as a result was able to win a patent infringement suit against Isaac Singer; Howe became rich by licensing a group of sewing machine manufacturers. Of these, Singer went on to become the czar of a sewing machine sales empire, with his name becoming synonymous with the product.[29]

Then there is the looming question: was the steamboat really invented, or was it just a normal, gradual, and inevitable evolution of small steps of technology leading to success? S. C. Gilfillan, a student of the social nature of new ideas, argues, "Probably no one has ever been able to invent anything of large importance, in the history of the ship. Everywhere we have seen, or been able to guess at a gradual evolution [i.e., innovation], or accumulation of quite little steps, never a revolutionary one, such as the 'invention of the steamship' would have been, had it ever taken place. That is understanding the word 'invention' to mean a catastrophic, sudden beginning."[30] To this Dorothy Gregg, in her PhD dissertation on John Stevens, observes, "The more knowledge we gather about the social nature of invention, the less impressive are the claims to inventive originality of such individuals as John Fitch, James Rumsey, Oliver Evans, Robert Fulton, and John Stevens."[31]

Since the invention of a complicated machine can only flow from advancements in science and technology, which usually take place by means of a gradual accumulation of knowledge over a period of time, if one were to follow Gilfillan's argument there would be no invention—only innovation of varying sizes. The line of reasoning, however, is wrong: technologists, as well as the framers of the Constitution and today's worldwide patent offices, recognize that invention does exist, and economists compute the role it plays in the economic development of a nation. Historians and philosophers may argue otherwise, but they are not inventors, who know the secret joy and the exuberant thrill of being the first person *ever* in the world to create something truly new. When that moment arrives, it is an overpowering rush that only the inventor can ever know (see fig. 2.10). Some may stand on the sideline and watch or

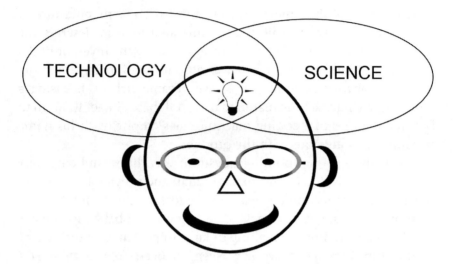

Figure 2.10. "Eureka!" or "Aha!" The mental manipulation of technology with persistence aided by the laws of science can sometimes produce a new and useful Idea. (Illustration by Jack L. Shagena.)

even claim that invention does not exist, but the individual who has created the new device certainly knows otherwise.

Historian Brook Hindle unambiguously states, "The steamboat was indeed something new under the sun,"[32] and that premise is adopted here. Prior to Fulton's successful demonstration, the steamboat had been prophesied by a number of individuals and had been worked on in England, Scotland, France, and perhaps Germany and Italy, but it was in America, isolated from the world's leading science and technology, that it first came into use. Most likely, one or more of these American steamboat pioneers deserve credit for the invention—hence the subject of this work.

Therefore the title inventor may be recognized by the *government* for a viable, technology-based, scientifically sound, new, and novel idea, as well as by *society* for a device that captures the public's imagination and serves a social need or want.

Claimants to the title of inventor of the steamboat have been made either by or for several individuals, with Robert Fulton the popular public choice. Regarding this, historian Robert C. Post states that

"Robert Fulton didn't invent the steamboat. . . . Mythology has often been taken too literally. Mythmakers have been potent determinants of the historical record, and it is often futile if not actually misguided to try to divorce what people *believe* to have been so from what in some 'objective' sense *was* so."[33] Such is the perhaps insurmountable hurdle facing this author. Should this effort not result in supporting Fulton as the steamboat inventor, it might well be futile to come logically to another conclusion, as it would be at odds with popular belief. However, with a better understanding of the events leading up to the steamboat, and with the concept of invention itself being more fully understood, along with the passage of time, it may still be possible to change people's minds. However, as my dad used to skeptically say when one of my ideas seemed far-fetched, "We shall see."

Notes

1. James Flexner, *Steamboats Come True: American Inventors in Action* (1944; reprint, Boston: Little, Brown, 1978), p. 365.

2. In any organization or microcosm of society there is usually one or more individuals who instinctively know how to do things. So, when a question arises as to the procedure to accomplish a task, the general answer given will be, "Ask so-and-so."

3. The author is indebted to P. J. Woodall for this observation.

4. Brook Hindle, *Emulation and Invention* (New York: W. W. Norton, 1981), p. 117.

5. Although engineers who invent often have large egos, they do not try to deceive their peers, as they know instinctively or have learned through training or practical experience—you can't fool Mother Nature. Her laws must be obeyed, and although one might occasionally deceive a coworker, the truth will eventually come out. Hence the only rational and logical policy is one of complete honesty. Since most engineers work for large companies, a preemployment agreement assigns any patent to the employer, and although the patent is issued in the inventor's name, except for through prestige, it seldom brings significant personal monetary reward.

6. Please see J. D. Bernal, *A History of Classical Physics* (1972; reprint, New York: Barnes & Noble, 1977), p. 37. Table I on page 37 has an out-

standing summary of the development of science. Also refer to the chapter starting on page 240 for the development of the steam engine.

7. Ray Palmer Baker, *Engineering Education* (New York: John Wiley & Sons, 1928). The reader's attention is drawn to a series of essays collected in this volume. This nontechnical book is planned for students of engineering and provides a diverse and enlightening discussion of science, engineering, and invention. It is a source for some of the ideas contained in this chapter.

8. For a discussion of the interrelationship of science, technology, and invention see Milton Kerker, "Science and the Steam Engine," in *The Development of Western Technology since 1500*, ed. Thomas Parke Hughes (New York: Macmillan, 1964), pp. 66–77.

9. See Eugene S. Ferguson, *Engineering and the Mind's Eye* (Cambridge, MA: MIT Press, 1993). Mr. Ferguson graciously provided the author a copy of this insightful book.

10. The author is indebted to Thomas O. Flink, who described this clever technique he had observed being used by the "Resident Inventor" (believed to be Joseph J. Albert) at the Martin Company in Baltimore, Maryland, around 1956.

11. Travis Brown, *Historical First Patents: The First United States Patent for Many Everyday Things* (Metuchen, NJ: Scarecrow Press, 1994), p. 88.

12. U.S. Constitution, art. 1, sec. 8, cited in Gustavus A. Weber, *The Patent Office: Its History, Activities and Organization* (Baltimore: The Johns Hopkins University Press, 1924), p. 1n.

13. *Papers of Thomas Jefferson*, ed. Julian P. Boyd, 20 vols. (Princeton, NJ: Princeton University Press, 1952), 15:171.

14. Quoted in Weber, *The Patent Office*, p. 3.

15. See Archibald Douglass Turnbull, *John Stevens: An American Record* (New York: Century, 1928), p. 108; and Dorothy Gregg, "The Exploitation of the Steamboat: The Case of Colonel John Stevens" (PhD diss., Columbia University, 1951), p. 112n.

16. Weber, *The Patent Office*, p. 4.

17. James Rumsey, Letter to Levi Hollingsworth, June 30, 1790. The author is indebted to the American Philosophical Society for providing a copy.

18. Weber, *The Patent Office*, p. 4.

19. L. Sprague de Camp, *The Heroic Age of American Invention* (Garden City, NY: Doubleday, 1961), p. 26.

20. Ella May Turner, *James Rumsey, Pioneer in Steam Navigation* (Scott-

dale, PA: Mennonite Publishing House, 1930), p. 133. Neither Rumsey nor Fitch was happy with the award of steamboat patents on the same date. Rumsey's lawyer and brother-in-law, Joseph Barnes, made an attempt to have Congress produce a patent law more favorable to inventors in 1792. A committee appointed in October 1791 drafted an amendment to the "Act to Promote the Progress of Useful Arts" and reported the results in March 1792. Barnes produced an impressively written, thirty-four-page pamphlet for consideration by the members of Congress, titled *Treatise on the Justice, Policy, and Utility of Establishing an Effectual System for Promoting the Progress of Useful Arts, by Assuring Property in the Products of Genius.* He identified two types of property, *local* and *mental*, where local is what we call real estate today, and mental we call intellectual property rights. His efforts were not heeded, and it would not be until 1836 before Congress would adopt patent legislation embodying some of his proposals.

21. Weber, *The Patent Office*, pp. 4–5.

22. George Basalla, *The Evolution of Technology* (Cambridge: Cambridge University Press, 1990), pp. 6–7.

23. Oliver Evans, *To His Counsel, Who Are Engaged in Defense of His Patent Rights for the Improvements He Has Invented* (1817), Milton S. Eisenhower Library, The Johns Hopkins University, microfilm S40778, p. 37.

24. Russell Bourne, *Invention in America* (Golden, CO: Fulcrum, 1996), p. 3.

25. Evans, *To His Counsel*, p. 7.

26. Ibid.

27. Robert C. Post, "The American Genius," in *The Smithsonian Book of Invention,* ed. Smithsonian editors (New York: W. W. Norton, 1978), p. 24.

28. Robert Fulton, *A Treatise on the Improvement of Canal Navigation; Exhibiting the Numerous Advantages to Be Derived from Small Canals* (London: I. & J. Taylor, 1796), p. x.

29. Mitchell Wilson, *American Science and Invention: a Pictorial History* (New York: Bonanza Books, 1954), pp. 132–34.

30. S. C. Gilfillan, *Inventing the Ship* (Chicago: Follett, 1935), p. 103.

31. Gregg, "The Exploitation of the Steamboat," p. 7.

32. Hindle, *Emulation and Invention*, p. 55.

33. Post, "The American Genius," p. 23.

Figure 3.1. Fulton's *North River Steam Boat* is shown in the upper image; in the lower are depicted the vessel's steam engine, boiler, pumps, gearing, and paddle wheels. The satisfactory meshing and interaction of many parts, some complicated, was necessary to achieve harmony and acceptable performance in the ensemble. (From Robert H. Thurston, *History of the Growth of the Steam-Engine* [New York: D. Appleton, 1901], p. 258.)

3.

The Steamboat as a System

If a man can write a better book, preach a better sermon, or make a better mousetrap than his neighbor, though he builds his house in the woods, the world will make a beaten path to his door.

—Attributed to Ralph Waldo Emerson

Less than a decade following the Revolutionary War, the creative genius or "Yankee ingenuity" of Americans was unleashed, and individuals set out to build a steamboat. It was, they reasoned, just a matter of constructing and assembling the right components and connecting them together, and success would follow. They had no historical precedent to emulate, since development of the steamboat was the *first* time ever a fuel-carrying, self-sustained craft (or anything like it) would be attempted. The steamboat was to be the world's very first *transportation system,* and the inventors completely failed, even remotely, to comprehend the significant complexities of their undertakings.

Why was this task such a giant leap in difficulty? After all, the

English firm of Boulton and Watt had already developed a steam engine, and spinning mills were being built throughout the United Kingdom. Further shipbuilding was not a new but rather an ancient skill, so it appeared to these early inventors that it should not be that hard. Clearly this was not the case, though, as after the pioneering work of Rumsey and Fitch, it would be more than twenty years before Robert Fulton would demonstrate his first commercially successful steamboat in 1807. What was missing?

The answer is, quite simply, the concept of systems development and the understanding of the interaction of a multiplicity of components where alone each entailed considerable complexity. These had to be integrated into working order under the heretofore unencountered constraints of *efficiency*, *reliability*, and *weight*. Although efficiency and reliability had already been factors in the design of stationary devices, the added constraint of weight and overall system reliability would challenge the creative resourcefulness of American inventors and developers.

Engineering as such, much less *systems engineering*, hardly existed as a discipline: the underlying scientific foundation had only been partially formulated, and the application of this knowledge to components was in an embryonic stage, and to systems, was completely nonexistent. The most skilled metalworkers of the day in America were blacksmiths (see fig. 3.2), and the most advanced artisans, the clockmakers. America at the close of the Revolutionary War was technologically challenged and would remain so for many years.

To better comprehend the difficulties of the task and the state of technology at the time, it is instructive to jump forward a little over two decades to the year 1803. Benjamin H. Latrobe, a highly respected engineer (see fig. 3.3), presented a paper to the American Philosophical Society of Philadelphia on May 20 stating his objections to the use of steam for impelling boats.[1] His concerns fell into three broad areas:

- *Practicality.* The weight and volume of the engine, fuel, and equipment left little capacity to accommodate passengers or cargo.

Figure 3.2. During the latter part of the eighteenth century, blacksmiths were the most skilled metal fabrication artisans in America. (From *Blacksmith Shop & Iron Forging* [1906; reprint, Bradley, IL: Lindsay Publications, 1983], p. 57-17.)

- *Safety.* During periods of rough weather, resulting vessel motion, along with vibrations caused by the irregularity of machinery motion, portended an unsafe environment.
- *Reliability.* Equipment failures would be frequent, and the operation and maintenance cost-prohibitive.

It was Latrobe's "sincere and well founded belief" that steam navigation was "beyond the realm of possibility in any practical sense." He recognized that his opinion would offend "many very respected and ingenious men" but urged them to "turn their talents to other fields which

Figure 3.3. Benjamin Henry Latrobe, responding to an inquiry to the American Philosophical Society about the status of steamboats in the United States in 1803, wrote that steam navigation was not practical. He seriously misjudged the ingenuity of Americans. (Rendered by William T. Sisson from a portrait by Charles Wilson Peale.)

perhaps more greatly benefit the people at large."[2] Latrobe, of course, was wrong, but in realizing this view was presented by a respected engineer about twenty years after work had started on the American steamboat, we can see that the difficulties and ridicule the pioneering steamboat engineers faced was certainly real and significant.

How could anyone have expected to succeed? Part of the answer lies in the perceived pressing national need; part in the willingness, despite frustrations, to undergo years of trial and error experimentation; but more importantly the innate belief of the inventors that they themselves, often misunderstood, could make a difference. They were driven individuals, envious of success, but not dismayed or deterred by failure. Frustration, however, was apparent, as evidenced by Fitch: "I confess the thought of a steamboat, which struck me by mere accident, about the middle of April 1785, has hitherto been very unfortunate to me; the perplexities and embarrassments through which it has caused me to wade, far exceed any thing, that the common course of life ever presented to my view."[3]

He was not alone, as Rumsey underwent a similar experience, described in his steamboat pamphlet:

> Those who have had the good fortune to discover a new machine, or to make any material improvement on such as have already discovered, must lay their account to encounter innumerable difficulties; they must arm themselves with patience to abide disappointments; to correct a thousand imperfections (which the trying hand of experience alone can point out) to endure the smarting shafts of wit, and what is perhaps more intolerable than all the rest put together (on the least failure of any experiments) to bear up against the heavy abuse and bitter scoffs of ill-tempered ignorance. . . . Happy for him if he escapes with so gentle an appellative as that of a madman.[4]

In addition, even James Watt in the early days of his steam engine work lamented, "Of all things in Life there is nothing more foolish than inventing."[5] Clearly the steamboat—and invention itself— were not easy.

Introduction to Systems Engineering

To understand the source of the failure and frustration of the early steam developers, it is instructive to understand the basic concept of *systems engineering* by examining a hypothetical mousetrap design problem. Thereafter, steam navigation can be revisited with a fresh and enlightened appreciation of the significant difficulties faced by the steamboat pioneers.

In the most basic sense, any system, be it simple or complex, can be represented by a single block diagram that has one or more inputs that undergo some process, from which one or more outputs are produced. A very easy-to-understand systems process is found in the preparation of food, using a recipe such as the baking of bread. The ingredients of flour, water, yeast, and shortening are the inputs; the mixing, kneading, waiting for the dough to rise, and baking is the process; and the loaf of bread, along with the never-mentioned dirty utensils, are the outputs.

Whereas cooking using a recipe and setting a mousetrap are such simple activities as not to require further explanation, development of a *new* recipe or design of a *new* mousetrap might. Therefore we will use the design of a mousetrap to illustrate the systems engineering process. The top-level block diagram is shown in figure 3.4. This is a relatively straightforward problem, but knowledge gained of the underlying systems process will yield insight into more complicated systems.

The most important basic results of a systems engineering approach is to break a complicated project into manageable elements that are dealt with in a step-wise and methodical process, then assembled (integrated) into a whole, working system. So with this in mind, we will proceed through the *essence* of the systems engineering approach using an easy-to-understand and humorous example. Afterward we will scrutinize the entire systems process with

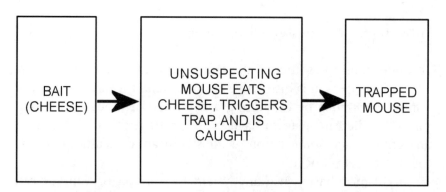

INPUT　　　　　　**PROCESS**　　　　　　**OUTPUT**

| BAIT (CHEESE) | → | UNSUSPECTING MOUSE EATS CHEESE, TRIGGERS TRAP, AND IS CAUGHT | → | TRAPPED MOUSE |

Figure 3.4. Top-level systems diagram for catching a mouse. Any engineering task, regardless of complexity, can be represented by such a diagram, which has inputs, a process, and outputs. Once this is done, the challenge is to break the task into manageable parts. (Diagram by Jack L. Shagena.)

enough (but hopefully not too much) detail, so as to establish the requisite background and understanding to appreciate the process being applied to a steamboat.

A Better Mousetrap

Let's say you the reader have an idea on how to build a better mouse-trap. You have thought about this for some time and plan to coat the trigger mechanism with the slippery compound called Teflon, and as a result you expect to catch a mouse more often saving money on the cost of cheese. You have heard the old saying "build a better mousetrap and the world will beat a path to your door," and you secretly anticipate that this improvement will bring you success and hoped-for riches. Also, you have been introduced to the systems engineering process that you plan to use in producing your new design. You sketch out the three-phase methodology shown in figure 3.5 and facetiously show the tripped mousetrap with the cheese gone, but you fully anticipate success.

Here is how the model works:

Concept phase. In the concept phase the *idea* of how to build a better mousetrap is germinated, that is, to coat the trigger mecha-nism with something very slippery like the material found on cooking utensils. Also as part of this phase, and derived from the existing mousetrap technology base, you break your design into a number of individual *pieces.* In engineering terminology this is called the *work breakdown structure.* When doing this it is good judg-ment to utilize proven approaches and techniques and not try to redesign the entire mousetrap from scratch. Experienced new-product designer engineers usually subscribe to the maxim "if it ain't broke, don't fix it"; the reasons for this will become apparent later. The work breakdown structure information generated in the concept phase is then passed along to the development phase.

Development phase. The pieces that comprise the design are then drawn or sketched, specifications prepared for the materials and processes to be used, and the drawings and instructions are sent to the model shop for fabrication. The shop, however, misunderstands

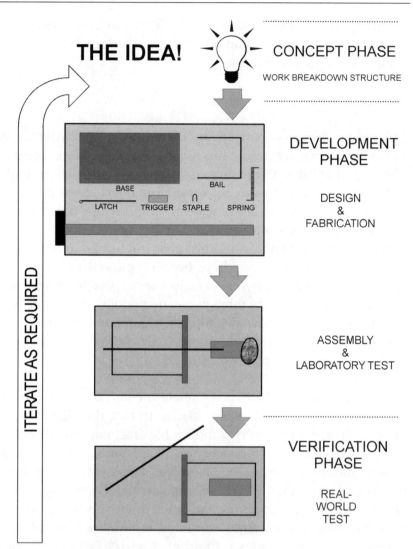

THE IDEA!   CONCEPT PHASE

WORK BREAKDOWN STRUCTURE

DEVELOPMENT
PHASE

BASE   BAIL

LATCH   TRIGGER STAPLE   SPRING

DESIGN
&
FABRICATION

ITERATE AS REQUIRED

ASSEMBLY
&
LABORATORY TEST

VERIFICATION
PHASE

REAL-
WORLD
TEST

Figure 3.5. Mousetrap Product Cycle. Adapted to a mousetrap, this is a simplified, logical, step-wise methodology with feedback that has proven effective in developing successful devices. (Diagram by Jack L. Shagena.)

one of the specifications and makes the spring too weak, so you have to modify the paperwork and have it made again. Also, they have a number of questions about other parts that several telephone calls and a visit finally resolve.

At last all the fabricated components are ready and they are assembled or *integrated* in the completed trap. Unfortunately in the process you discover the staples holding down the spring are not long enough and keep pulling out of the wooden base. You modify the drawing and have them remade, but you believe the shop should have made the base from a denser wood in order to avoid this problem. In any case the oversight is resolved with the new staples and the device is finally ready for *laboratory testing*. You set the trap and with a small wooden stick attached to a measuring gauge find the force to activate the trigger has been reduced by 50 percent over conventional mousetraps. After several more tests, you are convinced the device is working as planned and is ready to catch a mouse.

Verification phase. In the real-world test, or verification phase, the mousetrap, codenamed "Slippery Grabber," is baited with cheese, the bail is set, and the apparatus is positioned in a strategic place to await its unsuspecting victim. This is where, as engineers say, "the rubber meets the road." While contemplating positive results, at least for the designer, you attend to cleaning up the mess you made while constructing the mousetrap and put all of your tools away in full anticipation of success. There is a resounding snap, and you hurry back to the mousetrap's location just in time to see the mouse scampering away with the cheese. You are about to learn the most powerful concept in engineering. It is called *feedback*.

Iteration of process. If you have been at this for some time, you realize that seldom does success come on the first try, so the engineering methodology builds in a way to point you back to the beginning of the process, called iteration. In this way, lessons learned can be incorporated into future designs, which hopefully are more successful. Oh, you have already learned about iteration, haven't you? The spring, followed by the staples, needed modification, but that was the fault of the model shop, not the design—or was it? Now that the final product failed its first real-world test, this is clearly the responsibility of the designer—or is it?

Upon analysis you discover that the release mechanism has gotten stuck and that the trap did not spring as anticipated. Further investigation reveals that the Teflon had peeled off the latch, so your first thought is that the shop did not apply it correctly. . . . Perhaps. Remember, you used Teflon because it was very slippery and nothing would stick to it. So that being true, it follows that Teflon cannot stick to another surface, such as the metal trigger mechanism. Did the specification provide for any special surface treatment of the trigger before application of the Teflon? I thought so! But in any case, you are the responsible engineer, so as they say in your profession, it's "back to the drawing board" for another try. Such is the life of an engineer designing new products.

Another problem has arisen. There seem to be two different groups of protesters outside your work area. The word about your new mousetrap has leaked out, and one group, you presume a "bunch of liberals," is carrying signs reading BAN STEEL TRAPS and chanting, "Mice have rights, too." The other group, apparently "greens," appears worried that the wood base will impact the environment, because they are carrying a banner proclaiming, YOU ARE DESTROYING THE OZONE BY DEPLETING THE AMAZON FOREST. Nonsense! you think to yourself, they don't understand, and you decide to ignore them and open your mail. There is one letter from an individual who claims that applying Teflon to the mousetrap's trigger mechanism violates a patent he has for door latches and wants your patent attorney to call his lawyer. No one ever said it would be easy!

Why couldn't the mousetrap design have been done right the first time? Engineering managers have been asking that question for decades, if not centuries, and despite analysis, simulations, material process control, and testing, Mother Nature always has a few surprises. So engineers strive for perfection, but if success doesn't happen on the first try, they don't fight it. They simply learn that from failure comes success, and that's just the way it is!

The Real-World Systems Engineering Process

In the mousetrap example, three distinct engineering phases were identified: concept, development (includes design and integration), and verification, where the rubber meets the road. We see these again in the more complete systems (product) engineering cycle, shown in figure 3.6. The model is generic, but the descriptive terminology has been tailored to reflect the development of a complex hardware system such as a steamboat versus a product such as a mousetrap. Let us look at each phase again, but this time with a view of a much more complex system, and interject the approaches used by the early steamboat proponents.

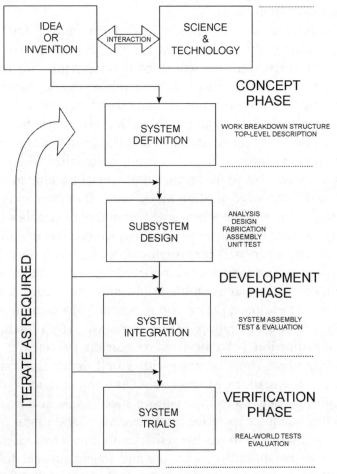

Figure 3.6. Systems (product) engineering cycle. This more generic and complete engineering development methodology, though overly complex for a simple device, is essential for complicated projects developed today, such as a new ship. (Diagram by Jack L. Shagena.)

Concept Phase

Intellectual horsepower, concentrated thought, or perhaps an inspiration or hunch initiates this phase, when an invention or new idea germinates. To be practical as previously discussed, the idea must have foundation in both *science* and *technology*. For if it cannot be done because of the laws of nature, or if the materials or processes do not exist to support the effort, it is just wishful thinking, not an invention or useful idea.

Sometimes a new idea dies at the concept point, as further reflection concludes that it would strain the technology base and is deemed too risky or impractical for development. This reasoning, although true at the time of early steamboat development, did not deter American inventors but certainly dampened English efforts. In other situations, the capital to proceed simply cannot be organized, or is available in limited amounts and not forthcoming when optimistic performance predictions do not materialize. The steamboat inventors were able to find sufficient financial backing to move forward to the systems definition block, but still in the concept phase.

Here the steamboat is broken down into manageable subsystems through a work breakdown structure, which consists of a number of work elements necessary for system success. Each is further described at a top level. To produce a steamboat, the inventor had to envision, design, construct, and assemble or integrate the following ten components: (1) a furnace, (2) a steam generator, (3) a vacuum generator, (4) an air (vacuum) pump, (5) motive power, (6) a forcing pump, (7) a transmission, (8) a means of propulsion, (9) steering, and (10) a floating vessel (boat or ship) with a hull design that can handle weight and vibrations of the equipment and minimize resistance through the water so as to maximize speed while at the same time providing adequate space for crew, passengers, and cargo.

All of these had to work together as an integrated system and be capable of reliably moving people and freight on schedule over a prescribed course at a speed and cost competitive with other modes of transportation. When viewed in this context, the task starts to look daunting, and after understanding the state of the contemporary science, technology, and component development, it starts to appear impossible. Indeed, for the American steamboat pioneers a practical solution was not to be introduced for another two decades. The top-level description of each of the work breakdown structure elements can be found in the paragraphs that follow.

1. *Furnace.* To generate steam, a source of energy had to be consumed in a metal or masonry vessel so as not to endanger the ship's wooden hull. Fitch initially used bricks for heat insulation for the furnace, but they were found to be too bulky and weigh too much. Rumsey and later Fitch employed furnaces and smoke-stacks made of iron.

Both wood and coal were used for fuel, but coal had the advantage of producing about twice the amount of heat for a given weight; hence, it was the better choice, as the saved weight could support passengers or cargo. Fulton used coal on his first speed evaluation trip to Albany, then switched to wood. In the latter part of the eighteenth century, wood was generally more readily available, and since the steamboats made frequent stops on inland waterways for passengers and freight, fuel could also be boarded at the same time.

2. *Steam generator.* Existing boilers employed for stationary steam engines (see fig. 3.7) were generally metal containers (pot boilers) encased with bricks sitting on a sturdy masonry foundation. They require a great deal of fuel to generate a moderate amount of steam; hence, the efficiency of these devices was low, their weight was heavy, and their integration onto a wooden vessel difficult because of the danger of setting the boat on fire. One of Fitch's boats actually caught fire during the night when the boiler fire was thought to be extinguished, but was not. Deliberate sinking into the river saved the vessel.

Because pot boilers were large and heavy, it was desirable to develop a new type of steam generator. Rumsey, recognizing this, invented the water tube boiler, which was later incorporated by Fitch and Stevens; this device eventually became the standard for the steam generating industry. It was the key to introducing high-pressure steam engines, but boiler technology moved slowly, and their explosions were the fear of steamboat owners, operators, and passengers well into the mid-nineteenth century.

3. *Vacuum generator.* Early steam engines worked by first moving

Figure 3.7. A Newcomen mine pump steam engine shown on Plate 37 in J. T. Desaguliers's 1744 book, *A Course of Experimental Philosophy*. James Rumsey had a copy of this book, which no doubt stimulated his fertile mind on steam engine construction. Note the man standing beside the brick structure to appreciate the size of the engine and supporting frame. (This copy of the illustration is from Frank P. Bachman, *Great Inventors and Their Inventions* [New York: American Book, 1918], p. 14.)

a piston in a cylinder one direction by the force of steam, then causing the steam to condense by cooling the cylinder, whereby a vacuum would be produced, allowing the atmospheric pressure to push the piston back to its starting point. Steam engine developer James Watt recognized the inherent inefficiency in this alternately heating and cooling of the cylinder itself, and in 1769 invented a separate vessel, called a condenser, in which to achieve the vacuum (see fig. 3.8).[6] Later, in 1781, Watt would patent the double-acting steam engine, which used steam to push the piston in both directions, but it would be three more years before the required slide-valve mechanism was developed and installed on a steam engine. American steamboat pioneers were aware of the separate condenser, and this approach was initially used, but achieving a proper design was frustrating, especially to Fitch.

4. *Air (vacuum) pump.* To get the steam quickly from the cylinder of the engine to the vacuum generator, it was necessary to use an air pump to create partial pressure in the condenser. The power to drive the pump was mechanically coupled from the engine. Although

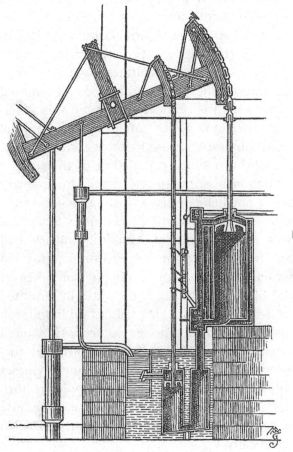

Figure 3.8. James Watt made a major improvement in steam engine efficiency and performance with his invention of the separate condenser. The condenser is shown in a cold-water bath at the bottom, adjacent to the right-side structure, with the air pump to the left. For an explanation of how his engine worked see Robert H. Thurston, *A History of the Growth of the Steam-Engine* (New York: D. Appleton, 1901), pp. 98–99. (From Bachman, *Great Inventors and Their Inventions*, p. 18.)

pumps had been used since the early part of the eighteenth century to remove water from mines, vacuum pump technology was not as well understood. As its capacity was related to the size of the steam cylinder and condenser, it did not yield immediately to an intuitive solution. The steamboat builders therefore solved the design problem through a lengthy trial-and-error approach.

5. *Motive power.* The earliest and only source of mobile, fueled power for one hundred years was the steam engine. It propelled the steamboats, the railroad locomotive, and the early automobiles. Like the pot boilers, existing engines were stationary, large, and heavy, and they produced limited horsepower. The weight-to-horse-power ratio (a figure of merit for mobility efficiency) of these was not suitable for steamboats, and improved, lighter-weight engines were required (see appendix B for more on this problem). As no suitable engine existed in America, the steamboat pioneers had to invent their own and were frustrated by the same development delays and failures that James Watt had encountered years earlier.

Though simple in concept, the mechanical precision required for effective operation between the cylinder and piston was generally beyond the available technology for American builders. The English company of Boulton and Watt used almost exclusively the firm of John Wilkerson Ironworks for production of their steam engine cylinders for a period of some twenty years. This eminent caster was able to achieve adequate precision by employing a technique he had developed for boring military cannons.[7] Since this milling procedure was not available in America, Rumsey had approximately thirteen-inch, lightweight copper cylinders formed around wooden mandrels—one for steam and one for water jet propulsion. He dripped water onto the top of the vertically mounted pistons to keep the leather or oakum packing tight to aid in the seal. The cylinders and imperfect water tube boiler worked well enough for two public demonstrations in December 1787. Fitch was successful in having a twelve-inch and later an eighteen-inch cylinder cast and bored. Initially he mounted the twelve-inch cylinder horizontally to save space, but he discovered the pistons leaked because of poor seals aggravated by gravity and found it necessary to remount the cylinder vertically.

6. *Forcing pump.* To provide boiler water to replace that expended through loss of steam, a forcing pump was employed. The action of the steam cylinder's piston was mechanically coupled to the pump to provide the work necessary to inject the water. This kind of device was fairly well known, as similar pumps had been used in the earliest steam engines to remove water from mines. In addition, a second forcing pump was required to deliver cold water for effective condenser operation. Rumsey cleverly diverted a portion of his jet propulsion water through the condenser, exhausting the warmed water overboard (see appendix A for details).

7. *Transmission.* Getting the output of the steam engine efficiently to the vessel's propulsion component also represented a difficult engineering challenge for many of the steamboat designers. Fitch noted the significant loss in his boatside paddle mechanism and later reported to the stockholders that his partner, Henry Voight, had made a significant improvement in reducing friction. Fulton's coupling mechanism appears overly complex and may have contributed to the boat's relatively poor speed performance for its weight and motive power (again, see appendix B).

Although this step lends itself to a visual, intuitive approach, the resulting mechanism should not rob the engine of power. It was in this area that Rumsey's genius excelled: the transmission was simply a single rod connecting the steam cylinder to the water-pump cylinder.

8. *Propulsion.* This was the area where steamboat inventors most diverged. Taking examples from nature and extant techniques, some went with a replica of a goose's or duck's foot; others, such as Nicholas J. Roosevelt, insisted on "throwing wheels over the side"[8]; and Fitch experimented with chain and paddles, side paddles, and stern paddles. Rumsey was initially alone in employing the water jet principle, although Fitch later considered but abandoned the approach, which had been suggested by Benjamin Franklin.[9] Rumsey most likely used jet propulsion because of his incorrect assumption that the paddle wheel was very inefficient; he may have been influenced by Franklin but claims not to have been. Although the jet principle was sound, and during the latter part of the twentieth century was to be employed on personal watercraft (jet skis)

and ferry shuttles, it was an idea that could not be supported by the technology of the day and as such was the weakest element in Rumsey's design.

9. *Steering.* Here a rudder with a tiller was employed, as it had been proven over years of practical experience to be effective (see fig. 3.9). Despite this, on one of Fitch's boats the rudder proved inadequate, resulting in the steamboat running aground. In general, however, this approach worked satisfactorily.

10. *Vessel.* Shipbuilders were to be found along the coast and inland waterways, as many had come from England to America initially to construct ships for the mother country. Since there was a good supply of wood and capable shipbuilders, it was not difficult to have a hull made, and except for the weight, this component posed the least problem to the steamboat builders.

The early machinery did produce significant stress on the ship; unfortunately for Fulton, the hull of his docked 1803 steamboat,

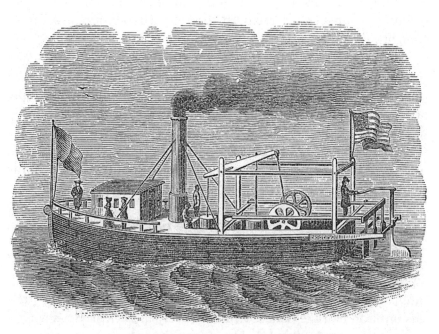

Figure 3.9. The rudder steered by a tiller was the device of choice to control the direction of small ships as well as steamboats. Shown is Fitch's 1788 steamboat. (From Thurston, *History of the Growth of the Steam-Engine,* p. 239.)

constructed in France, broke in half during a storm. The equipment was recovered and a stronger boat was built. Fulton particularly was aware of the drag of an improperly designed vessel and chose to make his *North River Steam Boat* very narrow and long to minimize the power necessary for a given speed. In addition, both Rumsey and Fitch had some knowledge of efficient hull design.

Now that the top-level requirements of the subsystems have been described, these are passed on to the development phase.

Development Phase

Using the work breakdown structure flowed down along with the overall requirements from system definition, each element or subsystem must undergo *analysis, design, fabrication, assembly,* and *unit testing* before proceeding to systems integration. This is where engineering really begins, so each of the steps will be examined.

Analysis. The analysis task is to apply the existing knowledge of science against both the system and each subsystem to ascertain from a theoretical standpoint the best possible performance that can be expected, accounting for losses and inefficiencies. After completion this information will serve as the baseline against which to measure actual versus predicted performance. It is in this area that steamboat pioneer Rumsey was far ahead of his contemporaries. He was familiar with some of the basic science and had, for his time, a unique ability to visualize the final product.

Design. In this step, drawings and specifications are prepared for the components of each subsystem and include the materials, processes, and procedures the fabricator is to use to produce the parts. It is interesting that successful early designers such as Fulton and Latrobe were also trained as artists and thus had a superior ability to produce accurate drawings. This provided better communication to the fabricator, leading to fewer errors in the produced parts.

Fabrication. At this point the producer of the designed components starts "bending metal." Since it is likely the designer has specified one or more steps in terms unfamiliar to the fabricator, communication is essential to ensure that the final parts meet the requirements. During the late part of the eighteenth century, nearly all communication with fabricators was verbal. Fulton's order from Boulton and Watt for his steam engine was a notable exception.

Assembly. Upon completion of fabrication, parts are put together, creating the completed subassembly. As it is likely that one or more parts will need adjustment, this step usually takes longer than initially anticipated. The engineer already knows this will happen, but the engineering manager or financial backer is eternally optimistic that everything will "go right the first time around."

Unit testing. Here the completed subassembly is tested against the predicted results of the analysis step above. Often, this requires that test fixtures be made and that instrumentation be connected. If the results cannot be explained and will to some degree adversely affect the performance of the overall system, redesign will be required, resulting in retesting. This is a critical and necessary step that was nearly always bypassed by the steamboat pioneers. Still in the development phase, we now proceed to *system integration,* which consists of assembly as well as test and evaluation.

System assembly (integration). Upon satisfactory completion of all of the subsystem tests, the units are ready for integration into the final system. The process is one of successively putting the units together and retesting them to ensure they function as intended. As the developer has most likely uncovered problems in each subsystem, it is almost certain there will be surprises at the integration stage. In some cases this will require redesign of one or more of the subsystems, with resulting retesting.

Test and evaluation. In the final stage of the development phase, the final integrated product is tested prior to public demonstration. For the steamboat pioneers this was usually the first time many of the subsystems were tested; not understanding the process, they simply bypassed much of the preparatory work. Hence, the anticipated performance did not occur, causing the developers to feel dejected and to lose the confidence of the financial backers. Eventually, however, through luck and pluck, both Rumsey and Fitch did succeed to some degree.

Once the integration is complete and sufficient testing has been done, the developer is ready to proceed to system trials in the *verification phase.*

Verification Phase

At this point the system is subjected to evaluation in the environment for which it was designed and scrutinized by the marketplace. It becomes almost trivial to predict that there will be more surprises, resulting in a reiteration of the process. These iterations are repeated with improvements being made until the product is satisfactory or, when the money runs out, it is abandoned.

Fulton had proceeded very carefully with the design and integration of his *North River Steam Boat* and made a successful introduction of steamboat service on the Hudson River on September 3, 1807. He later found it necessary to protect the paddle wheel with cages to keep other boats from inflicting damage, and because of steering difficulties he improved the rudder by substituting a wheel with ropes for the tiller. This, however, was only a minor part of the boat's verification phase story.

When the *North River Steam Boat* was taken out of service in November 1807 because of ice, Fulton notified Livingston that many improvements were necessary to correct what he described as a "cranky" boat. His partner reluctantly went along, and over the winter the vessel was completely rebuilt, adding sixteen feet to the length and five feet to the width; while strengthening the structure, the weight increased by a whopping 80 percent! This new design necessitated a re-registration, with Fulton now including the names of both owners and calling their boat the *North River of Clermont*. Neither biographers nor historians have recognized this rebuilding and restructuring of the first boat as direct fallout from verification phase testing. These improvements, however, stand as testimony of Fulton's determination to be successful.

In a real sense, the verification phase testing continues throughout the life cycle of the system. As problems are uncovered, solutions are implemented and the system becomes more robust and capable of sustaining the rigors of the real-world environment.

When writing his autobiography, John Fitch included a prologue to the steamboat section, in which the complexity of the steamboat development task led him to question his own sanity. He concluded that he was not a "lunatic" and in a backhanded way, more or less, spiritually came to grips with the concept of systems engineering.

With emphasis added, he wrote, "What I call Lunacy is a train of deranged unconnected Ideas. It is well known that a Steam Engine is a Complicated Machine, and to make that and connect it with the works for propelling a boat must take *a long train of Ideas and them all connected, and no one part of them disjointed,* for the Laws of God are so positive that the greatest favorite of heaven would not succeed contrary to the fixed Laws of Nature, no sooner than the most pro-fain [*sic*] sinner."[10]

It is also unfortunately true that as soon as the development is complete, the system is technologically obsolete. Invariably, engineers and artisans will find better ways to accomplish each function, but most likely these improvements will have to await the next new development cycle before being adopted. Fulton was thoroughly aware of this, and each new steamboat he built was continually improved using past lessons learned. He and Livingston had the financial resources required to launch the first steamboat, and with profits garnered through the New York state monopoly, he was able "plow the money" into new and better designs. The financial factor, so critical to development success but often overlooked, is discussed below.

Development Capital

Though generally not considered part of the systems engineering process, at least by the engineer, one of the perplexing problems for all inventors since the onset of development has been obtaining the necessary capital to convey a concept to practical reality, thereby garnering profits. James Watt eventually found a wealthy and understanding partner in Matthew Boulton; for John Fitch, a reluctant consortium of middle-class investors provided money and at times interference; for James Rumsey, it was Dr. James McMechen, followed by the Rumseian Society and lastly unreliable investors in London; and for Robert Fulton, the wealthy and politically connected Chancellor Livingston.

At all stages of the development of the steamboat, however, there were setbacks, delays, and disappointments, and without an understanding that a new and complicated system must progress through an often difficult and lengthy development process, the financial backers quickly got cold feet. Fitch was certainly a victim of this. Although today the process is infinitely better understood, often the individuals supplying the money, so-called venture capitalists, sometimes thrust their ideas onto the inventor.

Engineers and inventors have always had to live by an unwritten principle, which they utterly detest, called the *Golden Rule* of development. It simply states, "He who has the gold makes the rules." That's just the way it is!

Notes

1. George Calvin Carter, *Samuel Morey: The Edison of His Day* (Concord, NH: Rumford Press, 1945), pp. 60–61; and J. Franklin Reigart, *The Life of Robert Fulton* (Philadelphia: C. G. Henderson, 1856), p. 161.

2. Carter, *Samuel Morey*, pp. 60–61; Reigart, *The Life of Robert Fulton*, p. 161.

3. John Fitch, *The Original Steam-Boat Supported; or, A Reply to James Rumsey's Pamphlet Shewing the True Priority of John Fitch and the False Datings, &c. of James Rumsey* (Philadelphia: Zachariah Poulson, 1788), para. 3.

4. James Rumsey, *A Short Treatise on the Application of Steam, Whereby Is Clearly Shewn from Actual Experiments, That Steam May Be Applied to Propel Boats or Vessels of Any Burthen against Rapid Currents with Great Velocity* (Philadelphia: Joseph James, 1799), para. 1.

5. Ivor B. Hart, *James Watt and the History of Steam Power* (New York: Collier Books, 1949), p. 130.

6. Ibid., p. 117. Watt invented the separate condenser in 1769, but it was not until 1774 that the first really successful condensing engine was constructed.

7. Ibid., pp. 137–39.

8. Archer B. Hulbert, *The Paths of Inland Commerce: A Chronicle of Trail, Road, and Waterways* (New Haven, CT: Yale University Press, 1920), p. 110.

9. David Read, *Nathan Read: His Invention of the Multi-tubular Boiler . . .* (New York: Hurd and Houghton, 1870), p. 38.

10. Fitch, *Autobiography of John Fitch*, ed. Frank D. Prager (Philadelphia: American Philosophical Society, 1976), p. 144.

Figure 4.1. A very ornate Greek variation of Hero's Aeolipile, where fire, augmented by human breath, produces steam in the lion-headed base, causing the hollow ball to rotate. Although a horse is pictured on the sphere, the device did not produce enough power to accomplish useful work. (From Robert H. Thurston, *History of the Growth of the Steam-Engine* [New York: D. Appleton and Company, 1901], title-page frontispiece.)

4.

Early Steamboat Experimenters

The great achievements of the past were the adventures of adventurers of the past. Only the adventurous can understand the greatness of the past.

—Alfred North Whitehead

As the steamboat was born in America, the product of American ingenuity to serve a uniquely American need, it follows that American developers are the principal subjects of this history. They, however, are not the only steam navigation pioneers considered; work in England and Europe provided some of the enabling technology for successful steam navigation. It is therefore important to trace early steamboat history, which provides a bridge from which American contributions can be better understood.[1] Much of this early steamboat work suffered from undeveloped science and inadequate technology, and it generally occurred over short periods of time. It achieved limited success but provides us with a baseline for understanding that which followed.

A Precursor of Steam Power

One of the earliest applications of steam to produce motion, although without a useful application, was the ancient steam engine called an Aeolipile produced by Hero of Alexandria in 200 BCE.[2] It was similar to a thin metal basketball with small bearing holes on either side. The sphere was held in place by tubes inserted into each hole but allowed to rotate. Around the center circumference, on opposite sides, were vents in the shape of pipe elbows. Steam generated in the base of the structure was fed to the hollow sphere through the two mounting tubes. As the steam escaped through the vents it caused the sphere to spin. It did not, however, produce any useful work, but simply demonstrated the power of steam and the principle of reaction (see fig. 4.2).

In 1629 an Italian, Giovanni Branca, contemplated extracting useful work from steam,[3] and one of his devices is shown in figure 4.3. As the left-hand side of the driving axle, I, is not supported, the device could not work as pictured; hence, the sketch was probably conceptual, and the device never built.

Prophecy

Was it possible that Homer foresaw in the *Odyssey* the possibility of steam moving a vessel when he wrote, "So shalt thou reach the realm assigned, / In wondrous ships, self-moved, instinct with mind"?[4] Or perhaps

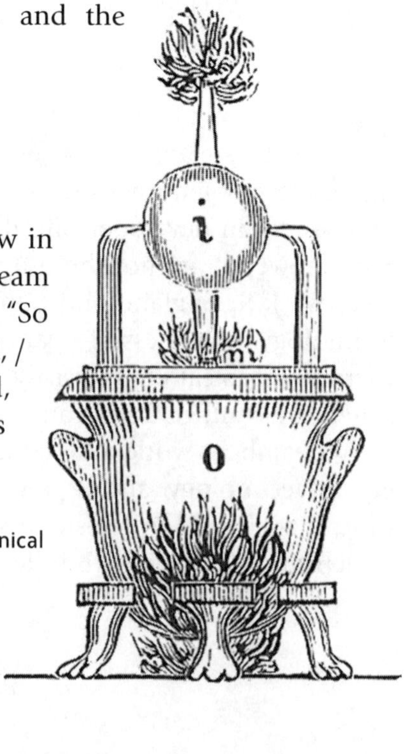

Figure 4.2. Hero's Aeolipile is one of the earliest uses of steam to demonstrate mechanical motion. (From David Read, *Nathan Read, His Invention of the Multi-tubular Boiler . . .* [New York: Hurd and Houghton, 1870], between pp. 14 and 15.)

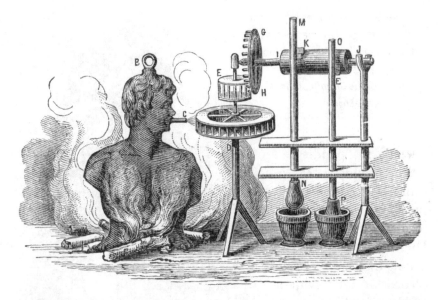

Figure 4.3. In a sketch by Giovanni Branca, the steam from a pot boiler, in the shape of the upper part of a man, was ejected onto a horizontal paddle wheel, which through gearing raised mortars that pounded grain or another substance in a pestle. (From Thurston, *History of the Growth of the Steam-Engine*, p. 17.)

Roger Bacon, the Franciscan monk, writing more than six hundred years ago may have anticipated steam navigation with the words, "I will now mention some wonderful works of art and nature, in which there is nothing magic, and which magic could not perform. Instruments may be made by which the largest ships, with only one man guiding them, will be carried with greater velocity than if they were full of sailors."[5]

A Plan by Papin, 1690

In any case, it was not until 1690 that powering a boat by steam was first explicitly proposed, when a Frenchman, Denis (also spelled Denys) Papin (see fig. 4.4), provided such a plan. He proposed to raise water by using a steam pump of his design with the water being discharged onto a water wheel. This wheel would in turn be con-

Figure 4.4. Denis Papin was one of the
most distinguished men of his time,
devoting much effort to the study
of physics. (From Thurston,
*History of the Growth of the
Steam-Engine*, p. 46.)

nected to paddle
wheels, which by
turning would cause
the boat to move
through the water.
There are some ac-
counts that say he suc-
ceeded in 1707 with a
paddle wheel boat on the
river Fulda, but his boat was
moved by human force or, as
described by H. Phillip Spratt, "par la
force humaine."[6] It is believed that his plan remained only a proposal, as an actual trial was never reported.

The Pumping Engine of Savery, 1698

One of the first practical uses of steam to perform work was that of
Thomas Savery (see fig. 4.5), who around the close of the seventeenth century developed the steam pump to remove water from
mines. In a 1702 London publication with the abbreviated title *The
Miner's Friend; or an Engine to Raise Water by Fire, Described, and of the
Manner of fixing it in Mines*, Savery described how the steam engine
worked and some other tasks it might be employed to perform. He
contemplated the use of steam to propel vessels but, fearing a hostile response by the British Admiralty, wrote, "I believe it may be very
useful to ships, but I dare not meddle with that matter."[7]

Figure 4.5. Thomas Savery, born about 1650, became a military engineer and is best known for his steam powered pulseometer water pump for mines. (From Thurston, *History of the Growth of the Steam-Engine*, p. 31.)

The Newcomen Engine, 1712

Probably inspired by the work of Papin and others, Thomas Newcomen, an iron-monger, blacksmith, and minister, along with partner and assistant John Crowley, a plumber, started working on a steam pump at the close of the seventeenth or the beginning of the eighteenth century. In 1712, they installed a steam pump engine near Wolverhampton to remove water from a mine.[8]

Unlike the Savery pulseometer pump, it worked on the principle of using steam pressure alternating with atmospheric pressure to move a piston in a cylinder. The steam raised the piston to the top, at which point water was injected into the cylinder to condense the steam, creating a vacuum whereby the pressure of the atmosphere (about fifteen pounds per square inch) forced the piston to the bottom.

By mechanically coupling the piston to a walking beam, sort of an elevated seesaw, it was possible to operate a pump to force water from the mine. This engine would be the basis for future steamboat work and the precursor for James Watt's later improvements.

The Patent by Hulls, 1736

Jonathan Hulls proposed the use of a Newcomen engine to power a steam vessel, and in 1736 obtained an English patent described as a "machine for carrying ships and vessels out of or into any harbor or river against wind and tide."[9] He planned to use the steam-powered vessel for towing (see fig. 4.6). One year later he published a pamphlet describing the invention that included a detailed description of the operation as well as the use of a stern paddle wheel. There is no evidence, however, that he pursued the concept to completion, as he wrote "That the Scheme I now offer is Practicable, and if encouraged will be Useful."[10] The time was not right, though, and he was scorned with the following doggerel rhyme:

> Jonathan Hulls,
> With his patent skulls,
> Invented a machine
> To go against wind and stream;
> But he, being an ass,
> Couldn't bring it to pass,
> And so was ashamed to be seen.[11]

Figure 4.6. The fact that Jonathan Hulls never ran a steamboat did not discourage steam navigation writers from giving him credit for producing a steam-propelled craft. (From Thomas W. Knox, *Life of Robert Fulton and a History of Steam Navigation* [New York: G. P. Putnam's Sons, 1886], p. 75.)

Hulls was a man significantly ahead of his time, as the steam engine was not yet sufficiently developed to produce enough horsepower for a given size and weight to be useful for steam navigation.

A Proposal from Bernoulli, 1753

In response to a prize offered in 1752 by the French Academy of Science for an essay on propelling a boat without wind, Daniel Bernoulli proposed filling a vertical tube with water, which would then be exhausted out the stern, propelling the craft forward (see fig. 4.7).[12] Also he advocated using an early form of the screw propeller, which could be powered by animal or steam power.

Watt's Separate Condenser, 1774

During the class year of 1763–64 at the University of Glasgow, James Watt, instrument maker for the school, was asked to repair a model Newcomen steam engine in the laboratory of the philosophy class (see fig. 4.8). After an examination of the model, Watt began studying the properties of steam and experimentally determined the latent heat of steam (energy required to transform water at 212 degrees Fahrenheit into vapor), a property he did not understand. He consulted his friend Dr. Joseph Black, professor of chemistry at the university, who provided him with the explanation.

With this new understanding, Watt quickly realized that the inef-

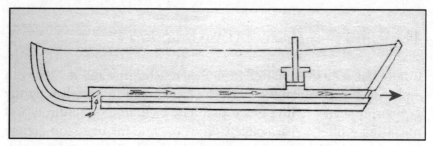

Figure 4.7. Bernoulli's water jet propulsion proposed in 1753. (From Read, *Nathan Read, His Invention of the Multi-tubular Boiler* . . . , between pp. 38 and 39.)

Figure 4.8. The Newcomen steam engine model was used by University of Glasgow instructors to demonstrate the harnessing of steam power. As it did not operate satisfactorily, the school's instrument maker, James Watt, was asked to repair the model and his investigation led to his invention of a separate condenser for the steam engine. (From Thurston, *History of the Growth of the Steam-Engine*, p. 84.)

ficiency of the Newcomen engine was due to the alternate heating and cooling of the cylinder. This happened when the hot steam was cooled to produce the vacuum to allow the atmospheric pressure to move the piston back to its original position. Watt reasoned correctly that the use of a separate vessel for the condensing of steam, aptly named a condenser, would eliminate this problem, leading to a more efficient steam engine. Eventually Watt partnered with the Soho manufacturing firm of Matthew Boulton to produce his first single-acting condenser steam engine in 1774. It was a much later variation of this basic engine that eventually powered Fulton's first commercially successful steamboat.

The Failed Steamboat of D'Auxiron, 1774

Comte J. B. d'Auxiron retired from the French Army about 1770 with a plan to build a steamboat. He was successful in establishing the D'Auxiron Society with several partners, including de Jouffroy, and obtaining a fifteen-year concession to operate his boats on French rivers, provided his plans were acceptable to the French Academy of Science.

D'Auxiron constructed a boat in 1774 on the Seine near Paris, using a two-cylinder atmospheric engine, but before the trial, a counterweight fell through the hull of the vessel, dooming his efforts. He was later taken ill and died in 1778, having achieved no success.

The First Successful Steamboat, by De Jouffroy, 1783

Having been a partner in the D'Auxiron experiment, the Marquis Claude de Jouffroy d'Abbans constructed a forty-three-foot boat on the river Doubs in 1778. It was fitted with two Newcomen-type engines, each having a cylinder diameter of 22.4 inches, with strokes of 64 inches. The cylinders were made of rolled copper sheets, a technique that would later be used by James Rumsey. "Duck foot" paddles or flaps provided propulsion, but the trial was not a success.

On his second craft, in 1783, a longer, 148-foot vessel called the *Pyroscaphe* was built with a double-acting, larger cylinder 25.6 inches in diameter and with a stroke of 77 inches (see fig. 4.9). The vessel was tested on the river Saône, near Lyon, on July 15, 1783, and it went against the current for about fifteen minutes. Unfortunately, De Jouffroy failed to secure the support of the French government, and when the Revolution started, his work was interrupted.[13]

Figure 4.9. De Jouffroy's 1783 steamboat on the Saône in France, as envisioned by one steam navigation writer. (From William McDowell, *The Shape of Ships* [New York: Roy, n.d., ca. 1948], p. 93.)

The Triple Hulls of Miller, 1786

Patrick Miller of Dalswinton in Scotland strapped together three hulls with two paddle wheels in between to create a wide, 22.5-by-77.3-foot boat called the *Edinburgh*. The paddle wheels, six feet in diameter and four feet wide, were driven by human power; hence the *Edinburgh* was not a steamboat. It was, however, the forerunner of a two-hull vessel built later by William Symington, a steam engine designer.

Paddle Wheels and a Poet, 1788

The tutor of Patrick Miller's son recommended the use of a steam engine to drive the paddle wheels, as it was too exhausting for the crew. The tutor introduced Miller to William Symington, who produced an engine for a double-hulled boat. The space between the hulls was used for the paddle wheel, and on October 14, 1788, the twenty-five-foot vessel achieved five miles per hour on Dalswinton Lake, near Dumfries (see fig. 4.10). Several people were on board, including the poet Robert Burns.

Figure 4.10. Patrick Miller's doubled-hulled steamboat of 1788, shown with the paddle wheel between the hulls. (From McDowell, *The Shape of Ships*, p. 93.)

Lord Dundas and the *Charlotte Dundas*, 1802

At Grangemouth Dockyard, Alexander Hart constructed the *Charlotte Dundas* for Lord Dundas, who named it for his daughter. William Symington supplied the ten-horsepower engine, with a 22-inch cylinder and 48-inch stroke. The engine drove a stern paddle wheel, and the tug successfully towed two ships totaling 140 tons for a distance of 19.5 miles in six hours on the Clyde and Forth Canal in 1802. Although the performance was satisfactory, the canal owners did not feel that the advantage provided by steam propulsion outweighed the possible damage to the canal banks, so the project was dropped. In his 1958 book, *The Birth of the Steamboat*, Spratt cites this effort as the world's first practical steamboat.[14]

Summary of the Early Steamboat Pioneers

To gather data that can later be used to summarize the contributions of English, French, and American steamboat developers, Table 1 has been compiled. It highlights the individual, concept date, and trial date(s), and it provides notes regarding the accomplishment.

Figure 4.11. Lord Dundas's twin-hulled *Charlotte Dundas*. The successful steamboat used two rudders, one on each side, strapped together to one tiller. (From McDowell, *The Shape of Ships*, p. 93.)

Table 4.1. Summary of the efforts of steamboat pioneers

Steamboat pioneer	Concept date	Steamboat trial date(s)	Notes
Papin	1690	1707	Human, not steam power
Hulls	1736	None	Concept only for steam tug
D'Auxiron	1770	None	In 1774 boat accidentally sunk
De Jouffroy	ca. 1771	1778 Unsuccessful	Some success on 1783 trial, but work
		1783 *Pyroscaphe*	interrupted by the French Revolution
Miller	ca. 1786	1787 *Edinburgh*	Human, not steam power
		1788 Steamboat	Engine by Wm. Symington
Dundas	ca. 1798	1802 Tugboat	Hull by Alexander Hart with engine by
		Charlotte Dundas	Wm. Symington; towed 140 tons on a canal

Every several chapters, this table will be updated with additional developers; in the last chapter, it will be presented again, where *the* steamboat inventor will be identified.

On to America

There simply did not exist in England or elsewhere in Europe the pressing need for the establishment of steamboat travel—it was not until steam navigation became entrenched in America that the rest of the world would take serious notice. To understand this, it is worthwhile to examine the premise advanced by the author that America led the world in steam navigation. The graph in figure 4.12, which shows the number of steamboats built in the United States over the period of 1807 to 1880, provides part of the answer. Note that the vertical axis is logarithmic, with each major division increasing by a factor of ten.

The curve is derived from data found in George Preble's *A Chronological History of the Origin and Development of Steam Navigation*, published in 1883.[15] The number of steamboats built each year was averaged over ten years, and the resultant value was plotted at the end of the decade. The year 1823 was the starting point, with fifteen steamboats. Between this point and Fulton's first *North River Steam Boat* in 1807, a period of sixteen years, a segmented line is drawn, connecting at the value of seven at the midpoint year of 1815.

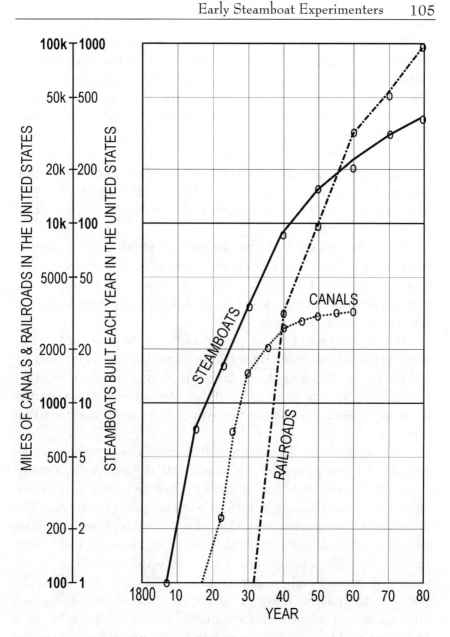

Figure 4.12. Graph showing the number of steamboats produced each year, miles of canals, and miles of railroad in the United States from 1800 to 1880. The canal-building craze lagged steamboats by little over one decade, but ten years later railroad construction would accelerate dramatically and eventually cause the demise of most canals. (Graph by Jack L. Shagena.)

To validate the appropriateness of the extrapolation from 1807 to 1823 it is necessary to examine other data. Jean Baptiste Marestier visited the United States in the latter part of the first quarter of the nineteenth century and for nearly two years collected data on steamboats on the East Coast and on the Mississippi. In 1824, he published his *Memoir on Steamboats of the United States of America*, providing the steamboat names along with technical data on the country's emerging fleet. His data begins with Fulton's efforts and appears to go through some time in 1820.[16] On January 27, 1823, Marestier reported his observations to the Institute of France, Royal Academy of Science.

From the extrapolated line, the average number of steamboats built between 1807 and 1823 is seven per year; and over a sixteen-year period, this totals 112. Marestier reported 95 by 1822,[17] so the graphed line represents a good approximation of America's steamboat construction during these years.

The significance of the graph lies in the rapid growth of steamboats in the United States once Fulton demonstrated their commercial feasibility. This growth validated the pent-up need to unite the young republic, as over one-half of the steamboats Marestier reported in 1824 were on the Mississippi River or on the rivers that flow into it.

Of interest also is the rapid growth of canals in America that paralleled steamboat construction a decade later. These data reflect canals constructed in the mid-Atlantic states but can be viewed as representative of the entire country.[18] In 1800, there were only twenty-six miles of canals on the eastern seaboard; however, canal fever was spreading rapidly, and thirty years later the mileage had jumped to fourteen hundred. By 1839, the investment in canals was $13.2 million, representing 10 percent of the country's total construction. The end of the boom came quickly, however, as by 1859 canal investment had dropped to 0.4 percent, with the number of miles leveling off to little over three thousand. By the early part of the twentieth century, the total length had dipped to just under two thousand miles.[19]

Most of the canals were bypassed by the fast-track growth of the railroad during the 1830s, and the watery ditches and leaking locks were quietly abandoned. Today only a few of the canals remain, with the Chesapeake and Delaware, dredged to sea level in 1927, the busiest canal in the United States and the second busiest in the world.

The United States, as shown in figure 4.13, initially led the world in steamboat tonnage, with nearly all ships built before 1830 being placed into domestic service. Other nations did take notice of America's leadership; Great Britain, a decade later, started to challenge the lead and overtook the United States by 1865. Still later, France and Holland produced steamboats, but production grew slowly by comparison.[20]

In the development of steam navigation, the U.S. lead resulted from a pressing national requirement. Entrepreneurs quickly

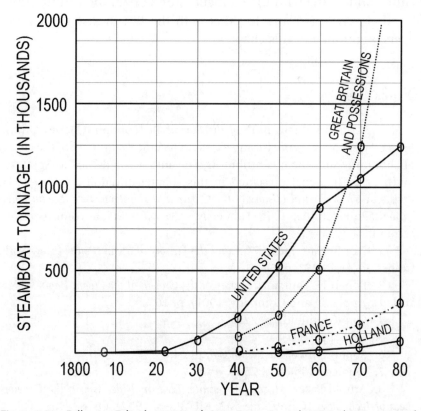

Figure 4.13. Following Fulton's 1807 *North River Steam Boat*, the United States jumped ahead of the rest of the world in tonnage; however, by about 1865 it was overtaken by Great Britain and its possessions. (Graph by Jack L. Shagena.)

evolved steamboat designs that addressed regional operational environments, adding credence to the oft-quoted saying "necessity is the mother of invention."

Looking Ahead

The following eight chapters address the contributions of American steamboat developers who pioneered steam navigation in the United States. These short biographies, or what the author prefers to call "technographies," examine each pioneer's efforts. In addition, the last chapter briefly addresses the work of three lesser-known efforts and, along with the epilogue, summarizes the contributions of all eleven Americans, together with the British and French pioneers mentioned in this chapter.

Notes

1. H. Philip Spratt, *Birth of the Steamboat* (London: Charles Griffin, 1958). This work is a concise and excellent technical summary of the efforts of about fifty steamboat contributors. The author is indebted to Spratt for much of the material presented in this chapter. In addition, George H. Preble, *A Chronological History of the Origin and Development of Steam Navigation* (Philadelphia: L. R. Hamersly, 1883), offers a comprehensive account.

2. Ivor B. Hart, *James Watt and the History of Steam Power* (New York: Collier Books, 1949), p. 64.

3. Robert H. Thurston, *History of the Growth of the Steam-Engine* (New York: D. Appleton and Company, 1901), p. 16.

4. Quoted in ibid., p. 223.

5. Ibid., pp. 223–24.

6. Spratt, *Birth of the Steamboat*, p. 24.

7. Quoted in ibid., p. 25.

8. Arnold Pacey, *Maze of Ingenuity: Ideas and Idealism in the Development of Technology* (Cambridge, MA: MIT Press, 1992), pp. 89–90.

9. Preble, *A Chronological History*, p. 7.

10. Thurston, *History of the Growth of the Steam-Engine*, p. 228.

11. Preble, *A Chronological History*, p. 9.

12. Benjamin Franklin was the U.S. representative to France from 1779 to 1785 and was familiar with Bernoulli's work. Upon his return to the United States he authored a paper that was presented to the American Philosophical Society that included Bernoulli's proposal to propel a boat by water jet.

13. Spratt, *Birth of the Steamboat*, pp. 38–39.

14. Ibid., pp. 8, 61.

15. Preble, *A Chronological History*, appendix, Tables I, II, pp. 424–25.

16. Jean Baptiste Marestier, *Memoir on Steamboats of the United States of America* (1824; reprint, Mystic, CT: Marine Historical Association, 1957), no. 31, p. 61. The last dated steamboat entry observed by the author was March 1820 regarding the *Volcano*, whose boiler burst near St. Francisville, Louisiana.

17. Ibid. Of the ninety-five steamboats identified, sixty-four were on western waters, with the remainder scattered across the eastern seaboard of the United States.

18. Christopher T. Baer, ed., *Canals and Railroads of the Mid-Atlantic States, 1800–1860* (Wilmington, DE: Eleutherian Mills-Hagley Foundation, 1981), p. 59.

19. A. Barton Hepburn, *Artificial Waterways of the World* (New York: Macmillan, 1914), appendix I, p. 147.

20. Preble, *A Chronological History*, appendix, Tables I, II, pp. 424–25.

Part 2.

The Candidates Considered

Figure 5.1. William Henry. (Rendered by William T. Sisson from a portrait in Francis Jordan Jr., *Life of William Henry of Lancaster, Pennsylvania* [Lancaster, PA: New Era Printing, 1910], title-page frontispiece.)

5.

William Henry
1729–1786

I have frequently heard of Mr. Henry applying steam as a means to urge boats through the water by the force of it, and that he had proposed laying a model of a machine before the Philosophical Society.

—Ann Henry, wife of William Henry, quoted in Jordan, *Life of William Henry of Lancaster, Pennsylvania* (1910)

William Henry was born on May 19, 1729, in Chester County, near the town of Lancaster—not far from the birthplace and boyhood home of Robert Fulton. In Francis Jordan Jr.'s 1910 biography of Henry, he is described on the title page as a "Patriot, Military Officer, [and] Inventor of the Steamboat."[1] In a paper by Herbert H. Beck, read before the Lancaster County Historical Society in 1950, he is identified as the "Progenitor of the Steamboat, Riflemaker, [and] Patriot."[2]

Henry was indeed a great American with significant creative skills. Having been a highly successful rifle manufacturer in the early part of his life, he went on to serve as the armorer for Pennsylvania, treasurer of Chester County, and delegate to Congress; he was also the initial benefactor to the well-known American artist Benjamin West. Having a keen interest in the sciences, he was a member of the American Philosophical Society and maintained a long friendship with David Rittenhouse, a famous instrument maker of his day.

William Henry was the inventor of the screw auger (see fig. 5.2), a spiral metal device that became a commonplace tool for drilling holes in wood.[3] He also invented a heat regulator called the sentinel register, but it is his contribution to the steamboat that is the focus here. Therefore his life up to the time of his steamboat work will be addressed.

Henry's grandparents, Robert and Mary Ann Henry, had found their way to Pennsylvania after arriving in New Castle, Delaware, in

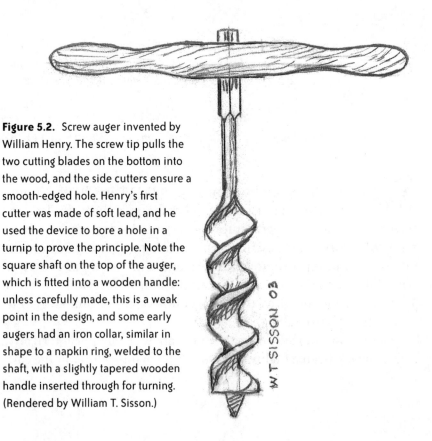

Figure 5.2. Screw auger invented by William Henry. The screw tip pulls the two cutting blades on the bottom into the wood, and the side cutters ensure a smooth-edged hole. Henry's first cutter was made of soft lead, and he used the device to bore a hole in a turnip to prove the principle. Note the square shaft on the top of the auger, which is fitted into a wooden handle: unless carefully made, this is a weak point in the design, and some early augers had an iron collar, similar in shape to a napkin ring, welded to the shaft, with a slightly tapered wooden handle inserted through for turning. (Rendered by William T. Sisson.)

1722. Of Scottish ancestry, they had sailed to America via Coleraine in the northern part of Ireland with their three adult sons Robert, James, and William's father John Henry.

Robert married Sarah Davis, and his brother James married her sister Mary Ann. Between the marriages, eight children were born, and later the two families moved to Virginia. One of the two boys born to Robert and his wife was Patrick Henry, the famous patriot, orator, governor of Virginia, and delegate to the Continental congresses.[4]

William's father, John Henry, remained in Pennsylvania and in 1728 married Elizabeth (Jenkins) DeVinne, of Huguenot descent. They settled in the area, and this union produced two boys and three girls, of which William was the oldest. The girls married into the families of Postlethwait, Bickham, and Carson. The progenitors of John Henry's family, Robert and Mary Ann, both died on the same day in 1735.

In 1744, at the age of fifteen, William was apprenticed to Matthew Roesser, a gunsmith located in Lancaster, which at the time was the colony's largest inland town. William's mechanical aptitude served him well (see rifle in fig. 5.3 and sentinel register in fig. 5.4), and six years later he left his apprenticeship to form a rifle manufacturing partnership with Joseph Simon. The business prospered.

In a short time he was able to acquire a desirable property on the public square of Lancaster, whereupon he made improvements to an existing brick dwelling (see map in fig. 5.5). It was here, according to biographer Mathew S. Henry, that his young neighbor Robert Fulton first visited him.[5]

As William Henry's father had passed away in 1747, and one of his siblings, Mary, had been widowed, his mother and sister moved

Figure 5.3. A Pennsylvania rifle that was most likely similar to the ones produced by William Henry. The rifle, with a long, spiraled barrel, had a far greater useful range than British muskets and has been called the American secret weapon of the Revolutionary War. (Rendered by William T. Sisson.)

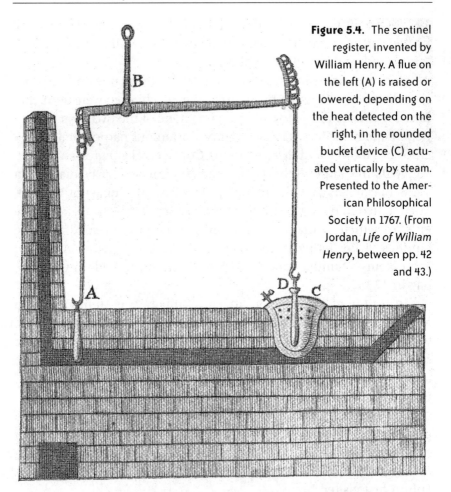

Figure 5.4. The sentinel register, invented by William Henry. A flue on the left (A) is raised or lowered, depending on the heat detected on the right, in the rounded bucket device (C) actuated vertically by steam. Presented to the American Philosophical Society in 1767. (From Jordan, *Life of William Henry*, between pp. 42 and 43.)

in to help manage the residence. It was at a tea party given by Mary that he was to first meet the clever and admirable Ann Wood, whom he would marry in 1755. From this union was born on March 3, 1757, William Henry Jr., who later distinguished himself as an officer and a judge; John Joseph Henry; and Benjamin West Henry, named in honor of the artist.

In 1759, the rifle manufacturing partnership with Joseph Simon was dissolved, and subsequently William decided to go to Europe to establish connections with makers of iron and steel. He sailed from Philadelphia in December 1760, after some delay because of England's war with France, and arrived in London in 1761.

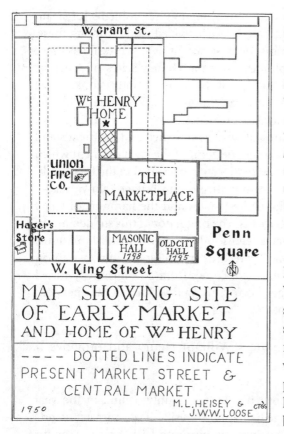

Figure 5.5. William Henry's home, at the location of the star, was centrally located in Lancaster and visited by aspiring portrait painter Benjamin West as well as by Robert Fulton. (From Hebert R. Beck, "William Henry," *Papers of the Lancaster Historical Society* 54, no. 4 [1950]: 83. Reprint permission by Lancaster County Historical Society.)

He is reported to have met with James Watt, inventor of the single-acting condenser steam engine; this meeting, in which he was cordially received, provided the spark for his subsequent steamboat work (more about this reported meeting later). After completing his business he sailed for America in November 1761, arriving in Philadelphia before the end of the year.

Regarding the period when Henry became involved in his steamboat work, biographer Jordan relies heavily on a written account provided by Robert H. Thurston. In 1871 Thurston became the chairman of the Mechanical Engineering Department at the Stevens Institute of Technology in Hoboken, New Jersey, and later in 1886 taught at Cornell University.[6] In 1891 he wrote *Robert Fulton: His Life and Its Results*, which contained an account of Henry's work. It is from this source that Jordan extensively quoted, and it is a reprint thereof from which the following is extracted. Thurston first starts with an overview of the steamboat's status at the turn of the eighteenth century:

Many other inventors were now studying the problems [of steam navigation] in different parts of the civilized world. Among these, none were more ingenious or as persistent or as successful as those of the then British colonies, later the United States of America. Among these was a group of New York and Pennsylvania mechanics, who, seemingly each more or less familiar with the work of the other, struggled on persistently, and finally successfully. A nucleus consisting of one of these men and his friends and coadjutors, became, ere long, the germ of the great moment which in the early art of the nineteenth century resulted in the final application of the powers of steam to the propulsion of steam-vessels,— first on the rivers of the United States and the harbours of Great Britain, then on all the oceans. The originator of this sudden movement in the United States seems to have been a man of unknown fame, and one of whom few records are preserved. Our own information, hitherto unpublished, comes from an indistinctly traced source; but its facts have been fairly well verified by independent historical investigation.

Professor Thurston introduces William Henry and provides background information:

> William Henry was born in Chester County, Penn., in the year 1729. . . . At an early age he became a resident of Lancaster, Penn., where he learned the business of gunsmith [*sic*]. After serving his apprenticeship he began business on his own account, and in a few years became the principle gunsmith in the province. . . .
>
> In the year 1760 Mr. Henry went to England on business with his vocation, and there he remained for some time. Having a mechanical turn of mind, the invention and the applications of steam by Watt being then much discussed, the idea of its application to the propelling of boats, vehicles, etc., so engrossed in his mind that on return to his home in Lancaster he began the construction of a machine, the motive power of which was steam. In 1763 Mr. Henry completed the machine, which was attached to a boat with paddles, and with it he experimented on the Canastoga [*sic*] River, near Lancaster, but the boat was by some accident sunk.
>
> This was the first attempt ever that had ever been made to apply steam to the propelling of boats. Notwithstanding the ill

luck that attended the first attempt in an undertaking of the practicality of which he had not the least doubt, he constructed a second model, with improvements on the first; and among the records of the Pennsylvania Philosophical Society to be found a design, presented by him in 1782, of a machine, the motive power of which was steam.

Professor Thurston then elaborates on Henry's steamboat work with an account provided by a German traveler:

An intelligent German traveller named Shoepff, who travelled through the United States in 1783–1784, whilst staying for a time in Lancaster, became acquainted with Mr. Henry. He says: "I was shown a machine by Mr. Henry, intended for the propelling of boats, etc., 'but', said Mr. Henry, 'I am doubtful whether such a machine would find favor with the public, as every one considers it impractical to make a boat move against wind and tide;' but such a boat will come into use, and navigate on the waters of the Ohio and Mississippi he had not the least doubt, though the time had not yet arrived of its being appreciated and applied." A sketch of the machine, with boilers, etc., made by Mr. Henry in 1779, is said to be still in the possession of his heirs.

John Fitch (for whom his biographers claimed the honour of the invention of the application of steam to the propulsion of boats) was a frequent visitor at Mr. Henry's house, and according to the belief of his friends obtained from him the idea of the steamboat. Fulton, then a young lad, also visited Mr. Henry's to examine the paintings of Benjamin West; and the germ that subsequently ripened into the construction of the "Folly" was probably due to those visits. Mr. Henry's deceased [sic] occurred on the 15th of December, 1786.[7]

From the foregoing quotation by Thurston and the claim by Jordan, it would be fairly easy to conclude that indeed William Henry was the inventor or originator of the steamboat (at least in America) as early as 1763. However, there are several facts that indicate otherwise.

When Henry went to London in 1760, he could not have learned of the steam engine from Watt, because the timing was wrong. It was not until the college session of 1763–64 at the University of Glasgow that the instrument maker Watt was asked by Professor Anderson to examine the laboratory's Newcomen model steam engine with the request to make it operable (see chap. 4).[8] From this Watt started his study of steam that led to his invention of the separate condenser, obtaining his first patent on the device in 1769. It would not be, however, until 1774 that Watt, then in partnership with Matthew Boulton, produced a working model of the single-acting condenser steam engine that was offered for sale. Therefore, for Henry to have obtained any information on Watt's steam engine in 1761 seems highly unlikely.

In order for Henry to have produced a steamboat, the skill of many different individuals would have been required for the fabrication, construction, and assembly of the many parts involved in the project, which would have spanned at least one or two years. Since no evidence has been uncovered as to who these individuals might have been, and no log or written account of an actual trial exists, most likely one never took place.[9]

This position would appear to be confirmed by John Fitch in his autobiography of 1785:

> On my way to Virginia I called on the ingenious Mr. Wm. Henry of Lancaster, as I wished the opinion of every man of science [regarding my steamboat]. And knowing the scheme to be far before my abilities, thought I could not be too cautious how I acted, and to take every mans [sic] opinion who could throw any light on it.
>
> To my surprise Mr. Henry told me that he had produced it many years ago, and had made a Draft of it which he intended to lay before the Philosophical Society, but that he had neglected it. And altho I gave no hint of the kind, I believe he thought that I suspected him, and he went and hunted his papers and produced a Draft for me, which was to propel a Boat by a Steam Wheel.[10]

Fitch's initial reaction was one of suspicion, as he believed the draft by Henry was made subsequent to his own earlier presentation before the American Philosophical Society (see fig. 5.6). This, how-

Figure 5.6. Philosophical Hall on Independence Square. Henry was a member of the American Philosophical Society, which met in the hall, and this was the location where John Fitch presented his steamboat ideas in 1785. (From Allen C. Thomas, *History of the United States* [Boston: D. C. Heath, 1901], p. 138.)

ever, turned to embarrassment when Henry graciously relinquished any claim, saying that his idea was different and that he had not brought it to public view, whereas Fitch was the first to publish his idea. Furthermore, Henry was confident of the success of the plan, and an elated and relieved Fitch departed for Frederick Town, Maryland, for a visit with Thomas Johnson, who had served as governor of Maryland from 1777 to 1779.

It appears, therefore, that although Henry did discuss a steamboat with Andrew Ellicott as early as 1775 and with Thomas Paine in 1778,[11] and although he prepared a sketch of a plan to propel a boat with a steam turbine in 1779, he did not at any time conduct actual trials. Other researchers who have studied his papers share this conclusion.[12]

A concluding remark is appropriate. Although William Henry did not make a significant direct contribution to the development of the steamboat, it does appear that he was the first to have conceived and documented the idea in America and, according to Herb Beck, deserving of the title *Progenitor of the Steamboat*. In addition, he made many valuable contributions to his adopted country that may be best summed up by John Galt, author of *Life of Benjamin West*, when he wrote in the preface, "Henry was indeed in several respects an extraordinary man, and possessed the power generally attended upon genius under all circumstances, that of interesting to the imagination of those with whom he conversed."[13]

Indeed, Galt's words appear to be a fitting assessment of the ingenious Mr. William Henry, Patriot, Military Officer, and great American.

Notes

1. Francis Jordan Jr., *Life of William Henry of Lancaster, Pennsylvania* (Lancaster, PA: New Era Printing, 1910), p. i. The author drew on this account for much of the information in this chapter. (Note: Ann Henry's statement above was made in May 1788, after her husband's death and is taken from Jordan, p. 53.)

2. Herbert H. Beck, "William Henry, Progenitor of the Steamboat, Riflemaker, Patriot," *Papers of the Lancaster County Historical Society* 54, no. 4 (1950): 65.

3. Mathew S. Henry, "Life of William Henry" (manuscript, American Philosophical Society, 1860), p. 33.

4. Ibid., p. 69.

5. Ibid., p. 59.

6. *International Cyclopedia*, ed. Richard Gleason Green, 15 vols. (New York: Dodd, Mead, 1887), 14:397.

7. The block quotations above are from William Graham Sumner and Robert H. Thurston, *Makers of American History: Alexander Hamilton and Robert Fulton* (New York: University Society, 1904), pp. 30–33.

8. Ivor B. Hart, *James Watt and the History of Steam Power* (New York: Colliers Books, 1949), p. 105.

9. James Flexner, *Steamboats Come True: American Inventors in Action* (1944; reprint, Boston: Little, Brown, 1978), pp. 383–84, points out that the erroneous story about Henry actually building a steamboat seems to stem from an article in the *Lancaster Daily Express* of December 10, 1872.

10. Fitch, *Autobiography of John Fitch*, ed. Frank D. Prager (Philadelphia: American Philosophical Society, 1976), pp. 154–55.

11. John Fitch, *The Original Steam-Boat Supported; or, A Reply to James Rumsey's Pamphlet Shewing the True Priority of John Fitch and the False Datings, &c. of James Rumsey* (Philadelphia: Zachariah Poulson, 1788), para. 3.

12. Fitch, *Autobiography of John Fitch*, p. 155n; Philip H. Spratt, *Birth of the Steamboat* (London: Charles Griffin, 1958), p. 30; and Flexner, *Steamboats Come True*, pp. 48–49.

13. John Galt, *Life of Benjamin West*, quoted in Jordan, *Life of William Henry*, p. 26.

Figure 6.1. James Rumsey. (Rendered by William T. Sisson from a portrait believed to have been painted by Benjamin West. From Ella May Turner, *James Rumsey, Pioneer in Steam Navigation* [Scottdale, PA: Mennonite Publishing House, 1930], title-page frontispiece.)

6.

James Rumsey
1743–1792

Tomorrow morning, I throw myself upon the wide world In pursuit of my [steamboat] plans, being no longer Able to proceed upon my Own foundations, I Shall bend my Course for Philadelphia where I hope to have it in my power to Convince a Franklin and a Rittenhouse of their utility by actual Experiment. . . . Conquer or fall *was my motto!*

—James Rumsey, March 1788

James Rumsey was born in March 1743 on a wedge of land known as Middle Neck (see fig. 6.2), between the Great and Little Bohemia creeks in Cecil County, on Maryland's upper eastern shore. Rumsey's ancestors had operated an ordinary (nowadays called a tavern) and grain mill in the area for many years, and it was here he received his early education and training as a blacksmith, millwright, and innkeeper. It is likely he left the area after growing up, and he is reported to have served in the Revolutionary War. Sometime around 1782 he settled near Bath, Virginia, now Berkeley Springs, West Virginia. It was there that Rumsey initiated his work

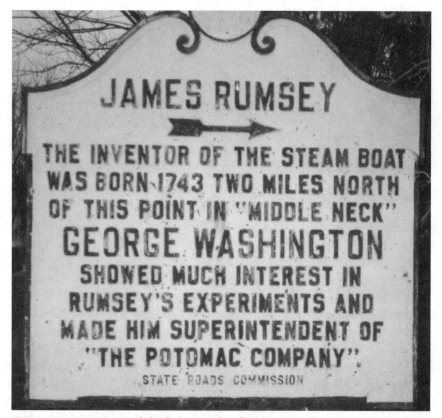

Figure 6.2. James Rumsey's birthplace is identified two miles from Warwick, Maryland, in this historical roadside marker. "Middle Neck" is a triangle-shaped section of land bounded by the Great Bohemia and Little Bohemia creeks on the northeast and southeast respectively and the Delaware state line on the east. (Photo by Jack L. Shagena.)

on a mechanically powered boat, followed by his work on the steamboat, and this passion for invention and success consumed much of his life until his untimely death in December 1792.

The progenitor of the Maryland family was Col. Charles Rumsey of the English army, who distinguished himself in Portugal under Cromwell. Later he was implicated in the Rye House plot against the reign of Charles II and was banished to St. Nicholas Isle near Plymouth for the rest of his life. It is said that his two sons, one of whom was also named Charles, were taken from him at that time.

Along with a relative, Charles Rumsey emigrated from Wales to Charleston, South Carolina, about 1665. In 1676 they then moved to New York, then to Philadelphia, and sometime later Charles struck out on his own, arriving in Maryland sometime before 1678.[1] Finding the Bohemia River off the Elk River at the upper part of the Chesapeake Bay, it is likely he explored the area carefully, looking for a place to establish a mill, build a house, marry, and raise a family.

About six miles from the mouth of the Bohemia, the wide river splits into a north branch, called Great Bohemia Creek, and a lower tributary known as the Little Bohemia Creek. Probably examining both branches, he found the upper fork to be navigable further east toward what was then Maryland, later Pennsylvania, and now Delaware.

About seven miles up the Great Bohemia Creek he located the limit of navigable water at a point where the creek separates into two much smaller streams. Calling this the "Head of Bohemia" (see fig. 6.3) he there constructed a dam and mill while establishing a dwelling about one mile down the Great Bohemia Creek on a bluff of land.

The location for the grain mill was strategic.[2] Running east and west a short distance north was the area's first cart road connecting

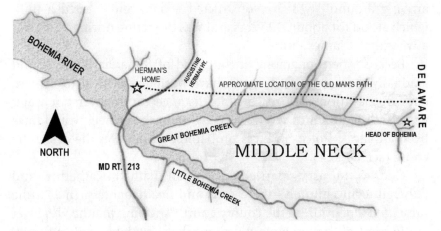

Figure 6.3. Map showing the "Head of Bohemia," near the location of the Rumsey family gristmill and tavern, and Bohemia Manor, the home of Augustine Herrman, who established the early trail known as the "Old Man's Path" across the Delmarva peninsula. The Bohemia River flows into the Elk River, which then empties into the Chesapeake Bay. (Map by Jack L. Shagena.)

the home of Augustine Herrman, Bohemia Manor, on the north bank of the Bohemia River with the shipping port of New Castle, Delaware. This route, which was really just a path through the woods, was nevertheless the principal cross-peninsula artery for the transport of goods for many years. It would later by eclipsed by the completion of the Chesapeake and Delaware Canal in 1829. The road, known for years as the "Old Man's Path," had been constructed about 1670 by Herrman, north of and parallel to where he had anticipated, as early as 1661, a canal connecting the head of the Bohemia with Appoquinmink River, which emptied into Delaware Bay.[3]

It was common at that time to identify locations by geographically descriptive terms, and the two streams that joined to form the Great Bohemia Creek could, with reasonable accuracy, be called the Head of Bohemia. As it was the most easterly navigable point, it was a logical place for a port of entry and subsequently became the location of America's first custom house.

With an overland trail nearby, navigable water to the Chesapeake Bay and Atlantic Ocean, adequate flow of water for a mill, and a shallow rocky stream bottom for fording, the location must have seemed to Charles Rumsey geographically ideal. Shortly after arriving around 1678, he constructed a dam and erected a mill, which stood for about 110 years and was operating during the formative years of James Rumsey.

Because the topography at the Head of Bohemia is gentle and gradually sloping, it is likely to have been impractical to achieve enough elevation in a reasonable distance for an overshot water wheel. With a dam of several feet, an undershot wheel would have been a logical choice, and it is probable that this was the design he chose (see fig. 6.4).

Charles Rumsey married a woman named Catherine and fathered eight children—three sons and five daughters. In 1710 he presented a petition to the county court: "shewing that he was Liver at the head of the Bohemia River and that he a had a wife and several small children to maintain, which to him were very chargeable, and continual passengers coming to his house, travellers from the province for Pennsylvania and from Pennsylvania to this province,

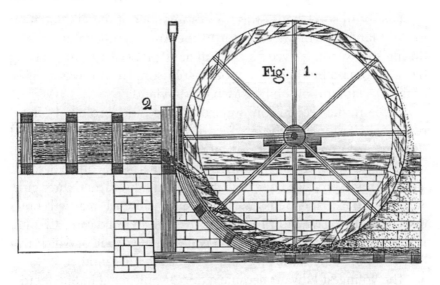

Figure 6.4. Schematic of an undershot water wheel that is probably similar to the one the Rumsey family constructed in Maryland and most likely like the wheels James Rumsey built in Berkeley County, Virginia. (From Oliver Evans, *The Young Mill-Wright & Miller's Guide* [1795; reprint, Wallingford, PA: Oliver Evans Press, 1990], pl. II.)

and to whom he in modesty gives entertainment and lodging, victuals, &c., without pay, which in time may amount to considerable sums of money"[4]; therefore he sought to be licensed to keep an ordinary. Because of his strategic location it was inevitable that waterborne travelers as well as those traversing the Old Man's Path would pass by his home looking for accommodations. The license from the county affixing fees for the food, drink, and lodging would provide a legal way to supplement his income.

It is likely that Charles Rumsey died in 1717, as that was the year his will was probated. He left to his son Charles tracts of land called New Hall, Concord, and Mill Pond totaling about eight hundred acres near the head of the Elk River. To William he bequeathed his plantation of about six hundred acres with a mill, as well as lots and wharves in Fredericktown; and to Edward, a tract of land consisting of one hundred acres, called "Adventure." Little did Charles realize that his great-great-grandson James would in three generations provide new meaning to the tract's name.

First-born son Charles was the commander of the Elk Battalion of the Maryland Volunteers prior to the American Revolution, and during the war he served as a Lieutenant of Cecil County, having been appointed by Governor Thomas Johnson. He also was a signer of the Declaration of the Free Men of Maryland.

William, the second son, went on to become a colonial officer, collector of customs for the province of Maryland, and a distinguished surveyor, acquiring a significant amount of land in Cecil County.

Edward, the third son, married a Miss Douglas of Scottish descent, and the union produced one boy, also named Edward, and three girls, Susanna, Mary, and Terissa. The younger Edward married Anna Cowman, who gave birth to at least five children: Edward; Charles; James, the steamboat inventor; and two daughters, one of which was Mary, and the other probably named Anna after her mother.

The youngest Edward became a doctor practicing in Shepherdstown, and sometime afterward in Lexington, Virginia, he operated a store with partner Leonard Robinson. He later moved to Christian County, Kentucky, and was named in 1788 by James Rumsey as executor of his will.

Other members of the family also left the Maryland homestead, with brother Charles marrying Betsy Maple of Sleepy Creek, and his sister Mary becoming the wife of Charles Morrow. The other sister, Anna, married Joseph Barnes, and it appears that all of them resided in the Shepherdstown area. This is probably the reason James eventually found his way to Sleepy Creek, near present-day Berkeley Springs, West Virginia, and was to benefit from his house and mill construction knowledge.

It is likely that while still in Maryland, James was married, as his daughter, Susanna, was not by his later marriage to Charles Morrow's sister, Mary. This latter union, however, did produce two children: a son, James, who because of scarlet fever was deaf and mute, and a daughter, Clarissa. It is worthwhile to note that brother and sister Charles and Mary Morrow married sister and brother Mary and James Rumsey. Later, when Rumsey went to England to construct his second steamboat, it was Morrow who looked after his sister and Rumsey's family.

On June 17, 1782, James Rumsey purchased a tract of land on Sleepy Creek called "World's End."[5] He was no doubt attracted to the site because of available waterpower, but it is interesting, and probably coincidental, that within a few miles of Rumsey's Maryland birthplace was a tobacco plantation known as Mt. Harmon at World's End. On Sleepy Creek, below a stone outcropping known as Butt-on Rock, Rumsey constructed his first sawmill.

Maj. George Michael Bedinger attested to Rumsey's service in the Revolutionary War, although there appear to be no records to confirm this. During this postwar period, Major Bedinger was living in the Bath area and decided to join Rumsey in a partnership in the milling business. To this end, Rumsey constructed the Bedinger Grist Mill on Sleepy Creek but found the day-to-day tedium of operations not to his liking. This prompted Bedinger to dissolve the partnership, and his biographer Danske Dandridge would later write about Rumsey, "A most unsatisfactory partner, Bedinger found the dreamy, absent-minded inventor, so much so, indeed, that Michael soon became disgusted with the whole concern, threw up the partnership, and, in 1784, went off to Kentucky to continue his career as a surveyor."[6] After this Rumsey constructed an early form of iron operation called a bloomery on nearby Meadow Branch for his brother Charles. At this site ore from Sleepy Creek Mountain was smelted into pig iron, with the same millrace being used to power both a gristmill and the bloomery.

In 1783 land developer Raleigh Colsten had Rumsey construct a mill that was uniquely designed to operate satisfactorily with low water. It was probably an undershot wheel similar to what his ancestor had constructed on the Great Bohemia Creek, and from this he would study and gather information of the relative efficiency of undershot versus overshot waterwheels. Use of this data would be a determinant in his later selection of the best means by which to propel his steamboat.

During 1783 Rumsey's vivid imagination had been examining and analyzing ways to propel a boat against the current. By the middle of the year he had revealed to John Wilson of Philadelphia some of his initial thoughts and plans to employ the power of steam. Before doing this, however, Rumsey became convinced that a far simpler mechanical approach was possible. This consisted of a paddle or flutter wheel in front of a boat (or a pair of boats) where the vessel was held against the current by setting poles from the stern to the stream's bottom. As the current of the stream rotated the wheel, the poles would be successively activated to push the boat forward (see fig. 6.5). This was not a new idea, but it is not known if Rumsey conceived of it independently.

Around this time Rumsey met James McMechen, who agreed to provide financial support to the always cash poor inventor, with the understanding of sharing in the profits derived from his inventions. This arrangement was formalized in a written agreement on November 10, 1784, with McMechen receiving a 25 percent share of the future profits. Sometime before the middle of 1784 they approached the Virginia General Assembly, and on June 28, 1784, the *Journal of the House of Delegates* reported,

> Whereas, it has been represented to the assembly by James Rumsey and James McMechen that they have discovered the art of constructing a boat of the burthen of ten tons, which shall sail or be propelled by the influence of certain mechanical powers thereto applied, the distance of between twenty-five and forty miles per day, against the current of a fresh water river, notwithstanding the

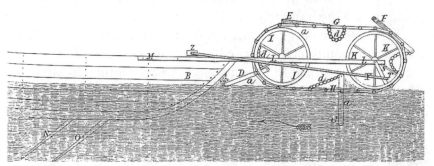

Figure 6.5. Rumsey's so-called *Stream* Boat. The current of the stream water pressed against the bow-mounted paddle or float boards, causing the wheels to rotate. Through mechanical coupling the setting poles, labeled N and O, pushed the boat upstream. (From James Rumsey, *Applying Water and Steam Power to Machinery, to the Propulsion of Vessels, &c.*, British Patent No. 1738, May 22, 1790 [London: George Eyre and Wm. Spottiswoode, 1855], fig. 21.)

velocity of the water, with the said burden of ten tons on board to be wrought at no greater expense than that of three hands. This Assembly being desirous of encouraging every effort of genius and industry which may tend to public good Resolved, That upon the said James Rumsey and James McMechen perfecting such a boat as above described this Assembly will grant them or their heirs a sum adequate to the importance of the discovery.

Ordered, That Mr. Alexander White do carry the resolution to the Senate and desire their concurrence.[7]

This provided for a possible reward but did not address an exclusive right to operate his invention in the waters of the state, a consideration he would later come to believe as being more important.

Looking for a business opportunity to sustain himself and provide funds for his mechanical boat and steamboat experiments, Rumsey entered into a partnership with Nicholas Orrick in 1783. They operated a mercantile business in Bath and, according to Orrick in an October 18, 1816, statement made under oath, "we lived together and were very intimate as partners and friends."[8] The length of the partnership probably did not last more than one year, as by June 1784, Rumsey and Robert Throgmorton had opened a commodious boarding house in Bath called "at the sign of the Liberty Pole and Flag."

Demand for accommodations in the town was high, as the warm springs had developed a reputation as a place for healing. Methodist bishop Francis Asbury, however, described it otherwise, calling the town Sodom and Gomorrah in miniature. Rumsey had not yet stumbled into his life's work, but the inn was to provide a unique opportunity for this aloof, absent-minded genius that would prove to be extremely fortuitous. He was finally in the right place at the right time with the right idea.

On September 6, 1784, Gen. George Washington was on his way west examining the possibility of navigable water communications with the East Coast (see fig. 6.6). Having previously acquired some building lots in Bath, he spent the night at the new inn owned by Throgmorton and Rumsey. Meeting Rumsey, the general was impressed by his manner and unassuming confidence and agreed to

Figure 6.6. George Washington's certificate provided to James Rumsey for a *stream* boat, not a steamboat, gave the beleaguered inventor some much-needed credibility. Shortly after receiving the endorsement, Rumsey informed the general of his plans to also build a steamboat. (From Alexander Johnston, *History of the United States* [New York: Henry Holt and Company, 1889], p. 149.)

witness a demonstration of his mechanical pole boat that would walk itself up the stream. It had been reported that they ventured to nearby Sir Johns Run for the demonstration, but it may well have taken place in the small stream at the rear of the inn.[9] In any case, Washington was impressed and the following day provided Rumsey a certificate:

Town of Bath, County of Berkeley in the state of Virga.,
September 7, 1784

I have seen the model of Mr. Rumsey's Boats constructed to work against stream; have examined the power upon which it acts; have been eye witness to an actual experiment in running water of some rapidity; and do give it as my opinion (altho' I had little faith before) that he has discovered the Art of propelling Boats, by mechanism and small manual assistance, against rapid currents: that the discovery is of vast importance, may be of the greatest usefulness in our inland navigation, and, if it succeeds, of which I have no doubt, that the value of it is greatly enhanced by the simplicity of the works which when seen and explained to, might be executed by the most common Mechanics.[10]

Washington, having a favorable impression of Rumsey's mechanical abilities, hired him to construct a house, a kitchen, and a stable on his property in Bath. After some delays occasioned by bad weather and a fire at the sawmill that destroyed Washington's lumber, the preoccupied Rumsey finally completed the structures. Probably dreaming about his next invention, however, his work was not completely satisfactory; nevertheless, this did not seem to injure the general's confidence in him.[11]

As the federal Articles of Confederation did not have a patent statute, Washington urged the secretive Rumsey to obtain protection from the states. So, in early November 1784, Rumsey traveled to Richmond, where the Virginia Assembly was in session. There he met Washington, divulged to him his plans for a boat propelled by steam, and petitioned the assembly for "the sole and exclusive right of constructing and navigating certain [mechanical pole or stream] boats for a limited time."[12] He was successful in obtaining from Virginia a ten-year patent starting January 1, 1785, and later from Maryland a similar agreement dated January 22, 1785; from South Carolina, encouragement in March of the same year.

Rumsey had conceived the idea of a steamboat shortly after moving to Bath, as in his 1788 pamphlet *Explanation of a Steam Engine*, he states, "The idea of applying steam to the purpose of navigation, occurred to me as early as 1783."[13] This is borne out by John Wilson and Juliana Stewart of Philadelphia. When Wilson was visiting Bath, also called Warm Springs, in late July or early August of 1783, Rumsey told him "he intended to construct a boat to go by the power of steam, and pointed out the great expenses it would save in water carriage." When Wilson returned to Philadelphia he related this story to Juliana Stewart, and later they both signed certificates to that effect.[14]

Although much of his work was done in secrecy, Rumsey did reveal his plans to several friends. In addition to George Washington, his mercantile business partner, Nicholas Orrick, and his steamboat partner, James McMechen, he confided his steamboat ideas in March 1784 to his former mill partner, George Michael Bedinger, before he left for Kentucky; to George Rootes in 1784; and Charles Morrow in 1785.[15]

Rumsey's concept of a steam engine was probably derived from a 1744 description and drawing of a Newcomen mine pump engine in

J. T. Desaguliers's book, *A Course of Experimental Philosophy* (see fig. 6.7).[16] Although the book is not mentioned in Rumsey's writings, it is known that he possessed a copy, as it was used to illustrate a point with a Baltimore inventor, Englehart Cruse.[17] This slow-operating, heavy, stationary design of wood and iron set on a masonry foundation was entirely unsuited for a mobile application; but the possibility of using a steam engine to power a boat set Rumsey's fertile mind to work on how to adapt the basic concept to steam navigation.

A highly visual thinker, it is likely that he was able to examine each individual component that would make up a steamboat and then conceptualize how they would be integrated together to achieve a working system. He had to consider each of the subsystems (first introduced in chapter 3), including the vessel, means of

A

C O U R S E

O F

Experimental Philofophy.

B Y

J. T. DESAGULIERS, LL. D. F. R. S.

Chaplain to HIS ROYAL HIGHNESS
FREDERICK, PRINCE of WALES, &c.

Formerly of *Hart-Hall* (now *Hertford-College*) in *Oxford*.

VOL. II.
Adorn'd with Forty-fix COPPER-PLATES.

L O N D O N:
Printed for W. INNYS, at the Weft End of *St Paul's*; M. SENEX,
in *Fleet-ftreet*; and T. LONGMAN, in *Pater-nofter-Row*.
M. DCC. XLIV.

Figure 6.7. In 1744 Desaguliers published volume 2 of his book on "experimental philosophy" (i.e., physics) that contained text, drawings, formulas, and examples that could be understood without knowledge of calculus. It is known that James Rumsey had a copy and was familiar with a Newcomen steam engine described therein. (From J. T. Desaguliers, *Course on Experimental Philosophy*, vol. 2 [London: printed for W. Innys, M. Senex, and T. Longman, 1744], title page.)

propulsion, motive power, a transmission, a steam generator, a vacuum generator, and a forcing pump. Working through the list, it is possible to glean from his writings a few of his thoughts.

Vessel. The steamboat starts with a vessel, and in May 1785, Rumsey engaged his brother-in-law Joseph Barnes to construct a boat of six tons' burden. It was built at Sir Johns Run, near Bath, and was completed in the summer of 1785, when it was then floated down the Potomac to Shepherdstown, where it awaited the machinery. Barnes, however, later said he did not know until after the boat was completed that Rumsey intended to power it by steam.[18]

Rumsey was keenly aware that the shape of the boat would have a significant impact on its resistance through the water, so it is likely that he provided his brother-in-law with some specifics on the hull design. Rumsey learned about a vessel's resistance through the water from experiments on his mechanical boat that he reported to George Washington in a letter, dated March 10, 1785: "I find the Resistance of water against Boats Increases Exactly as the Squares of the Velocity of the Boats against it,[19] Nither Can their be a general Rule to give the Resistance that Boats of the Same Burthen and Velosity meets with If their forms is Different for I find that Bad Shaped Boats meets with nearly three times the Resistance that good ones Do of the same Burthen." Summing up his progress, he provided a rule-of-thumb formula that three square feet of paddleboard is required for each ton taken up by his mechanical boat. He goes on to write about his steamboat,

> I have taken pains to afect another kind of Boats upon the princeples I mentioned to you at Richmond [i.e., a steam-propelled boat] I have the pleasure to inform you that I have Brought it to the greatest perfection[.] It is true it will Cost Sum more that the other way But when Done is more manageable and Can be worked by a few hands, the power is amence and I am quite Convinced that Boats of pasage may be made to go against the Current of the Mesisipia or ohio River, or in the gulf Stream from the Leward to the the Windward Isslands from sixty to one hundred miles per Day. . . . [It is] Stirckly agreeable To philosophy, The principles of this Last kind of Boat I am Very Cautious not to Explain to any

person, as it is easy performed and the method would Come Very natural to a Rittenhouse or an Elieott.[20]

From this account it is clear that at this point Rumsey had thought out the steamboat design and had a fairly clear image of how he planned to proceed. In addition, his comment regarding being strictly in agreement with "philosophy" implies that he had some understanding of the underlying scientific principles.

Propulsion. Rumsey had probably gathered information on the efficiency of undershot water wheels from the low-water mill he constructed for Raleigh Colsten in 1783 and from his mechanical boat with the flutter wheel in the front. In his later British Patent No. 1738 of May 22, 1790, he states the efficiency of an undershot wheel is not more than one-eighth or 12.5 percent, a figure he probably derived from experimentation.[21] As Edwin Layton points out, the figure is low: John Smeaton later found by experiment that the efficiency is closer to one-third.[22]

From Rumsey's knowledge of the resistance of a boat's hull through the water, he no doubt realized the resistance would be the same for the boat moving through the water or the boat remaining stationary and the water moving past. Understanding this engineering principle of reciprocity and believing the efficiency of an undershot to be only 12.5 percent, he likely concluded the paddle wheel to be too inefficient for use on a steamboat. This conclusion, however, was wrong.

Rumsey's underestimation of paddlewheel efficiency led him in January 1785 to experiment with water jet propulsion (see fig. 6.8). Nicholas Orrick provides a description of the apparatus he used:

> He caused to be made a hollow square tube (made of pine boards) about eight feet long, and about one inch and a half diameter in the cavity of the tube, which he suspended upon fine cords, and having a common gimlet hole bored at the end of said tube, and on the one side of the same, and hanging some weights over a small pulley, by a thread tied to the end of said tube, whereupon pouring water by hand into the tube, it drew up a certain weight; which being repeatedly tried, I asked Mr. Rumsey what he meant

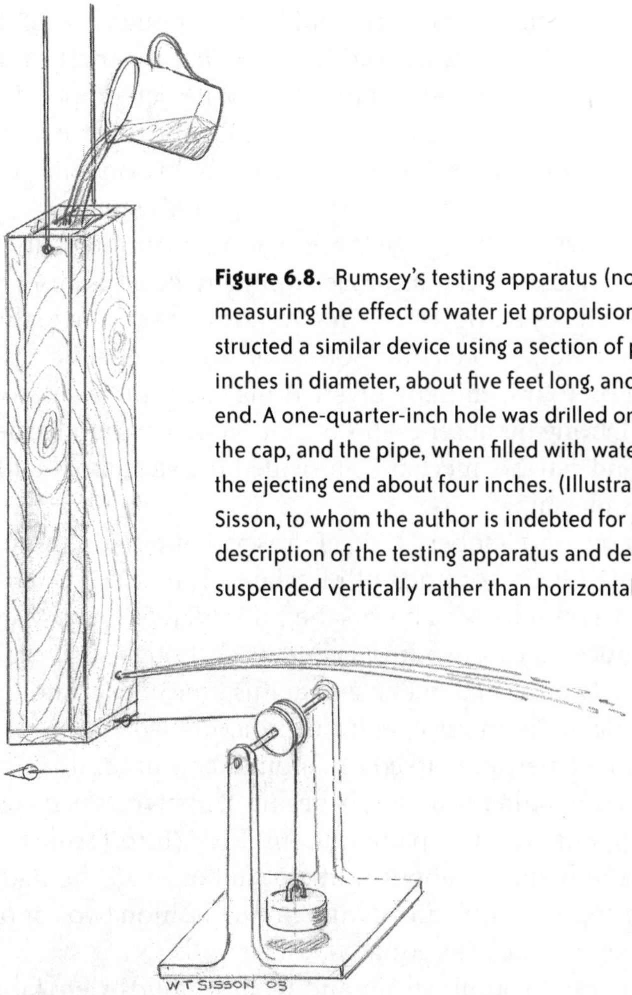

Figure 6.8. Rumsey's testing apparatus (not to scale) for measuring the effect of water jet propulsion. The author constructed a similar device using a section of plastic pipe two inches in diameter, about five feet long, and capped on one end. A one-quarter-inch hole was drilled on the side through the cap, and the pipe, when filled with water, would deflect the ejecting end about four inches. (Illustration by William T. Sisson, to whom the author is indebted for analyzing the description of the testing apparatus and determining it was suspended vertically rather than horizontally.)

or intended by the experiment, to which he made answer, that by that principle he would make the boat go, and after some small matter of conversation, he took his pen and ink, and retired to make calculations, as was his custom after experiments made, many of which he made during the winter.[23]

Apparently based on these results and his estimation, albeit low, of the efficiency of the paddle wheel, Rumsey decided to use water jet propulsion for his steamboat. It has been suggested that Rumsey learned the concept from Benjamin Franklin, as in August 1785 he returned to America from France and in December reported to the American Philosophical Society Bernoulli's water jet proposal as well as many other things he had learned while abroad.[24] Rumsey claimed, however, that water jet propulsion was his original idea and indeed he experimented with it before Franklin's return.

In July 1785, acting on George Washington's recommendation, the Board of Directors of the Potomac Company hired Rumsey to superintend their operations of making the river navigable. Although this provided a welcome source of income for the cash-strapped inventor, it also initially diverted his attention away from his boat work. During his absence, his brother-in-law Joseph Barnes, an intelligent and capable mechanic, provided the steamboat effort the necessary leadership.

Motive Power. In October 1785 Rumsey contacted Thomas Johnson, who had been governor of Maryland from 1777 to 1779 and who owned a foundry with his brother in Frederick Town, Maryland, with a request to cast two iron cylinders. Although such cylinders were routinely produced in England at this time, the technology was not yet available in America, and the project failed.

Rumsey immediately embarked on another approach, this time having a wooden mandrel thirteen inches in diameter turned on a lathe and employing this as a pattern around which to form sheet copper. Using the mandrel, about mid-November 1785, he had a Frederick Town coppersmith named Mr. Zimmer fashion two copper cylinders each about three feet long.

Four brass cocks to admit steam and regulate fluids were fabricated by Christopher Raburg of Baltimore, and these were delivered by the end of October. Thereupon Conrad Byers fitted two of the cocks with handles and springs. Byers also made "two pistons of about thirteen inch diameter, to which he brazed flanches, about one and a half inches broad" to be used inside the cylinders.[25] Most likely the pistons were fitted with leather or oakum for a good fit in the copper cylinder (see fig. 6.9) as he mentions these sealing materials in his British patents. In Rumsey's 1788 description of his steam engine and boat, he showed water being dribbled onto the

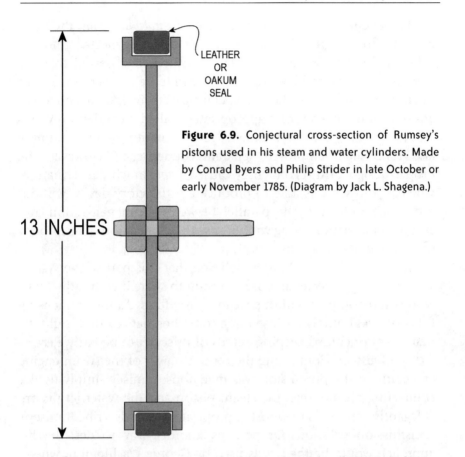

LEATHER
OR
OAKUM
SEAL

13 INCHES

Figure 6.9. Conjectural cross-section of Rumsey's pistons used in his steam and water cylinders. Made by Conrad Byers and Philip Strider in late October or early November 1785. (Diagram by Jack L. Shagena.)

tops of the pistons to keep the seal wet, thereby achieving a better fit.[26] The pistons were taken to Zimmer's workshop, where two of them were fitted into the cylinders, with the job being completed before Christmas of 1785.[27]

Vacuum generator. Curiously, nothing in the writings of Barnes or Rumsey has been found to describe how the vacuum generator or condenser was constructed. In his 1788 *Explanation of a Steam Engine,* on plate 3, Rumsey barely shows the barrel-like condenser hidden behind the furnace (see appendix A). The description only mentions that steam is passed off "for condensation in the condensing vessel." He also shows a pipe leading from the vessel to the input water chamber that allowed the warm condenser water to be expelled overboard.

In Desaguliers's 1744 *Course of Experimental Philosophy*, the New-comen "fire" engine described used cold water injected into the cylinder for the vacuum cycle. James Watt had observed that the alternate heating and cooling of the cylinder led to very inefficient operation, and by 1774 he was operating a steam engine with a separate vacuum generator he appropriately called a condenser. Watt's work was fairly common knowledge in America, as both Rumsey and John Fitch employed separate condensers.[28] However, the proper design of the condenser caused Fitch much consternation, whereas Rumsey was able to produce a satisfactory device without too much difficulty. The parallel between steam generation and steam condensation along with Rumsey's appreciation of reciprocity likely accounts for his success.

It is possible that Rumsey believed that this part of the system should be kept a secret and did not want to share it with individuals who could compete with his steamboat efforts. As the condensing function was entirely enclosed in a container, hence out of sight, he may have concluded keeping details of its secret to be fairly easy.

Transmission. Connecting the output power of the steam engine efficiently to the propulsion was ingenious: Rumsey simply used a connecting rod between the steam piston and the water jet piston. "Elegantly simple" is the most appropriate way to describe Rumsey's transmission solution. Or perhaps another way to describe his approach would be the words used by George Washington, where-upon learning Rumsey had moved from a stream boat to a steamboat said, "[It is] the ebullition of his genius."

Steam generator. Up to this time Rumsey planned to use a pot boiler; however, he would quickly learn of its inherent capacity limitations. In the winter of 1785–86 he invented his pipe or tube boiler, which was destined to become the standard for the steam-generating industry. This also was a critical step in achieving a steamboat with equipment small and light enough to leave sufficient room for passengers and cargo.

In his 1788 British patent titled "Boilers for Steam Engines," Rumsey established a well-thought-out understanding of boiler design, presenting numerous variations. One of these was the bent-pipe, zigzag configuration he had previously published in the May 1788 issue of *Columbian Magazine* (see fig. 6.10). In his patent he even proposed preheating the injected water by "employing the heated air and smoak [*sic*] after it has been applied to the boiler to warm the vapor or fluid intended to be evaporated previous to their entering the boiler."[29]

Assembly. Although no account appears to exist as to the fabrication of pumps, they are reported as complete by December 1785:

Mr. Joseph Barnes and James McMechen brought down the river Potowmac, to the Shannandoah [*sic*] Falls [i.e., Harpers Ferry], a boat of about six ton burthen, with a variety of machinery on board, amongst which were two cylinders of copper, about thirteen inches in diameter, and near three feet long, a copper boiler,[30] four large brass or copper cocks, *pumps,* &c., where the said Barnes and McMechen, under the direction of Mr. James Rumsey, continued adapting and suitably fixing the said machinery to said

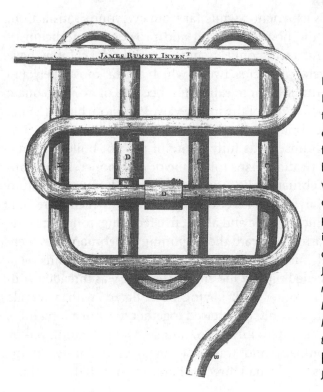

Figure 6.10. Rumsey's bent-pipe tube boiler, described by him to form pockets of a bottle case. Water was injected via a check valve into the preheated pipe that immediately produced steam. (From James Rumsey, *Explanation of a Steam Engine, and the Method of Applying it to Propel a Boat* [Philadelphia: Joseph James, 1788], pl. 1.)

boat, until the seventh of January, 1786, when the ice driving in the river, obliged them to desist.[31] (emphasis added)

Trial 1. The equipment was removed from the boat and stored in the cellar of Francis Hamilton for the winter. As soon as the weather permitted, the team was back at work, and on March 14, 1786, the first trial was made.[32] Hamilton reports, "She moved against the current some distance, though not to much satisfaction, owing to the imperfections of the machinery."[33]

According to Barnes, the heat of the steam rendered some parts useless, necessitating repairs, primarily consisting of replacing some of the soft solder with hard solder, which has a higher melting point.

Trial 2. After some days of work, another trial was attempted, and Barnes reported, "and she was again tried, but failed in the repair work, though it made many powerful strokes before it failed, and sent the boat forward with such power that one man was not able to hold her."[34]

No doubt this experience, while far from everyone's satisfaction, lent credence to the power of steam and provided a modicum of encouragement. Achieving only "many powerful strokes" may have resulted from several problems, two of which appear most likely: (1) the water in the trunk became exhausted because of an inadequate design or improper action of the intake valves, or more likely (2) the capacity of the pot boiler to produce steam was overcome.

By this time Rumsey was fully aware of the pot boiler's limitations and had fabrication plans for a tube or pipe boiler. In late January and early February 1786, Rumsey obtained from Anteatum Iron Works in Maryland six so-called gun scalps, which were flat-iron bars several inches wide and about five feet long, rolled into the shape of a gun barrel.[35] Toward the beginning of February they were taken to Michael Entler in Shepherdstown, where the seams were welded to form sealed pipes. One end of the pipe was threaded with a male thread; the other end, having been flared, with a female thread so that several could be screwed together to form a long tube.

At this point, and up to July 1786, Rumsey was still employed by the Potomac Company and was away most of the time. In his absence it appears that he had directed Barnes to install the mechanical pole equipment on the boat, as the pipes lay on the floor at the welder's facility for six months or longer.[36] In September or October 1786, Barnes returned to the welder (blacksmith), and we have this account from Entler: "Mr. Joseph Barnes came to me, and told me he wanted them put together; said Barnes then prepared a block [i.e., cylinder], and assisted by me to screw them together and bend them, they were bent in the shape of a worm of a still [i.e., tight spiral], with this difference, that the rounds were placed so close, as nearly to touch each other . . ." (see fig. 6.11).[37]

Trial 3. The pipe or tube boiler was fitted into the furnace and connected to the other parts of the steam engine. The following is Barnes's account of the trial:

> The next experiment was attempted in December [1786] with the new constructed boiler, but the violence of the heat was so great, from the steam, that it melted the soft solder that the great part of

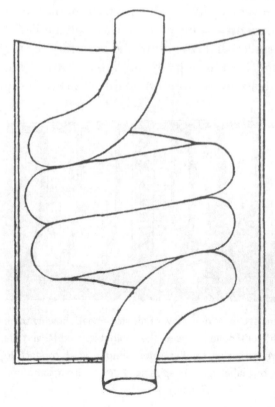

Figure 6.11. Rumsey's tight-spiral tube boiler was placed in a furnace to generate steam for his Potomac River steamboat, but leaks developed where the pipe sections were screwed together. (From James Rumsey, *Boilers for Steam Engines, for Distillation, &c.,* British Patent No. 1673, December 6, 1788 [London: George Eyre and Wm. Spottiswoode, 1854], fig. 12.)

the machines was put together with, and rendered it entirely useless, until repaired with hard solder; about this time, the ice drifting, carried off the boat which the machinery was made for, and destroyed her in such a manner, that repairing her was equal to one half of the expense of building a new one.[38]

Hard luck once again had fallen to the hapless Rumsey, who since July 1786 had devoted all of his energies to his steamboat, to the detriment of other responsibilities. It was now necessary to wait until spring 1787 to initiate the repairs to the badly damaged boat and equipment.

Trial 4. Rumsey went to work repairing his newly constructed pipe boiler and equipment and, probably before the boat was made seaworthy, had the boiler operating by July 1787. When it was tested, Charles Morrow was very impressed with its performance, saying he conceived it "to be the most capital contrivance to make steam that can ever be invented, for when the machine is not at work, the whistling of steam may be heard for at least half a mile."[39]

Trial 5. Finally, by September 1787, all of the necessary repairs had been made, and the boat was ready for another trial (see schematic, fig. 6.12). With about two tons on board, exclusive of the

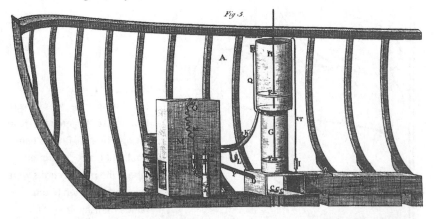

Figure 6.12. Rumsey's isometric sectional bow-view of his steamboat, showing some of the equipment installed. The tube boiler is inside the cabinet labeled M, and the steam cylinder is stacked above the slightly smaller water cylinder, which connects to the water jet trunk via a valve box labeled D. (From James Rumsey, *Explanation of a Steam Engine*, fig. 5.)

machinery, Rumsey was finally able to enjoy some small measure of success. The puffing craft steamed two miles per hour on the Potomac, "but the new boiler was so badly made that it opened at several joints, which let great quantities of steam escape."[40] Despite this serious problem, however, Rumsey and his team could nevertheless taste success by realizing upon solving the boiler problem, both the pressure and volume of steam would increase, thereby achieving a proportional increase, in vessel speed. They were nearly ready for a public trial.

Trial 6, for the public, on December 3, 1787. It was time for the work of the so-called crazy Rumsey to be examined by the public. Rumsey had in the days and weeks beforehand made sure a suitable number of known and respected dignitaries were present on the date to certify the boat's performance. Included in the anxious crowd was retired Revolutionary War major general Horatio Gates (see fig. 3.13); Robert Stubbs, a teacher at the academy in Shepherdstown; and about a dozen other local residents who were well respected. Invited on board were a group of about six ladies, the very first women ever to ride on a steamboat, and Rumsey thrilled the skeptical crowd by propelling his boat upstream at three miles per hour, continuing the demonstration for three hours.

Trial No. 7, for the public, on December 11, 1787. After

Figure 3.13. Retired Maj. Gen. Horatio Gates served in the Revolutionary War with Washington and was invited by Rumsey to his public demonstration, as he was judged to be a creditable witness. (From E. Benjamin Andrews, *History of the United States* [New York: Charles Scribner's Sons, 1915], p. 97.)

making some repairs to the boiler, another public demonstration was given on December 11, this time, a speed of four miles per hour was achieved. Rumsey was a local hero and was proclaimed the inventor of the steamboat. His work, however, was far from finished. He understood completely that his boat was imperfect in many respects and that much money would be required to develop it into a practical commercial entity. He was also fully aware that rival John Fitch, who was working on a steamboat in Philadelphia, had made several trials and was already claiming to be the inventor of steam navigation.

Rumsey could lose no time. On December 17, 1788, while he was in Annapolis, Maryland, visiting the governor and delivering his certificates to the General Assembly, he wrote George Washington and enclosed two certificates that attested to these successful steamboat trials. Rumsey feared, however, that the general would be displeased, as he had failed to carry through his pole boat work, for which Washington has issued a certificate. He therefore spent a significant part of the letter recapping events that had led him to abort the pole boat plan in favor of the steamboat. He was searching for a way to acknowledge the general's support of him but was uncertain as to how to proceed: "I would wish to Say Something to the public about it, on your account. But Doubt my own Abelityes [abilities] to give that Satesfacteon I would wish, It has gave me much uneasyness Especially as I have By a train of Unforseen Events So Often apeared to you as a parson acting Inconsistantly and I can Say in truth however unfortunate I have been in the attempt that my greatest ambetion is & has been to Deserve your Esteem."[41] Rumsey did not mention the pamphlet he was preparing and no doubt hoped that Washington would write back providing him something that could be used to bolster his claim for priority in the steamboat work. However, the general is not known to have responded, as Rumsey continued writing his pamphlet and for the time being would leave out any reference to his March 10, 1785, letter to Washington, in which the steamboat was addressed.

During the latter part of December 1787, Rumsey completed his pamphlet and organized notarized copies of statements about his boat, which he called certificates. He located a printer in Berkeley

County who produced copies dated January 1, 1788, with the title *A Plan, Wherein the Power of Steam Is Fully Shewn, by a New Constructed Machine, for Propelling Boats or Vessels, of Any Burthen, against the Most Rapid Streams or Rivers, with Great Velocity.*[42] In the introduction he attests to the technological and fabrication difficulties he faced in his steamboat work:

> This is the fate of the unlucky projector, even in the cities of Europe, where every material is at his command, and every artificer at his service. A candid public will then consider my situation, thrown by hard fate beyond the mountains, and deprived of every advantage which the grand mover, money, produces. They will easily perceive how my difficulties have been multiplied, which is the only reason of my not exhibiting my long promised BOAT before this; and which I hope will be a sufficient apology. Even now, these difficulties render my machinery very incomplete.[43]

Rumsey made plans to travel to Philadelphia, providing Charles Morrow with his power of attorney. Still not hearing from George Washington, on March 25, 1788, Rumsey sent him five copies of the pamphlet and provided a cover letter apologizing for using the general's name and advising him of his travel plans:

> With this you will Receive five pamphlets Respecting My Boat and other plans, the Subject is not handled Quite to my wish as I was Obliged to get a person to correct my Coppy. In doing which my Ideas in Several places were new[ly] modled, but not so much as to injure the truth, I wished to Introduce, But has made Sum things rather Obscure.
>
> I hope sir, that the necessity there was of such an Explanation being made to the public, will plead my Excuse for Taking the Liberty of Introducing your name into my Concerns, and Shall Do my best Endeavors to Conduct myself In Such a mannor as in Sum degree to Deserve the Honor it Does me, you may Rest ashored that all that I have proposed is within my power to perform, I have proven them all Experimentilly, and have modles by me Sufisantly Large to Convince any Compitant Judge (that may Examom them) of the truth thereof—

Tomorrow morning, I throw myself upon the wide world In pursuit of my plans, being no longer Able to proceed upon my Own foundations, I Shall bend my Course for Philadelphia where I hope to have it in my power to Convince a Franklin and a Rittenhouse of their utility by actual Experiment, as Mr. Barnes is to Set out in about ten Days after me with all the machinery in a wagon.[44]

The die was cast. Rumsey had somewhat boldly sought to capitalize on Washington's early support and, not having received any objection from the general, decided to liberally use his name to support his claims of priority. Despite the positive and upbeat tone reflected in his letter, underneath this veneer was a significant amount of anxiety. In a melancholy letter to Charles Morrow written from London on the anniversary of his departure from Shepherdstown (see his home in fig 6.14), he reflected on the previous year:

I believe nothing Short of Desperate Circumstances Such as mine then was, could have Enabled me to persue the precarious undertaking I has Set out upon and leave the once peaceful Shades of Berkeley; you Can have no Idea my friend of the Various Sienes [sinews] of anxiety and care, that I have went through, attended with the agitations—both of body & mind that hope, fear, success, and disappointment in general occationed, yet not retreat, (except that of poverty and Contempt through a wilderness of want) was left for me to attempt and therefore a desperate persuit became necessary.[45]

Rumsey continues, "*Conquer or fall* was my motto," and this was squarely on the line as he was off to the scientific capital of America in search of financial support and help from more skilled mechanics. For sure, he had trepidations about laying his life's work in front of the American Philosophical Society and the most respected scientists in Philadelphia, but for one time in his inventive career he was not to be disappointed.

On April 18, 1788, the American Philosophical Society received a letter from Rumsey describing his water tube boiler, improvements to a saw and gristmill, and a proposal for pumping water by steam.

Figure 6.14. According to tradition, the Rumsey home in Shepherdstown. However, houses constructed during the latter part of the eighteenth century often had outside chimneys; inside chimneys became popular with the introduction of stoves about fifty years later. (From Turner, *James Rumsey*, between pp. 76 and 77.)

A subcommittee consisting of Dr. John Ewing, David Rittenhouse, and Dr. Robert Patterson was appointed to examine Rumsey's inventions, as well as a water tube boiler submitted by Henry Voight, partner of John Fitch. At stake was an annual premium offered by the society for new, useful ideas. In defense of the originality of his boiler, Rumsey provided the society on May 10, 1788, "The plan of a boiler before the Philosophical Society, by Mr. Voight . . . is different from mine in form but not in principle, and I have good reason to believe that he got his first idea of it from mine, which I shall endeavor to make appear—at any rate I shall incontestibly [*sic*] prove, that I was two years before him in that invention."[46]

This argument as to first priority would not stop with the boiler, as Fitch announced that he was planning to publish his own pamphlet refuting Rumsey's claim to having invented the steamboat. To counter this, Rumsey decided that his limited supply of pamphlets published in Berkeley County, Virginia, would not be adequate; hence, he decided to reprint it with some corrections and also include the letter to George Washington of March 10, 1785, identifying the date he had first disclosed steam to the general. The pamphlet was renamed *A Short Treatise on the Application of Steam Whereby Is Clearly Shewn from Actual Experiments That Steam May Be Applied to Propel Boats or Vessels of any Burthen against Rapid Currents with Great Velocity*, was printed by Joseph James of Philadelphia, and was dated May 7, 1788.

Fitch counterattacked on May 10, 1788, an overcast rainy afternoon, with his pamphlet, titled *The Original Steam-Boat Supported; or, A Reply to Mr. James Rumsey's Pamphlet Shewing the True Priority of John Fitch, and the False Datings, &c. of James Rumsey.*[47] The battle lines were drawn, with each attempting to hold the same ground. Fitch had obtained certificates from several individuals who supplied components for Rumsey's steamboat from which one could infer that Rumsey did not start work until about one year later than he claimed. If these were indeed true, it would allow Fitch to claim priority, hence maintain that he, not Rumsey, invented the steamboat.

The counterattack was to lead to another offense by Rumsey's supporters on July 7, 1788, when Joseph Barnes, now acting as Rumsey's attorney, published in Philadelphia a pamphlet with the concise title of *Remarks on Mr. John Fitch's Reply to Mr. James Rumsey's Pamphlet*. Barnes revisited the individuals who had provided Fitch certificates; he found that at the time the certificates were given to Fitch, the dates provided were from memory, but now the individuals had found receipts showing actual dates in consonance with Rumsey's pamphlet. Barnes accused Fitch of fraud, as Fitch, having been aware of at least one wrong date in one of his certificates, nevertheless printed the erroneous information.

Although the pamphlets did have some effect on public opinion, key decision makers of the day, most of whom were members of the American Philosophical Society, were quick to side with Rumsey. His inventions—not just the steamboat but also the tube boiler, his saw and gristmill improvements, and his plan to raise water by steam—signaled to them that he was a true genius. Some of the members, having remembered Fitch from a 1785 visit to the Society, had concluded that by comparison with Rumsey, the socially awkward Fitch was simply not as creditable.

On May 1, 1788, Rumsey advertised in a broadside a "Proposal To forming a Company, to enable James Rumsey to carry into execution, on a large and Extensive Plan, His Steam-Boat and Sundry other Machines herein after Mentioned." Rumsey would retain one-half interest in the company, with the other half being divided into fifty shares, to be sold at a cost of twenty Spanish-milled dollars each. Response was immediate, with Benjamin Franklin (see fig. 6.15) and a number of others signing up in little over one week. By May 9, 1788, the company was formed and named the Rumseian Society, and by May 15, the Society had nineteen members, with forty-four of the fifty shares being subscribed.

When it was learned (or perhaps rumored) shortly after the new group's organizational meeting that Fitch's company was planning on patenting Voight's water tube boiler in Europe, the members of the Rumseian Society were incensed. Wanting to protect their investment, they met on May 13 and subscribed another one thousand dollars with the express purpose of sending Rumsey to Europe for patents. Benjamin Franklin and others provided letters of introduction, and on May 14, 1788, with a letter of credit for two hundred pounds sterling, Rumsey set sail on a very fair day with a prevailing southwest wind heading for London.

Figure 6.15. Benjamin Franklin was the first to buy shares in the Rumseian Society, and his confidence in the inventor led others to quickly subscribe. (From John Clark Ridpath, *A Popular History of the United States of America* [New York: Phillips & Hunt, 1883], p. 330.)

One of Rumsey's first business meetings in London was with the firm of Boulton and Watt, makers of the world's finest steam engines. Many of the problems Rumsey and Fitch had faced in their steamboat work were directly related to developing a suitable engine to propel their vessel. If Rumsey could succeed in working out a business arrangement with Boulton and Watt for the steam engine, a great deal of his difficulties could be overcome.

Matthew Boulton and James Watt courteously received him and were so impressed with his ideas and ability that the precepts of a draft agreement for a business arrangement were discussed (see fig. 6.16). Although Rumsey believed that they had achieved "a meeting of the minds," he was nevertheless shocked upon receiving the written proposal, as he considered it one-sided in their favor.

What Rumsey failed to recognize, however, was that though he had a fertile inventive mind with great ideas, they were not of any real value until developed into and embodied in a marketable product. The process from concept to working model is aptly called

Figure 6.16. Matthew Boulton, left, and James Watt were so taken with James Rumsey and his ideas that they were willing to undertake the most unusual step of bringing him into the partnership. Rumsey, however, misguided by a London friend of Benjamin Franklin, demanded too much, and the partnership arrangement failed to materialize. (From Robert H. Thurston, *History of the Growth of the Steam-Engine* [New York: D. Appleton, 1901], pp. 94, 80.)

product development, and that was precisely what was missing from Rumsey's steamboat work—and exactly what the company of Boulton and Watt was capable of doing. It was a combination of technology, machines, skilled workers, capital, marketing, and time that would transform an idea into a product that people would be willing to buy. As Rumsey failed to understand the complexities of the development process, he underestimated its value and misread the proposal as being unfair.

Rumsey may have been able to work through the differences with Boulton and Watt had it not been for the well-meaning advice of a newly acquired London friend, Mr. Vaughan. Benjamin Franklin had provided Rumsey a letter of introduction to Vaughan, which included a request to guide Rumsey through the "Modes of proceeding in public affairs," in London. As counsel to Rumsey, Vaughan had urged him to seek better terms and even suggested that the Boulton and Watt steam engine might be made in Ireland with British workers, as their patents were not valid there. This suggestion incensed the fair-minded Matthew Boulton and was one of the factors that led to the ceasing negotiations with Rumsey. In Boulton's letter to Rumsey of August 14, 1788, just before negotiations were terminated, he wrote with perceptive wisdom, "Partnerships ought to be founded on equitable principles & like a pair of Scales be Balanced, either with Money, Time, knowledge abilities or by possession of a Market & upon these principles were our propositions grounded but it now appears to me that our Sentiments are not in unison & that you have mistaken your road to the goal in view."[48] Rumsey made one more attempt to salvage the proposed partnership, but to no avail. Two weeks later, on August 29, in a letter signed by both Matthew Boulton and James Watt, Rumsey was politely notified that negotiations should terminate but that they wished to remain on friendly terms. He turned his attention to obtaining a patent for his tube or pipe boiler and by November 6, 1788, submitted his application. The patent was awarded by the British government one month later, on December 6.

Indeed, Rumsey had mistaken the road to success, as his attempts over the next four years to raise money, find and keep skilled workmen, and finally build and demonstrate a steamboat would prove to be extremely challenging. He bravely pursued his plans but on more than one occasion was disappointed by financial backers who failed to provide the necessary capital for the perceived risky venture.

During his stay in England, Rumsey made an arrangement with a wealthy and respected Mr. Whiting to finance the construction of a one hundred-ton burden boat at a cost of six hundred guineas. In a March 7, 1789, letter to his brother Edward, he wrote, "A gentleman here has undertaken to furnish me with a vessel to try my experiment upon. She is now building at Dover, 72 miles from London. She is large enough to go to the East Indies. The engine is making for her and I expect to make the trial in May." At this point in his letter Rumsey seemingly realizes the significance of the experiment he is undertaking, continuing, "The eyes of many are upon me. . . . This may truly be called the crisis of my life, should I succeed, I shall gain more reputation than I ever thought possible to fall to the share of any one man. If I fail, I shall be ridiculed and abused in all of the public prints in Europe."[49]

The inventor, still on the honeymoon of his difficult steamboat development, was optimistic and a short time later headed off to Paris to secure a French patent. Rumsey was much encouraged by Thomas Jefferson, the American representative to the French court, who entertained and introduced him to his friends and took a keen interest in the steamboat project. He met Joel Barlow, future confidant of Robert Fulton, and even considered using Barlow as his agent in France.

By early June 1789, Rumsey was back in Dover checking on the progress of his vessel. Learning of current news about the organization of the new American government, he wrote Thomas Jefferson on June 6, passing along the information and also commenting about his concerns regarding the proposed patent legislation. About his steamboat activities, he included, "I meet with many delays In geting forward my Experiment. It will be ten days yet before I Can have the Vessel Launched, by the time She gets to London I Expect to have the machinery ready to put into her. What it may take to fix it is uncertain, but hope not long; I have a dread Comes on me as the day approaches on which I have So much at Stake, yet Every review I take of my plans Confirm me more and more in its Success."[50]

Rumsey continued to be cautiously optimistic yet was understandably anxious. After returning to London he learned that his benefactor and boat financier Whiting had gone bankrupt. With a vessel about to be delivered to a London dock, and with Whiting having been delivered to a London jail, the inventor's future was suddenly thrust into jeopardy. In a letter to Charles Morrow, Rumsey explained his predicament and advised his brother-in-law that he was finally successful in borrowing enough money to pay for the boat. He also noted, "The machinery is now on board the Vessel and is going slowly together. I remain quite sanguine. . . . I have called her *Columbian Maid* but think to change it to *The Rumseian Experiment* as soon as success is ascertained."[51]

On September 8, 1789, Rumsey wrote to Thomas Jefferson predicting a speed of six miles per hour and wished him well on his return trip to America, as his French assignment had been completed. As Jefferson's ship was delayed from sailing because of weather, there were follow-up letters with Jefferson, who was anxious to learn the results of the experiment before departing for America. Rumsey, however, experienced many, many delays, and it was not until late February 1790 that a trial was made.

Trial 8. In a letter to Charles Morrow, Rumsey presents his account of the February 25, 1790, trial: "Let it then Suffice to know that every posable disapointment attended my Experiment, when it was all put together it proved So imperfect that almost the whole of it has been to do over again, the great delay an anormous Expense attending it, made my freinds doubtful, and uneasy; and thereby put it out of my power to obtain money from them, to pay my bills, which dayly came upon me."[52]

Rumsey was concerned that the bailiffs would arrest him and that he would be held in a London jail, so for a fortnight he remained in hiding. He was finally able to form a partnership with Samuel Rogers and Daniel Parker, who provided some money to avert the immediate threat of imprisonment. However, the press of finances over the next two years became so great that in May 1792 Rumsey accepted a forty-day job in Ireland to consult on a canal and earn some much-needed cash. In addition, he later constructed some mills to add to his income until his partners finally decided to back the completion of the steamboat. By the fall of 1792, the *Columbian Maid* was finally ready for a test run.

Trial 9. In an unfinished letter to his friend John Brown Cutting, Rumsey described another setback:

> In September we put her into dock, much out of repair; about the middle of November we took her out, almost everything ready for experiment, for to try which we fell down the river, near to Greenwich, where we were making preparations, and tried our boiler, which answered to admiration, and in two or three days should have had the whole ready but for an accident that happened to the vessel on the 21st of November. She made fast, by consent of the captain, to a ship, (belonging to a charity society called the Marine Society;) a haughty member of the society came on board and ordered the captain to have her cast off. The order was put into execution upon ebb tide, and our careless, worthless master (worse than a tory) let her go to shore.[53]

As the tide went out, the ship's anchor and some pilings went though the ship's bottom, allowing eight feet of water over the deck at high tide. Once again, bad luck had followed the sometimes hapless Rumsey.

Trial 10. Continuing the previous letter to John Brown Cutting, Rumsey reported on December 18, 1792, some progress, with the anticipation of full success to follow:

> It is, my dear friend, six days since I laid by this letter; it should before now have been attended to, had any opportunity offered; as there has not, I need not apologize for the delay, but will now take up the subject where I left off. We did not get ready to try our engines till Saturday last, when it worked with great success. I think it was internally very perfect; a spring placed on the outside that opened the steam valve proved something too weak to perform its office handsomely; Lenaker had to help a little at each stroke; the

vessel went forward against the tide, and pulled hard to get from her moorings. Mr. Rogers was on board, and was highly delighted at the performance. Yesterday he met with Mr. Parker on the stairs, and gave him joy at what he was pleased to call an experiment, and wanted invitations sent off to the Duke of Clarence, Sir Joseph Banks, and others of the Royal Society, and, the people inflated with insanity, talked of instantly putting engines upon ships of the navy against France. . . . Between ourselves, my friend, I have little doubt of success.

Two days, later on December 20, James Rumsey appeared before the Committee of Mechanics of the Society of Arts to discuss the utility of a model he had previously provided for the equalization of water on water wheels. He delivered a lecture on hydrostatics "to the admiration and satisfaction of all present" but afterward complained of a violent head pain and died several hours later of a cerebral hemorrhage. Thus ended his brilliant career, and the fate of the steamboat, without his leadership, was doomed.

Trial 11. In February 1793, Mr. Daniel Parker, Rumsey's business partner, assembled a group of witnesses for a demonstration of the *Columbia Maid* on the Thames in London. The trial was reported to be successful, and the vessel achieved a speed of four knots during many times out. However, there was much still to be accomplished on the yet imperfect craft, and without the inspired leadership of James Rumsey, his steamboat efforts died as he had.

Robert Fulton was in London during the time Rumsey was working on the *Columbia Maid,* and it is said that the two met, as Rumsey "mentioned Fulton as if they were on terms of intimacy" in a letter to George West.[54] George Henry Preble reports that Rumsey received "frequent visits there [London] from a young American studying engineering, who showed a sympathetic and intelligent interest in Rumsey's labors. This young man was *Robert Fulton.*"[55] It is not known if Fulton witnessed the February trial, but he wrote in his notebook under the title of "Messrs. Parker & Rumsies experiment for moving boats" the following observations as reported by his biographer Sutcliffe. After a consideration of their several points, in the form of questions and answers, he avers, "It therefore appears that the Engine was not loaded to its full power, that the water was lifted four times too high and that the tube by which the water escaped was more than five times too small."[56] (An analysis of Fulton's observations is presented in chapter 11.)

Since Fulton later received credit for the invention of the steamboat, Rumsey's only real legacy is in the form of the British patents he obtained, which validate his true genius. His creative mind was constantly engaged and sometimes consumed in envisioning, in three dimensions, yet another new and useful device. On September 12, 1791, he had written to his friend George West, "I hope you will excuse me if the letter should not be so full as you expect or I intended especially when I tell you that I am in the middle of a Specification of a patent I expect to have more patents than any man in Europe."[57]

In four years in England, Rumsey obtained four British patents:

Patent No. 1673, December 6, 1788. Titled *Boilers for Steam Engines, for Distillation &c.,* this is the invention for which Rumsey was sent to England and is his most recognized contribution. This device found universal acceptance for use in steam generation. The detailed descriptions and presentation of various mechanical configurations in the patent attest to his understanding of steam boilers. One configuration, the tight spiral, was used in his Shepherdstown steamboat.

Patent No. 1738, May 22, 1790. Titled *Applying Water and Steam to Machinery, to the Propulsion of Vessels &c.,* this patent addresses a wide variety of ways of providing prime motion. Discussed are water jet propulsion, propulsion by the force of reaction, overcoming the force of inertia, an improvement to a mill originally proposed by Dr. Barker, and Rumsey's stream boat, using poles set against the stream's bottom to push the vessel forward.

Seldom attributed to Rumsey is another application of the setting poles—applied to the precursor of the automobile. The inventor notes, "The poles may be used to move carriages by themselves, as may the ratch & pinion wheel in good roads, yet it may be better to have them both fixed to same carriage, and only use the poles when the wheels get into gullies or are going up hill, at which time the friction of the wheels upon the ground may not be sufficient to cause the carriage to move forward."[58]

In this patent Rumsey describes a floating dock for moving vessels, where the water's depth is not otherwise sufficient for the vessel to traverse.

There are several elaborate drawings augmented by written descriptions of extracting power from a head of water, albeit slowly, by using water pistons, bellows, or a pendulum. Although these techniques are inherently efficient, because power has a time component, they were too slow to be useful and were hence impractical. His thinking, however, was in the right direction, and he would subsequently discover better ways of producing efficient and useful hydraulic devices.

Patent No. 1825, September 24, 1791. Titled *Applications of Water Power to Mills and Machinery,* this patent identifies a number of useful ways to convert waterpower to rotary motion. In general Rumsey substitutes metal for wood and improves seals with packing of leather or oakum. Interesting is the application of intermeshed gears of the same size followed by intermeshed elliptical gears driven by waterpower for producing mechanical power (see fig. 6.17). The complement of this would be a pump. At some time, however, it is likely that Rumsey's teeming mind would have realized the pump

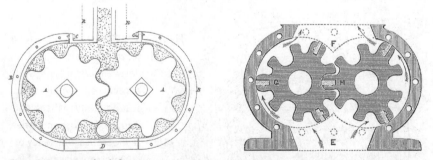

Figure 6.17. On the left is Rumsey's 1791 patented intermeshed-gear motor, designed for extracting rotary motion from a head of water; one hundred years later, on the right, rotated intermeshed gears were used to pump water in a steam-powered fire fighting engine. (Images respectively from James Rumsey, *Application of Steam and Water Power to Mills and Machinery,* British Patent No. 1825, September 24, 1791 [London: George Eyre and Wm. Spottiswoode, 1855], fig. 9; and Thurston, *History of the Growth of the Steam-Engine,* p. 366.)

possibility, and this continuous process would have been a better approach to water jet propulsion than his piston design.

A significant breakthrough came to Rumsey during the summer of 1792. After puzzling over how to extract both the potential energy in the head of water along with the kinetic energy from the moving water, he invented the precursor to the water turbine, which has a theoretical efficiency of 100 percent.

Patent No. 1903, August 23, 1792. Simply titled *Obtaining and Applying Water Power*, Rumsey at last hits upon the essence of the ultimate concept for extracting water power in the most efficient manner. He describes a hollow-tube rotary mill construction in which the water successively drives straight veins affixed to a wheel, instead of driving the whole wheel at once (see fig. 6.18). The final step to extracting the last bit of kinetic energy is to curve the veins, something he reported to his American business partner Joseph Barnes.[59] Unfortunately, Rumsey died before he was able to obtain a patent for this last improvement.

In this patent, Rumsey depicts two boats strapped together with a large spiral propeller fixed between (see fig. 6.19). He notes that by applying power to the propeller "by horse, by steam or by men . . . the boat will take motion with a power and velocity suitable to the force applied."[60] Had Rumsey lived and been able to give further

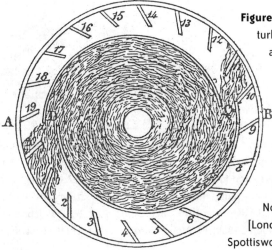

Figure 6.18. Rumsey's hollow-tube turbine, which ejects water against veins. By making the veins curved, something he subsequently realized and reported to Joseph Barnes, a theoretical efficiency of 100 percent was possible. (From James Rumsey, *Obtaining and Applying Water Power*, British Patent No. 1903, August 23, 1792 [London: George Eyre and Wm. Spottiswoode, 1855], fig. 5.)

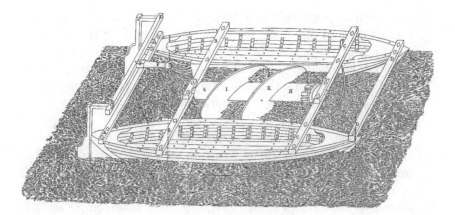

Figure 6.19. Rumsey's concept for utilizing a screw propeller to propel a boat, where the power came from man, horse, or steam. Note the use of two rudders, one behind each skiff, connected to a common tiller. (From James Rumsey, *Obtaining and Applying Water Power*, British Patent No. 1903, August 23, 1792, fig. 11.)

thought to this approach, it is likely he would have recognized its inherent advantages over his water jet approach and would have adapted it to a future steamboat.

Whereas these patents remain to attest to his many contributions to mechanics, in Shepherdstown, West Virginia, James Rumsey —rather than the textbook favorite Robert Fulton—is introduced to schoolchildren as the rightful inventor of the steamboat. Ask anyone in town about steam navigation, and he or she will point with pride to Rumsey's monument on a hill overlooking the Potomac where his public trials took place in December 1787. In Cecil County, Maryland, where Rumsey was born, Robert Fulton—who lived just over the northern Cecil County border in Pennsylvania—is recognized as the inventor of the steamboat.

Just who was this brilliant man that impressed George Washington, Benjamin Franklin, and many other leading scientific men of the day? The man that prompted Thomas Jefferson to proclaim him "the most original and the greatest mechanical genius I have ever seen"?[61] Perhaps his friend authored the best description just after his death, when John Brown Cutting wrote Jefferson on December 24, 1792, the day Rumsey was buried:

On Friday evening last died Mr Rumsay of Maryland. On the evening before while sitting with some members of the Society for encouraging british arts sciences and manufacture (who from the high opinion they entertained of his genius, had frequently consulted him at their meetings) he complained of a sudden pain in his head, and resting it on his hand on the table, in an instant became apparently lifeless. . . .

Unaided by education—unacquainted with those rudiments of science that are furnished in books or brought by masters—his mind seem'd to be as it were imbued in the elementary principles of mechanics. Of the most intricate combinations those he had almost an intuitive perception—Theories the most complicated sprung up spontaneously and correctly in his thoughts and by the dint of incessant research invention and experiment he wou'd reduce them to practice—and shape them to purposed of utility. . . .

In a word in Mr Rumsay we have lost a citizen the vigor and extent of whose intellect in this particular department of most useful science has seldom been surpass'd—and according to many of the best judges has very, very seldom equal'd.[62]

James Rumsey was interred in an unmarked grave at St. Margaret's Church in London. In the parish register the letters "G. D.," for Great Dues, were placed next to Rumsey's name, attesting to the fact that he was a person of consequence. Through the efforts Shepherdstown High School students on May 18, 1955, a plaque was unveiled with a closing inscription: "This Epitaph by Robert Herrick . . . was restored in memory of JAMES RUMSEY."[63] Upon his untimely death the world lost one of its most creative and talented engineers and inventors.

In August 1788, Rumsey had provided a fairly detailed description of the construction and operation of his Shepherdstown steamboat with the publication *Explanation of a Steam Engine and the Method of Applying It to Propel a Boat*. Using that description and discussions with members of the Rumseian Society of Shepherdstown, West Virginia, who in 1987 made a reproduction of Rumsey's steamboat, a schematic of his steam engine and apparatus has been developed. It can be found along with an explanation in appendix A.

Notes

1. George Johnston, *History of Cecil County, Maryland* (1818; reprint, Baltimore: Regional Publishing, 1989), p. 508; Ella May Turner, *James Rumsey, Pioneer in Steam Navigation* (Scottdale, PA: Mennonite Publishing House, 1930), p. 3; and "The Rumsey Family," *Sunday Sun* (Baltimore), January 9, 1904. (Note: The Rumsey quote above is found in Turner, *James Rumsey*, pp. 116, 161.)

2. G. Harry Davidson, "Birthplace of James Rumsey," *Cecil Whig* (Elkton, MD), June 3 and 17, 1938.

3. Ralph D. Gray, *The National Waterway: A History of the Chesapeake and Delaware Canal, 1769–1985* (Urbana: University of Illinois Press, 1989), p. 2. Here it is mentioned that Herman called the river Minquaskil, but a survey by the American Philosophical Society identified the river as the Appoquinmink, which is the name that appears on modern maps.

4. Johnston, *History of Cecil County*, p. 242.

5. Information on Rumsey's 1782–84 business ventures were obtained from the Rumsey display at the Museum of the Berkeley Springs, WV, curator Jeanne Mozier, 1986; Turner, *James Rumsey*, pp. 6–7; and Bertha Taylor Voorhorst, "The Cradle of the Steamboat," *Daughters of the American Revolution Magazine*, February 1937, p. 103.

6. Quoted in Turner, *James Rumsey*, pp. 6–7.

7. Ibid., pp. 11–12.

8. Ibid., p. 64.

9. Suggested to the author by Jeanne Mozier in the fall of 1988.

10. "Letter from George Washington to James Rumsey, September 7, 1784," ed. John C. Fitzpatrick, *Writings of George Washington from the Original Manuscript Sources, 1745–1799*, Library of Congress, American Memory, http://memory.loc.gov/.

11. "Letter from George Washington to James Rumsey, July 2, 1785," ed. John C. Fitzpatrick, *Writings of George Washington from the Original Manuscript Sources, 1745–1799*, Library of Congress, American Memory, http://memory.loc.gov/. In this letter, where Washington offered Rumsey the job of superintendent of the Potomac Navigation Company, he wrote, "As I have imbibed a very favorable opinion of your mechanical abilities, and have found no reason to distrust your fitness in any other respects; I took the liberty of mentioning your name to the Directors" for the job of superintendent.

12. Turner, *James Rumsey*, p. 15.

13. James Rumsey, *Explanation of a Steam Engine, and the Method of Applying It to Propel a Boat* (Philadelphia: Joseph James, 1788), last paragraph.

14. Joseph Barnes, *Remarks on Mr. John Fitch's Reply to Mr. James Rumsey's Pamphlet* (Philadelphia: Joseph James, 1788), certificate nos. 15, 16.

15. James Rumsey, *A Short Treatise on the Application of Steam, Whereby Is Clearly Shewn from Actual Experiments, That Steam May Be Applied to Propel Boats or Vessels of Any Burthen against Rapid Currents with Great Velocity* (Philadelphia: Joseph James, 1788), certificate nos. 15, 8, 11.

16. Brooke Hindle, *Emulation and Invention* (New York: W. W. Norton, 1981), p. 60; J. T. Desaguliers, *A Course of Experimental Philosophy*, vol. 2 (London: Printed for W. Innys, M. Senex, and T. Longman, 1744), pl. 36.

17. Englehart Cruse, *Projector Detected or, Some Strictures, on the Plan of Mr. James Rumsey's Steam Boat* (Baltimore: John Haynes, 1788), p. 5.

18. Rumsey, *A Short Treatise*, certificate no. 12.

19. The general explanation was that the number of particles of water impacted in a given time doubled with a twofold increase in speed, and the impact of each particle also doubled, hence a fourfold increase—the speed squared rule.

20. "Letters of James Rumsey," ed. James Padgett, *Maryland Historical Society Magazine* 32 (1937): 18.

21. James Rumsey, *Applying Water and Steam Power to Machinery, to the Propulsion of Vessels, &c.*, British Patent no. 1738, May 22, 1790 (London: George Edward Eyre and William Spottiswoode, 1855), p. 4.

22. Edwin T. Layton Jr., "James Rumsey: Pioneer Technologist," *West Virginia History* 48 (1989): 12n.

23. Barnes, *Remarks on Mr. John Fitch's Reply*, certificate no. 14.

24. David Read, *Nathan Read: His Invention of the Multi-tubular Boiler . . .* (New York: Hurd and Houghton, 1870), p. 38.

25. Barnes, *Remarks on Mr. John Fitch's Reply*, certificate no. 6.

26. Rumsey, *Explanation of a Steam Engine*, pl. 3.

27. Barnes, *Remarks on Mr. John Fitch's Reply*, certificate no. 10.

28. Ibid., para. 3: "Mr. Rumsey had in the year 1785, prepared a steam engine upon the plan used and improved in Europe to propel his boat."

29. James Rumsey, *Boilers for Steam Engines, for Distillation, &c.*, British Patent no. 1673, December 6, 1788 (London: George Edward Eyre and William Spottiswoode, 1854), p. 2.

30. Turner, *James Rumsey*, p. 66, says Rumsey used an iron pot boiler and even shows a sketch of the container opposite page 66. As Rumsey employed copper for both his steam and water cylinders, he was familiar with the metal

and its solderability. Thus it is not surprising that he also used copper for his initial boiler, later switching to iron pipe with water injection. A soldered copper tube would have been easier to fabricate, but in the furnace the solder would have immediately melted, rendering it useless.

31. Barnes, *Remarks on Mr. John Fitch's Reply*, certificate no. 8.

32. This is the date Francis Hamilton, who kept a daybook, reports and is the date used here, but in Barnes's certificate he states that April was the time of the first trial. See Rumsey, *A Short Treatise*, certificate no. 12.

33. Quoted in Barnes, *Remarks on Mr. John Fitch's Reply*, certificate no. 8.

34. Quoted in Rumsey, *A Short Treatise*, certificate no. 12.

35. Barnes, *Remarks on Mr. John Fitch's Reply*, certificate no. 1.

36. On September 19, 1786, Rumsey reported in a letter to George Washington that a trial of the mechanical boat was made on Saturday, September 9, 1786, but with little success. The wooden poles with iron tips would often slip on the bottom, causing the boat to rotate, with the resulting loss of power imparted to the paddle wheel. Although Rumsey maintained his continued belief in the mechanical boat concept, he was to abandon it and concentrate on his steamboat.

37. Quoted in Barnes, *Remarks on Mr. John Fitch's Reply*, certificate nos. 5, 7.

38. Quoted in Rumsey, *A Short Treatise*, certificate no. 12.

39. Ibid., certificate no. 11.

40. Ibid., certificate no. 12. It seems likely that when one end of the iron pipe was flared to accommodate the male thread, this weakened the joint, leading to a stress failure. When Rumsey later published his boiler schematic in *Columbian Magazine* in May 1788, he showed larger pipe couplings connecting the five-foot sections together. This is likely a good example of learning by doing—so important to engineering development.

41. "Letters of James Rumsey," ed. Padgett, p. 140.

42. Turner, *James Rumsey*, p. 111, and plate between pp. 112 and 113. In this original pamphlet, Rumsey wrote that Englehart Cruse of Baltimore visited him "In or about the Month of June 1787" and "begged" his opinion of a steam engine Cruse was planning. Rumsey discussed steam theory, pointed out the defects in Cruse's approach, and revealed his steamboat progress. When Cruse later constructed a steam engine and sought a state patent, this caused Rumsey to rail that he "had the audacity to petition the Maryland Assembly to give him exclusive rights for the emoluments of another's invention, so surreptitiously obtained; but he received the denial he so justly merited." Sometime after May 9, 1788, Cruse

responded with an unconvincing rebuttal titled *Projector Detected, or Some Strictures, on the Plan of Mr. James Rumsey's Steam Boat.* For an excellent account of the controversy, see John W. McGrain, "Englehart Cruse and Baltimore's First Steam Mill," *Maryland Historical Magazine* 71, no. 1 (Spring 1976). The quotations in this note are from this source, p. 66.

43. Rumsey, *A Short Treatise.* In May 1788, the January pamphlet, with minor changes, was reprinted; this later publication is the source of the quotations in this chapter.

44. Quoted in Turner, *James Rumsey,* p. 116. See also "Letters of James Rumsey," ed. Padgett, pp. 142–44, who has the date one day earlier on March 24, 1788, along with other minor changes in the transcription.

45. Quoted in Turner, *James Rumsey,* pp. 160–61.

46. James Rumsey, "Letter to John Ewing et al., American Philosophical Society Subcommittee," May 10, 1788, Milton S. Eisenhower Library, The Johns Hopkins University, Baltimore, microform 1240, no. 45218.

47. John Fitch, *The Original Steam-Boat Supported; or, A Reply to James Rumsey's Pamphlet Shewing the True Priority of John Fitch and the False Datings, &c. of James Rumsey* (Philadelphia: Zachariah Poulson, 1788).

48. Quoted in Turner, *James Rumsey,* p. 151.

49. Ibid., p. 157.

50. *The Papers of Thomas Jefferson,* ed. Julian P. Boyd, 60 vols. (Princeton, NJ: Princeton University Press, 1958) 15:172.

51. Quoted in Turner, *James Rumsey,* p. 169.

52. Ibid., p. 175.

53. Ibid., pp. 196–98.

54. Ibid., p. 209.

55. George H. Preble, *Chronological History of the Origin and Development of Steam Navigation* (Philadelphia: L. R. Hamersly, 1883), p. 12.

56. Quoted in Alice Crary Sutcliffe, *Robert Fulton and the* Clermont (New York: Century, 1909), p. 330.

57. Quoted in Turner, *James Rumsey,* p. 183.

58. Rumsey, *Applying Water and Steam Power to Machinery,* p. 10.

59. See Joseph Barnes, "Essay on Overshot & Undershot Wheels & on Rumsey's Mill" (manuscript, American Philosophical Society, 1793), fig. 4; Layton, "James Rumsey," p. 12, last para. of note 13.

60. James Rumsey, *Obtaining and Applying Water Power,* British Patent no. 1903, August 23, 1792 (London: George Edward Eyre and William Spottiswoode, 1855), p. 6.

61. *The Papers of Thomas Jefferson,* ed. Boyd, 14:699.

62. Turner, *James Rumsey*, pp. 199–201.

63. James L. Hupp, "West Virginians to the Rescue: James Rumsey's Memorial in London," *West Virginia History* 30, no. 2 (1969): 506–507.

Figure 7.1. John Fitch. (Rendered by William T. Sisson from a crude wood engraving found in James T. Lloyd, *Lloyd's Steamboat Directory and Disasters on Western Waters* [1856; reprint, Cincinnati: Young & Klein, 1979], p. 18.)

7.

John Fitch
1743–1798

*I know of nothing so perplexing and vexatious to a man of feeling,
as a turbulent Wife and Steam Boat building, I experienced the
former and quit in season, and had I been in my right senses I
should [have] undoubtedly treated the latter in the same manner.*
—John Fitch, *The Autobiography of John Fitch* (publ. 1976)

John Fitch was born on January 21, 1743, in Windsor, Connecticut, as the fifth child in his family. His mother died when he was about five years old, and after his father remarried, he was raised by a benevolent stepmother, who sent him to public school some distance from his home. At an early age, he demonstrated a remarkable capacity for absorbing knowledge and later, reflecting on his primary education, observed that he was "crazy after learning."[1]

Fitch left home at age eighteen to serve as an apprentice to a clockmaker; several years later he departed Connecticut to seek his career and fortune. He cleaned clocks; made brass and silver buttons; and became a silversmith, surveyor, land jobber, and mapmaker. Finally ending up in Bucks County, Pennsylvania, in 1785, the idea for a steamboat "struck" him quite by accident, and for the next several years, he doggedly pursued his steam navigation dream. Ultimately, however, success was not to be had, as he ended up disappointed, dejected, and drunk—dying by his own hand in Bardstown, Kentucky, in 1798.

The progenitor of the Fitch family was John's great-grandfather, Thomas Fitch, of Braintree, Essex, England.[2] After Thomas passed away, his five sons, along with their mother, moved to New England and purchased about one-twentieth of the township of Windsor, Connecticut.

Thomas's son Joseph was John's grandfather, whose first-born son, also named Joseph, was John's father. The younger Joseph Fitch married Sarah Shaler, and this union produced John; his two older brothers, Joseph and Augustine; two older sisters, Sarah and Ann; and a younger sister, Chloe.

John started school at the age of four, just before his mother died. Being very close to the Windsor-Hartford township border, he had to travel about 1½ miles to the schoolhouse but, according to his own account, "learned to spell pretty well" the first summer he was there.[3]

Until about age eight or nine, this small-framed boy remained in school full-time, adsorbing information, and when he was at home, he spent time reading his father's limited library of books instead of playing with other children. About age eleven he became aware of a geography book by Thomas Salmon and requested that his father buy it for him.[4] Having no success with his dad, he planted potatoes on some unused farmland, which in the fall he harvested and sold for ten shillings. As the book cost twelve shillings, he was obliged to borrow two shillings and asked a neighboring merchant to obtain the book for him from New York. After obtaining the book, the precocious young Fitch quickly learned geography, being able to recall the longitude, latitude, population, and religion of any nation in the world.

When Fitch was about eleven or twelve years old, his schoolmaster realized that the lad had learned all the arithmetic he was able to teach. He therefore introduced Fitch to surveying, which involved teaching him about plane geometry and the tools employed for measuring surface distance.[5] The basic linear unit for surveying at that time was the link, which was defined to be 7.92 inches, or equal to 0.66 feet. One hundred links, or 66 feet, was the length of a surveyor's chain, which was marked in four equal parts, each 16.5 feet, called rods.

The same unit used today, the acre, was defined to be ten square chains, which equaled 10 times 66 feet by 66 feet, or 43,560 square feet. If a surveyor were working in rods, by a straightforward conversion, it could be easily determined that an acre equaled 160 rods, which could be a rectangle 16 by 10 rods, 20 by 8 rods, 40 by 4 rods, or any combination where the product equaled 160.[6]

Governor Roger Wolcott owned farmland adjacent to Fitch's father and probably learned from the schoolmaster about John's mastery of surveying.[7] The governor requested permission to borrow young John to help him survey, to which his father readily agreed. This capable and somewhat forward young boy immediately ambled off to Wolcott's property to assist him in laying out some lots.

After Fitch had surveyed a straight line for some distance, the property turned to form a ninety-degree corner, and Wolcott asked John if he knew how this might be done. Recalling his basic geometry lessons, Fitch had Wolcott stand near a bush on the corner of the property as he trampled a segment of a circle in the grass at a distance the length of the 66-foot chain. He then had Wolcott stand two rods (33 feet) farther along the line they had been surveying and again trampled another segment of grass, this time crossing the first segment. A distance one rod below the intersection, announced Fitch, was the point that should be connected to the bush to establish the ninety-degree corner (see fig. 7.2).

Later that day, Wolcott asked Fitch how to survey out one acre of land in order to provide it to Isaac Morton. As one side of the parcel was 30 rods long, the question posed to Fitch was, how wide should it be to encompass exactly one acre? Fitch immediately replied 6 rods. Wolcott seemed to question the width, prompting John to explain that a parcel 8 by 20 rods was an acre, as was 4 by 40 rods. As 30 was halfway between 20 and 40, likewise 6 was halfway

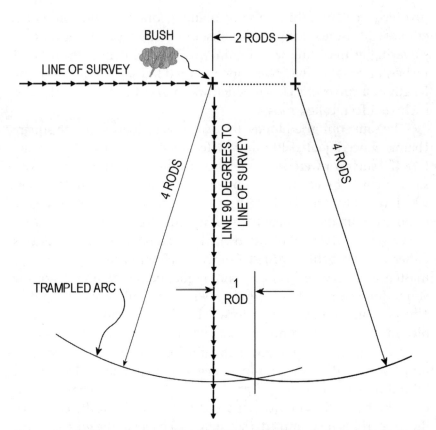

Figure 7.2. Fitch's surveying technique for turning a ninety-degree angle. At this moment the young John Fitch was justifiably proud of achieving a modicum of technical success. His knowledge of surveying would serve him well later, when he was exploring Kentucky and the Northwest Territory. (Diagram by Jack L. Shagena. Note: Not to scale.)

between 4 and 8; therefore 6 was the correct answer. Wolcott agreed and proceeded with the survey. But Fitch, of course, was wrong, as 6 times 30 equals 180, which is larger than the correct value of 160. Recalling the incident much later, Fitch said he had made a mistake, simply that—but the perpetually downtrodden Fitch may have unconsciously succumbed to the Robin Hood syndrome, taking from the rich and giving to the poor!

As the day drew to a close, it became too late to lay out lots for several other individuals. Wolcott instructed Fitch on what to do the following day and departed. The dutiful boy carried out the assigned task, and a few days later the governor came by his home to retrieve the surveyor's chain. John fully expected that he would be rewarded with a few coins, but Wolcott put the chain in his saddlebag and rode off. Unfortunately, this seemingly unfair treatment by others would be a pattern in Fitch's life that would eventually mold his character into a cynical knot.

Though still frail and relatively small for his age, John was nevertheless becoming more able to provide help on the farm, so his brother and father kept him busy with chores. He was worked so hard that, at one time, he compared his brother to a Virginia slave master. This unrelenting manual labor depressed him greatly, as he had no time to learn and lamented further for "the want of proper books." Despite this, he later wrote, "I seemed to be beloved by both old and young as I could speak rationally to the old and was always foremost amongst my playfellows."[8]

To avoid the physical labor for which he was unsuited, he started searching for a trade when he was seventeen and spent a week at sea but did not find it to his liking. He later wrote, "I came home more rich than when I went . . . [but] being as much at a loss how to dispose of myself as ever."[9] A short time later he became acquainted with a local clockmaker, Benjamin Cheney, and upon reaching the age of eighteen apprenticed himself to Cheney for several years with a bond secured by his father.

During the apprenticeship, however, Fitch was kept busy with Cheney's farm duties, as the master did not allow him into the shop to learn the business of clockmaking. As he kept a log of his farm hours and complained about the unfairness of his situation, it was decided to let him go to work for Cheney's brother, Timothy, also a clockmaker. There he was treated somewhat better, as eventually he learned about working with brass and even started building his own wooden clock. The ambitious and industrious Fitch, however, was still not satisfied with the progress of his learning and the fact that he seldom received enough food to eat, so he decided to strike out on his own. He had reached the age of twenty-one, but a few months remained on his apprenticeship, which was still bonded by his father. Agreeable terms were worked out with Cheney, and he departed.

Fitch wanted to start his own business but was penniless. At that time, Reuben Burnham was pursuing his sister Chloe with a "strong passion" and, seeking to win Fitch's favor, lent him twenty shillings to start a small brass business. In about two years, the industrious and hard-working young man was worth about fifty shillings, having repaid his debt to Burnham, now his brother-in-law. This was Fitch's first real business success and provided this now tall and somewhat gangly young man with the confidence that hard work would indeed reap rewards.

Having some ready cash, he was approached by a couple of individuals to go into the potash business. With no experience or knowledge in this field, he decided to go to work for a potash company for several months to learn; however, his business venture with his partners was not successful.

During this time, Fitch was lodging with a Mr. and Mrs. Beaman, who received a visit from Lucy Roberts, a sister of Mrs. Beaman. There she met Fitch, and after a courtship of six months, they were married on December 29, 1767, when Fitch was twenty-four and Lucy believed him to be several years older. A son was born to this turbulent union on November 3, 1768. The marriage was not a happy one, and after about 1½ years, Fitch started making plans to leave; thus on January 18, 1769, he abandoned Lucy, who was then expecting their second child, a daughter.

Although he subsequently had misgivings about his decision to leave, Fitch's dogged determination was one of the unflagging personal traits that at times during his life would produce limited success but would never secure long-term happiness for this tormented man. Later he would steadfastly pursue several passions: surviving capture and torture by the Indians to acquire and hold land in Kentucky; to survey the Northwest engrave; to print and sell his maps; to be recognized as the inventor of the steamboat; and to establish his own religious sect. But never again would he experience passion for another woman.

Fitch set his course for Albany and, passing through Pittsfield, remained there for three months, eking out a meager living. Upon reaching Albany, he found prospects no better for his future; thus he considered going down the Hudson River to New York and on to Jamaica to visit his well-to-do uncle Timothy Shaler, brother to his deceased mother.

This route, however, would not afford work opportunities, and being very short on money, he decided to go by land, cleaning clocks along the way. It is most likely that Fitch traveled south on the west side of the Hudson River, as upon arriving in New York City, he wrote, "And when I came there [I] inquired amongst the shipping but could not get a ready passage [to Jamaica]. Which determined me to pursue my journey into the Jerseys."[10] From the west side of the Hudson River it would have been necessary to cross the river, sail past Coney Island, and proceed up Rockaway Inlet into Jamaica Bay to reach his uncle's home.

His travels south from Albany would have most likely taken him through, or at least close to, Hackensack, Rutherford, Belleville, and Newark before he arrived at Elizabethtown Landing, now Elizabeth, New Jersey. The significance of this route lies in the fact that in 1753, a Newcomen steam pump was installed at Belleville by Philip Schuyler to remove water from a deep copper mine (see fig. 7.3).[11] The engine was in operation by 1755; however, a fire at the mine destroyed the engine about 1760. By July 1761, it had been rebuilt and was again back in operation. In 1768, a second fire put the engine out of operation, and it was not rebuilt until after the Revolution.[12] When Fitch passed through about 1769, the steam engine was not operating, but it is very likely that the man who was "crazy after learning" would have learned about it or would have seen it. (More about this subject later.)

At this point, Fitch was still heading south, looking for his opportunity in life as he passed through Elizabethtown Point and on to Rahway. One day he came upon the home of Benjamin Alford and, quite by accident, witnessed him being scolded by his wife. This incident made unshakeable his belief that he could not return to his wife and occasioned a poem that he sent to a friend in Windsor. The page-long, poorly written verse is strikingly unremark-

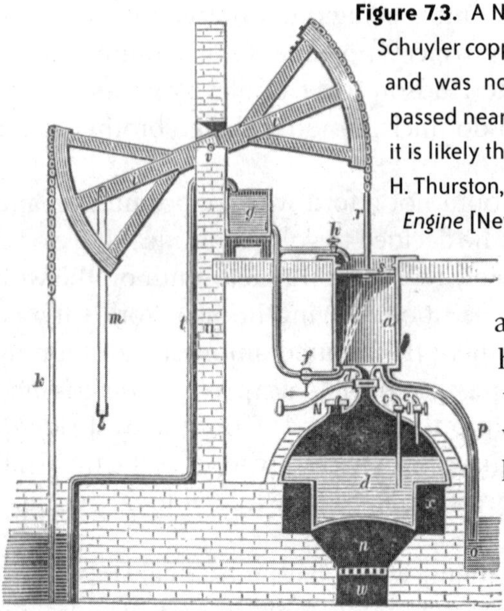

Figure 7.3. A Newcomen-type steam pump at the Schuyler copper mine had been damaged by fire and was not in operation when John Fitch passed near Belleville, New Jersey, in 1769, but it is likely that he saw the device. (From Robert H. Thurston, *History of the Growth of the Steam-Engine* [New York: D. Appleton, 1901], p. 59.)

able in its composition, but he no doubt kept a copy; two decades later when writing his autobiography, he quoted it.

Passing through Brunswick, he went into Trenton, where he became acquainted with Matthew Clum, a tin man, and subsequently met James Wilson, a silversmith. Wilson was an honorable man who had inherited his father's fortune but was given to drinking and seldom was sober. Fitch went to work for Wilson for a while repairing watches and also made about fifty to sixty pairs of brass buttons.

He ventured north to Springfield and south to Mansfield to clean a few clocks and peddle his buttons, which sold immediately. He then cast more brass buttons and this time some silver ones, which he sold in Monmouth and Raritan. Fitch was "growing at least one dollar a day richer," and when James Wilson got into difficulties with his creditors, Fitch was able to acquire his tools, described by him as "the best set of silversmiths tools in America."[13] With these he established his own silversmith shop in Trenton.

His business prospered, and in a short time Fitch was in need of help, hiring former employer James Wilson, along with several other individuals, as journeymen. To expand the operation, he occasionally borrowed money from a number of leading merchants in Trenton, but he was always careful to repay the loans on time. He was able to hire several journeymen, and by the time the Revolutionary War began, his worth had grown to eight hundred English pounds, a small fortune for his time.[14]

When the British Parliament enacted tax laws on the colonists, the fiercely independent Fitch violently objected, applying for and accepting a commission as first lieutenant in the New Jersey Militia. Being quick to learn the operation of mechanical things, and furthermore very capable with his hands, he was encouraged to go into the gunsmithing business to supply arms to the soldiers. Twenty guns were collected and in a few days Fitch had them operating with bayonets attached (see fig. 7.4).

When Fitch was ordered to take his first command, the troops under him were wary of his leadership. This concern resulted in an election of another lieutenant, with Fitch losing by two votes, and this lack of confidence by those "ignorant boys" distressed him greatly. He returned to his gunsmithing, pouring himself into the effort—working seven days per week and up to sixteen hours each day.

For a second time he was ordered to take command, but again the troops grumbled, and his superior officers decided to place another lieutenant in charge. Fitch was chagrined and after some consternation decided to desert his company. He was to regret this decision and would consider calling for his own court-martial, but he never did. For a while he served as a private under an officer in Bucks County, but his duty was short.

When British troops appeared in Trenton in November 1776, Fitch fled to Bucks County with his gold and silver, which he buried on the property of Charles Garrison. Unfortunately it was later found and stolen; however, Fitch eventually received partial restitution.

Learning of the needs of Washington's army, Fitch procured and supplied them with tobacco, beer, and whiskey, eventually earning about four thousand dollars in Continental currency. This small for-

Figure 7.4. Fitch had an excellent mechanical aptitude, being able to learn quickly how to fabricate the necessary parts, make necessary adjustments, and repair rifles. (Rendered by William T. Sisson.)

tune, however, was quickly deflated by the war's end to one hundred dollars. Attempting to salvage some of his money, Fitch formed a land company and headed down the Ohio River to Kentucky to purchase land warrants from the state of Virginia with his money and the funds of others. He surveyed some of the choicest land in the area of what is now Bardstown, recorded the warrants in Richmond, and returned to Pennsylvania in 1781.

As a result of this successful venture, Fitch went on a second land-jobbing trip in 1782. Going down the Ohio, he and a group of others were captured by Native Americans, and he barely avoided being killed. The prisoners suffered hostile treatment, lack of food, and cold-weather exposure while being marched toward Detroit in what was then the Northwest Territory. The British troops in the area were unaware of the defeat of Cornwallis at Yorktown in October 1781, which for practical purposes had ended the Revolutionary War. They accepted the prisoners from the Indians, sending Fitch and twenty others to the seventy-to-eighty-acre Prison Island on May 25, 1782.

Not wanting to idle away his time, the highly ingenious and industrious Fitch was able to scrounge various parts of old guns, barrel hoops, and other pieces of broken or discarded metal. With these he constructed a set of tools, including a forge, vise, and molds for making brass and silver buttons. He also made wooden clocks. Some raw materials and tools, such as borax for soldering and files for shaping metal, were obtained by bargaining with the guards, who ventured into nearby populated areas for supplies. His products were sold to the guards and bartered to the prisoners. As he employed several of his fellow inmates, he was popular and found his existence in this microcosm of society to be satisfactory.

He also established a small garden, planting seeds readily furnished by the guards, who wished to share in the harvest. He enlisted the help of another individual to assist him in this enterprise as well, but as his captors had encouraged the activity, it was met by many of his fellow prisoners with some suspicion.

When word of the end of the war finally reached the British on the island, Fitch was sent back east in a prisoner exchange, arriving back in Pennsylvania in the early part of 1783. He then organized another land company for the purpose of surveying the Northwest Territory, expecting that Congress would encourage settlement with land grants such as those provided by Virginia in Kentucky. He was able to find partners and raise money, and during the summers of 1783 and 1784 he surveyed some of the choicest land while spending the winters in Pennsylvania.

Having a very good knowledge of the territory, he petitioned Congress to name him surveyor for the Northwest and enclosed recommendations from several influential individuals, fully expecting his request to be granted. While waiting for Congress to act, the industrious Fitch decided to make a map of the territory that would be more accurate, smaller, and less expensive than the existing map, produced by Hutchins and Morrows. His first step was to produce a draft map using existing information augmented by the knowledge he had gathered from his own surveying.

Making such a map was not a trivial task and called into play the skills normally residing in a group of individuals, not just a single person. Fitch first obtained a sheet of copper and made it suitable for engraving by hammering and polishing its mill surface to a mirror finish. It was then necessary to engrave the draft map, in reverse, onto the copper, so that it would read properly when printed. Fitch had an engraving tool he used for marking brass and silver buttons and with a steady hand produced an astonishingly good image and graphics.

Fitch then fashioned a press and created a backing for the copper sheet to keep it perfectly flat during the printing operation. To make quality images, he had to learn about lithography, which he probably did from a local printer. Fitch then produced a number of copies (see fig. 7.5), selling them in Maryland, Virginia, and Pennsylvania, providing a meager source of revenue for his subsequent steamboat work.

To better appreciate the extraordinary talents of this man, it is worthwhile to summarize his map project. He organized and managed the land-jobbing trip to the Northwest; surveyed a portion of the territory; translated the results to an accurate map; hammered,

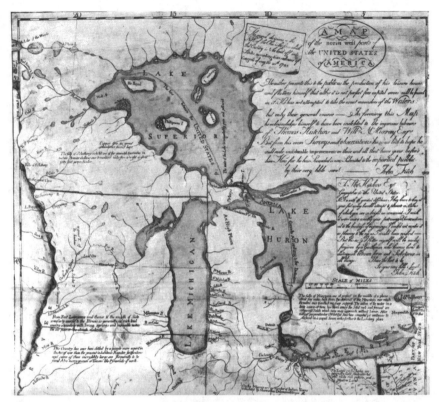

Figure 7.5. Portion of Fitch's *Map of the North West Parts of the United States of America*, showing the Great Lakes. This effort was a remarkable achievement for a self-taught individual, who surveyed the area, polished the copper engraving plate, engraved and printed the map, and finally sold copies, some hand-colored by a friend, to accumulate money for his later steamboat efforts. (From James T. Flexner, *Steamboats Come True* [1944; reprint, Boston: Little, Brown and Company, 1978], inside cover.)

polished, and engraved a copper plate; fabricated an offset press; and printed, distributed, and sold copies of his map. In terms of today's nomenclature, he researched, designed, produced, and marketed a new and better product, for the most part unaided by others—to find this range of capabilities in a single individual is indeed remarkable.

Of this map Fitch later humbly wrote, "It is true it was but Coarsely done [but] it was cheap Portable to any who wanted to go to the Woods and more to be relyed upon than any then Published."[15]

Once again, however, Fitch was to be disappointed, for he failed to receive the anticipated appointment from Congress as surveyor of the Northwest Territory. Furthermore, the federal government adopted a land grant policy different from that used in Kentucky, making his survey work effectively worthless. He was not defeated, however, for his steamboat plans were still to come.

While returning from a church service led by Rev. Nathaniel Irwin one Sunday in April 1785, Fitch professes in his later writing that when a carriage passed him, "A thought struck me that it would be a noble idea if I could have such a carriage without the expense of keeping a hors[e]. . . . I soon thought that there might be a force procured by steam and set [out] to make a draft [and later Rev. Irwin] shewed me Martins Philosophy with a steam engine laid down in it. Till then I did not know that there was such a thing in nature as a Steam Engine."[16] To believe this account, that the idea of a steam-propelled vehicle struck him for the first time at that moment, would require a serious stretch of the imagination. As Prager observed in a footnote in Fitch's autobiography, "This would make him more naive than he was."[17] Carroll Pursell notes, "This story has the ring of myth, and it is probably safe to assume that Fitch had previously seen the force of steam demonstrated in some form."[18]

Thomas Jefferson observed that one idea leads to another to another, and this is the history of innovation and invention. To make the leap from "a force procured by steam" to the propulsion of a carriage, and later a steamboat, is the rambling of a dreamer, not an inventor. As Fitch jumped intellectually rather quickly from the force of steam to a steam engine to propulsion of a boat, he must have had some of the steps already worked out in his mind. Most likely he had already envisioned how to construct the steam engine and generate the required steam long before he saw the book. It is highly probable that when Fitch passed through Belleville, New Jersey, some fifteen years earlier, he became familiar with the steam mine pump; else his leap in thinking would have not been possible.

In the later part of 1874, James Rumsey, through his partner, Dr. James McMechen, had been able to obtain the promise of a large tract of land, provided Rumsey could successfully demonstrate his stream or so-called pole boat. Prager points out that William C. Houston of Trenton was a member of the congressional committee proposing the award and was also associated with Fitch's land-jobbing company; hence, he was the likely conduit to Fitch of Rumsey's stream boat work.[19]

During the summer of 1785, Fitch constructed a small model of his steamboat showing a chain with small paddles at the side for propelling the boat (see fig. 7.6). At times, however, he even questioned his own sanity for attempting such a project, but in a determined fashion he staunchly pursued his dream. He obtained from Dr. John Ewing a certificate as to the feasibility of his plan,[20] along with an endorsement from William C. Houston, who at the time was no longer a member of Congress. Fitch presented a petition to Congress on August 29, 1785, asking for encouragement in his endeavor; however, none was forthcoming, leading him to be more contrite than ever and to "prove them to be but Ignorant Boys, I determined to pursue my scheme as long as I could strain a single Nerve, to forward it."[21]

About one month later, on September 27, 1785, Fitch laid his plans before the American Philosophical Society (see fig. 7.7), but was not invited to join—as Rumsey would be three years later—and did not receive any significant constructive feedback. He next visited Benjamin Franklin, who received him and spoke falteringly of his plan but would not endorse it in writing. Fitch was to continue to attempt to win over Franklin with letters and visits but to no avail, as Franklin was a proponent of the water jet propulsion technique being pursued by Rumsey.

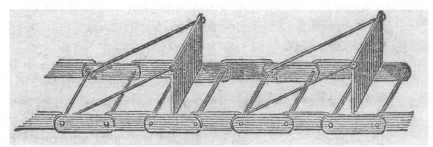

Figure 7.6. A section of the chain and paddle boards above the end pulleys is shown. To keep the boat from turning it would also be necessary to likewise equip the opposite side. (From Thomas Westcott, *Life of John Fitch* [Philadelphia: J. P. Lippincott, 1857], p. 131.)

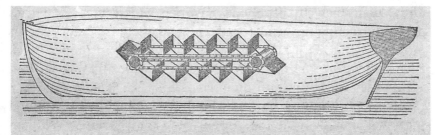

Figure 7.7. Sketch of John Fitch's model boat presented to the American Philosophical Society on September 27, 1785, that was propelled by the movement of the chain with paddle or so-called float boards. (From Westcott, *Life of John Fitch*, p. 131.)

Fitch set out to Kentucky to check on his land patents, and on his way he visited William Henry in Lancaster, a member of the American Philosophical Society, as Fitch "wished the opinion of every man of Science."[22] Henry informed him that a number of years earlier he had made a draft of a steamboat but never laid it before the Society. The initially suspicious Fitch was much relieved when Henry graciously agreed that Fitch should deserve the credit, for he was the first to make his plan public. (For a more detailed summary of this meeting, refer to chapter 5.) Quite satisfied, Fitch set off to visit former Maryland governor Johnson in Frederick Town.

Fitch showed Johnson his plan, and he seemed to be very interested, recommending Fitch present it to the Virginia Assembly, which was currently in session in Richmond. The governor then suggested he pay a visit to George Washington at Mount Vernon. Fitch, knowing of Rumsey's work and of Washington's certificate to Rumsey, silently reacted to the suggestion, thinking "his Excellency was heart sick of Boat projects."[23] Nevertheless Fitch, not wanting to ignore the governor's advice, did visit Washington in November 1785 on his way to Richmond and learned that in addition to a stream boat, Rumsey had also confided in the general his plans for a steamboat. Washington was unwilling to give Fitch a certificate; since Rumsey's steamboat plans predated his, this was to give Fitch much consternation.

Proceeding to Richmond, Fitch met with Governor Patrick Henry and James Madison, who were pleased with his steamboat ideas; Henry agreed to provide a written endorsement of Fitch's map, anticipating this would facilitate sales and raise money for his project (see fig. 7.8). Finding his Kentucky deeds in proper order, Fitch decided to return to Bucks County and on his way stopped again to see Governor Johnson in Frederick.

Johnson was ailing and in bed but was willing to see Fitch and suggested that he visit the current Maryland governor, William Smallwood (served 1785–88), in Annapolis as well as the General Assembly. To this end, Johnson wrote Fitch a very complimentary letter of introduction that in part reads, "[John Fitch] has spent much thought on an improvement of the Steam Engine, by which to gain a first [rate] power applicable to a variety of Uses, amongst others to force Vessels forward in any kind of Water. If this Engine

Figure 7.8. Patrick Henry, governor of Virginia, met with John Fitch late in 1785 and supported his steamboat efforts with words but no money. Henry endorsed Fitch's Northwest map with the anticipation that others would buy it, providing funds for the steamboat project. (Print from a painting by Alonzo Chappel [New York: Johnson, Fry & Company, 1862].)

can be simplified, constructed and made to work at a small expense, there is no doubt but it will be very useful in most great Works, amongst them in Shipbuilding."[24]

This letter, and in particular this paragraph, became pivotal for Fitch's subsequent argument for priority in the invention of the steamboat. Fitch continues in his autobiography, "The Governour in this letter alludes to our discourse on Rumseys plan, where he says amongst others to force Vessels forward in any kind of water, which convinces me that he was sincear at the time, and had no Idea that Rumseys [boat] would make way any otherwise than against the streams [i.e., a stream boat]." There is, however, a much more plausible interpretation of the words "amongst others": they simply stated that of the many uses for a steam engine, the steamboat was simply one. We will return to this later during the discussion of the pamphlets controversy.

Fitch then returned to Pennsylvania, where he petitioned the Assembly for support that was given—verbally, not financially. He then went to Annapolis and petitioned the Maryland Assembly for money to procure a steam engine from Europe, but to no avail. Hearing the Delaware Assembly was sitting in Dover, he ventured there and received "many flattering assurances of their indeavours to serve me" but little else.[25]

On March 18, 1786, the tenacious and unrelenting Fitch visited the New Jersey Assembly and submitted a petition signed by fourteen individuals, including John Stevens (see chap. 12) and Stevens's father-in-law, Col. John Cox, for "protection & encouragement of the Legislature."[26] Within three days, without any opposition, they provided him an exclusive right to operate his steamboat on the waters of the state. With this, Fitch was able to organize the Steamboat Company, raise money, and finally proceed with his plans to construct a steam-powered vessel.

His first priority was to find an individual with experience in building a steam engine. Quite by accident, he came upon the capable Henry Voight, described as "a plain spoken Dutchman, who fears no man," whom Fitch regarded as a mechanical genius with abilities superior to his own.[27] Before building a full-scale engine,

Voight suggested constructing a model, an approach that was approved by the company.

A one-inch cylinder model was constructed at little cost but proved unsuccessful because of friction. Sometime in June 1786, they began work on a three-inch cylinder, which worked well enough to convince them that a full-scale version would be successful.[28] They also envisioned the possibility of steam being applied to both sides of the piston, a concept that Watt had patented four years earlier in 1782.[29] These were heady times, as both Fitch and Voight called themselves "enginears."

Trial 1. Fitch had intended to use water jets to propel his boat, but Voight convinced him to experiment with other techniques (see fig. 7.9). Around the middle of 1786, before the work on the model engine was completed, they obtained a small boat or skiff and outfitted it "with a screw and Paddles and one or two other modes" of propulsion.[30] The craft was launched into the Delaware River, and Fitch and Voight climbed aboard, anxious to try out the new techniques. As they manually turned the mechanisms, the results were disastrous, and the embarrassed pair had to row to shore. Fitch reports, "[Voight] stole off from me and left me alone to take care of the Machinery and stand the scoffs and snears of those who awaited our arrival."[31]

The unsuccessful trial, however, set Fitch's creative mind to work, and very quickly he conceived the idea of using paddles and cranks. On each side of the boat he would mount several paddles, one

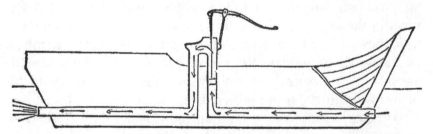

Figure 7.9. Benjamin Franklin had learned of water jet propulsion from Bernoulli and proposed this arrangement. Not shown are check valves in the pump's piston to allow it to rise without moving the water. Fitch was impressed with the concept of hydraulic propulsion, but his partner, Henry Voight, persuaded him otherwise. (From Westcott, *Life of John Fitch*, p. 135.)

behind the other, which, alternating, would descend vertically and move backward, propelling the boat forward, then ascend and move forward to repeat the cycle. He was, in effect, emulating the action of a number of humans paddling a large canoe.

Fitch's biographers, Westcott and Boyd, report that toward the end of July 1786, Voight and Fitch configured the skiff with side paddles, installed the three-inch model steam engine and boiler, and conducted another trial.[32] This, however, is unlikely, as no mention of such a trial can be found in Fitch's autobiography, and he states that his three-inch model was not finished until August, one month later. Furthermore, a three-inch steam cylinder moving the boat with side paddles does not appear to be possible, considering the inherent friction losses in the crank mechanism and the minuscule force generated.[33]

To build a practical boat of twenty tons' burden, Fitch deemed that a twelve-inch cylinder would be required for power. Unfortunately, the funds of the Steamboat Company were exhausted, and members were unwilling to contribute another time. Fitch petitioned the Pennsylvania legislature in September and narrowly failed to obtain £150. Recalling the enthusiasm that was generated upon receiving the monopoly from New Jersey, Fitch set out to obtain additional ones from several states. He went again to Dover and succeeded on February 3, 1787, in obtaining a fourteen-year monopoly in Delaware. From there he went into New York and petitioned the Assembly, which responded on March 19 with a monopoly on the waters of their state. After another success in Pennsylvania on March 28, the Steamboat Company was once again energized. The company was reorganized, with Fitch having no more authority than any other member, but again money was available.

A contract was given to the company of Brook and Wilson for a boat 45 feet long, with a 40-foot keel and an 11-foot beam drawing 3⅔ feet of water at a cost of fifty dollars.[34] The boat was delivered in April, and Voight began to construct a boiler consisting of 1,300 bricks weighing 7,000 pounds mortared together on the deck of the craft. The rest of the equipment was installed, and by early May the steamboat was ready for a trial (see fig. 7.10).

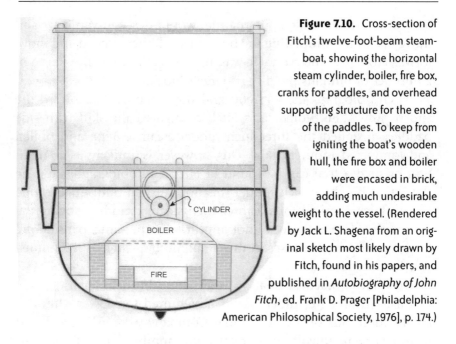

Figure 7.10. Cross-section of Fitch's twelve-foot-beam steamboat, showing the horizontal steam cylinder, boiler, fire box, cranks for paddles, and overhead supporting structure for the ends of the paddles. To keep from igniting the boat's wooden hull, the fire box and boiler were encased in brick, adding much undesirable weight to the vessel. (Rendered by Jack L. Shagena from an original sketch most likely drawn by Fitch, found in his papers, and published in *Autobiography of John Fitch*, ed. Frank D. Prager [Philadelphia: American Philosophical Society, 1976], p. 174.)

Trial 2. It appears that the second trial was a dockside test of the steam cylinder, which had been mounted horizontally. In this position, the weight of the piston compressed the seal on the bottom of the cylinder, causing it to leak. Fitch describes the event:

> And having everything ready to put on board we got our Works compleated about the beginning of May 1787. But fixing our Cylinder Horosontal we always had a Leaky Piston, of course we could not succeed. We had also wooden caps to our cylinders, which admitted Air, of course very injurious. When Mr. Voig[h]t and myself pointed out those defects to out company, they generiously advanced money to set them Right. Which was takeing the works from the foundation, and refitting them all again, a very tedious and expensive job. We also discovered that our Steam Valves were very imperfect, altho they was ingeneously contrived by my Friend Mr. Voig[h]t, and which every man of science would have approved until the defects were discovered by actual Experiment.[35]

In the process of correcting known problems, some additional ones were uncovered, as is often the case in the development of a new device. The condenser did not perform satisfactorily, leading Voight to develop an improved pipe or tube condenser. This improved engine performance significantly but pointed to an inadequacy in the boiler to produce a sufficient amount of steam. Fitch noted, "The Company haveing Run themselves to a vast expence [was] discouraged by such a continued scene of disasters."[36] To stimulate enthusiasm, Fitch drafted a lengthy article about the steamboat, which he proposed to have published. Showing it around to the members of the Steamboat Company had the desired effect, and over the next several months, money was found—and the steamboat was ready for another trial (see fig. 7.11).

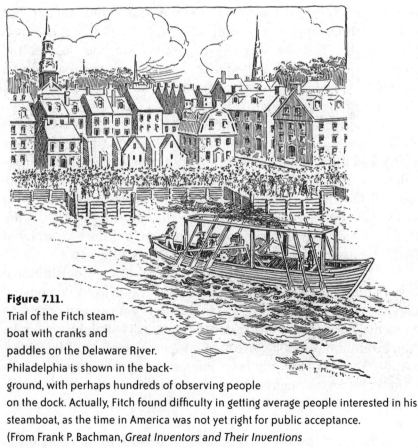

Figure 7.11.
Trial of the Fitch steamboat with cranks and paddles on the Delaware River. Philadelphia is shown in the background, with perhaps hundreds of observing people on the dock. Actually, Fitch found difficulty in getting average people interested in his steamboat, as the time in America was not yet right for public acceptance.
(From Frank P. Bachman, *Great Inventors and Their Inventions* [New York: American Book Company, 1918], p. 28.)

Trial 3. Here is Fitch's account of the event, which took place some time near the latter part of August 1787:

> When we had got the alterations made, we found ourselves more embarrassed than ever. We found our Engine to work exceedingly well, and plenty of steam, but [the steamboat was] not to go fast enough to answer a valuable purpose on the Delaware. Which threw me into the greatest consternation. We found that we must have a greater force than we had supposed, to carry it to answer the purpose of stages on the Delaware. We were convinced that we had got the whole force from our 12 inch cylinder, and the cylinder must be inlarged.[37]

How fast his steamboat went is not mentioned in this account, but in a September 5 letter to Governor Edmund Randolph, it is stated, "his boat attained the speed of two miles per hour against the strongest current of the Delaware."[38] Assuming the current to have been about two miles per hour, the steamboat most likely achieved about four miles per hour.

The stagecoach run from Philadelphia to Trenton, a distance of about thirty-eight miles, took about five hours (at eight miles per hour), and the steamboat was simply not competitive at about one-half the speed (see fig. 7.12). It was unfortunate for Fitch that one bank of the Delaware was populated with towns and had a road easily traversed. Such was not the case for many rivers, such as the Potomac and Fulton's rocky-banked Hudson, but the geography of Fitch's landscape did not favor his steamboat endeavors.

Dr. William Samuel Johnson—a member of the Constitutional Convention, then sitting in Philadelphia—was given a demonstration of the steamboat on the Delaware along with several others. In a letter to Fitch the following day, August 23, 1787, Johnson wrote, "the Exhibition yesterday gave the Gentlemen present much satisfaction."[39] This has generally been interpreted to mean that the boat went for a trial on August 22, but as no mention is made of such, it might have been that the delegation just came on board for a for a "show and tell."

As the speed had been too slow, it was decided by the company to

To the PUBLIC.

THE FLYING MACHINE, kept by John Mercereau, at the New-Blazing-Star-Ferry, near New-York, fets off from Powles-Hook every Monday, Wednefday, and Friday Mornings, for Philadelphia, and performs the Journey in a Day and a Half, for the Summer Seafon, till the 1ft of November; from that Time to go twice a Week till the firft of May, when they again perform it three Times a Week. When the Stages go only twice a Week, they fet off Mondays and Thurfdays. The Waggons in Philadelphia fet out from the Sign of the George, in Second-ftreet, the fame Morning. The Paffengers are defired to crofs the Ferry the Evening before, as the Stages muft fet off early the next Morning. The Price for each Paffenger is *Twenty Shillings,* Proc.* and Goods as ufual. Paffengers going Part of the Way to pay in Proportion.

As the Proprietor has made fuch Improvements upon the Machines, one of which is in Imitation of a Coach, he hopes to merit the Favour of the Publick.

                                    JOHN MERCEREAU.

New York Gazette 1771.

Figure 7.12. Stage lines, similar to this one advertised in the *New York Gazette* in 1771 by John Mercereau, ran regular service along the banks of the Delaware, moving passengers and goods from town to town, which made it difficult to introduce competitive steamboat service. (From D. H. Montgomery, *Student's American History* [Boston: Ginn, 1916], p. 167.)

procure an eighteen-inch cylinder, which Fitch predicted would increase the speed by a factor of 50 percent. By October 1787, the patterns for the cylinder were completed and delivered to the Warrick Furnace to be cast. The casting operation, however, was not successful, with the caster destroying his first attempt, leaving the company in a quandary as to how to proceed. It was resolved thus: "Finally it was concluded that if we could not get a cylinder to fit the Boat, that we would get a Boat to fit the cylinder. And as our old Boat was 45 feet Keel and 12 feet Beam, and we bespoke one of 60 feet Keel and 8 feet Beam, thinking a 12 inch cylinder might move that with the same velocity, as an 18 inch cylinder would the other, and we lightened the Boat about 3½ Tons and the velocity was more in proportion to the weight, than the length of the Beam" (see fig. 7.13).[40]

Although a narrower boat would incrementally decrease its resistance through the later, the elimination of weight was also crit-

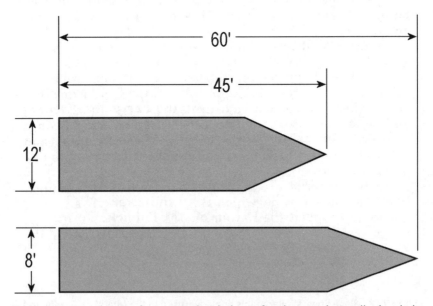

Figure 7.13. To achieve a faster speed with the 45-foot boat Fitch initially decided to replace the 12-inch steam cylinder with one 18 inches in diameter. Unable to successfully have the larger cylinder cast, he subsequently decided to construct a narrower and longer boat and use the 12-inch cylinder, correctly anticipating a higher speed. His new beam-to-length ratio of 0.133 was slightly larger than the 0.098, marginally stable ratio used by Fulton on his *North River Steam Boat*. (Diagram by Jack L. Shagena.)

ical. This would reduce the amount of water being displaced—hence the amount of water being moved by the hull when propelled.

Trial 4. The new hull was procured, and the old 12-inch cylinder steam engine installed. In addition, Voight made improvements to the mechanism by moving the paddles, reducing the work lost to friction, and, very importantly, a new, smaller, and lighter pipe boiler was installed. In July 1788, Fitch gives this account of its operation: "After experiencing innumerable difficulties more than any mortal man can conceive, except Mr. Voig[h]t and myself, we got the boat to work and set off [on] a Journey to Burlington. We went exceedingly well till we came quite opposite the Town, and in about 20 or 30 Poles [110 to 165 yards] of the upper Wharf where we intended to have come to, when our cussed pipe Boiler got such as leak we could not work the Engine any further, and where we came to anchor."[41]

With the help of the tide, Fitch and Voight got the boat back to Philadelphia and repaired the leaky boiler. Thereafter, with about thirty people on board, another twenty-mile run to Burlington was made in three hours and ten minutes. Factoring in the favorable current, caused by the tide, of two miles per hour, this represented a speed of slightly less than five miles per hour. Unfortunately, this was still not fast enough to compete with the stagecoach, which made the Philadelphia-Trenton run in less than five hours. The steamboat was also demonstrated to at least thirty other individuals during October, as attested in Fitch's pamphlet.[42]

The Steamboat Company was becoming disjointed, and Voight, who had been neglecting his family, decided to forsake the effort and "would have nothing more to do with it." Fitch was disappointed, dejected and even contemplated suicide.[43] With the faltering company threatening the project's success, Fitch was adamant to rekindle their interest. Once again, the doggedly determined inventor pulled himself up by his bootstraps, and on December 5, 1788, he presented a "Proposal of John Fitch to the Steamboat Company."[44] In the plan the number of shares would be increased from forty to eighty, with the current shareholders owning one-half and the other forty shares being sold at ten pounds each to raise enough money to continue the development of the steamboat. At such a time when the new company achieved success, they were to merge with the old company, which had spent the equivalent of sixteen hundred English pounds without producing a boat capable of competing with the stagecoach line.

To convince old and new investors this was prudent, he set out in his proposal the steps that would have to be taken to achieve success and an estimate of the costs involved. Most interestingly, he included a rare bit of theory, which illustrated that he had an understanding of some of the underlying principles involved in steamboat design. Fitch correctly pointed out that the resistance of a boat through the water varied according to the square of its velocity. By increasing the cylinder from twelve to eighteen inches, the power to move the boat would increase as the square of the ratios—specifically, a proportional increase in speed from four to six miles per hour.[45]

As this was still not fast enough to be competitive with the stagecoach line, he suggested running with the two-mile-per-hour tide to achieve a total of eight miles per hour. He also included a schedule of suggested fees, demonstrating that the boat would be economically viable. After some amendments, the proposal was adopted, and the company again had a full head of steam.

>━━\\\//━━<

During the early part of 1788, Fitch became embroiled in controversy with James Rumsey as to who had priority for the invention of the steamboat. After Rumsey had demonstrated his steamboat on the Potomac River in December 1787, he produced a pamphlet claiming priority, which contained, among many certificates, a letter from Governor Johnson that identified the date that Rumsey had advised him of his intention of applying steam to a boat as October or November 1785. Concurrent with this, Rumsey requested to have a cylinder cast at the foundry owned by the governor and his brother in Frederick Town, Maryland.[46]

As Fitch had obtained a very complimentary letter from Johnson on November 25, 1785, introducing him to Governor Smallwood, in which there was no mention of Rumsey's steamboat work, the combative and very suspicious Fitch believed that the governor had misrepresented by a full year Rumsey's plan to use steam; Fitch would try to exploit this, arguing that Rumsey only started working on his steamboat after he learned of Fitch's boat. The explanation given by the governor was that Rumsey had sworn him to secrecy and that he was simply encouraging both of them, as the steamboat was a worthwhile project.

Fitch set about interviewing individuals who had supplied Rumsey components for his steamboat with the belief that although Rumsey may have actually thought of applying steam before him, as he had learned from George Washington, he actually started work in 1786, not in 1785. To this end, Fitch was successful in obtaining certificates from several of Rumsey's suppliers that cast serious doubt on the actual dates of Rumsey's steamboat work. The information supplied, however, was based on the recollections of the suppliers, and when they later checked their actual written records, the dates on the supplied certificates to Fitch were found to be in error.

An angry and upset Fitch set about to discredit Rumsey's "wicked and invidious pamphlets," drafting a response and including certificates. One of the members of the Steamboat Company, Richard Wells, described by Fitch as a man with an able pen, undertook to rewrite and reorganize the draft into a presentable document. It was published in May 1788 with the long title of *The Original Steam-Boat Supported; or, A Reply to Mr. James Rumsey's Pamphlet Shewing the True Priority of John Fitch and the False Datings, &c. of James Rumsey.*

These charges were answered in another pamphlet by Joseph Barnes, Rumsey's brother-in-law, business partner, and now attorney, in July 1788; the pamphlet countercharged John Fitch with perjury, falsehood, want of memory of candor, and even bribery of one of the individuals from which he obtained a certificate.[47] The apparent discrepancies that Fitch claimed in the dates were, for the most part, explained by Barnes, and an unbiased reading of all three pamphlets, sometimes with convoluted arguments, appears to support the original Rumsey position.[48]

In an attempt to refute the statements in Barnes's pamphlet, Fitch obtained new affidavits and certificates that were published in a handbill.[49] This seemed to end the dispute; however, the controversy did consume a significant amount of his energies during the middle part of 1788. His fanaticism for steamboat priority damaged his credibility, but for a man driven by only one passion in life, his zealous pursuit seems quite understandable.

⋙～\ψ⁄～⋘

After completion of the organization of the new Steamboat Company near the end of 1788, Fitch was kept busy settling accounts from the old company that had accrued because of the low state of finances. The company had agreed to procure an eighteen-inch cylinder, which would, they anticipated, provide the power for the required increase in speed for the boat, so early in 1789, Fitch started looking for a supplier. He found Mr. Drinker of Atsion Furnace, and the casting was ordered in March 1789.[50] After several attempts, a satisfactory cylinder was produced.

Around August 1789, the new engine was about complete, but one of the new investors, Dr. William Thornton (see fig. 7.14), insisted on testing his own condenser, made of eight-pound copper sheet.[51] Fitch protested, arguing the metal was too weak to withstand the pressure of the atmosphere when it created a vacuum, but to no avail. The condenser was installed and "crushed together at once," setting

Figure 7.14. William Thornton was one of the partners in Fitch's steamboat company and in 1802 became the commissioner of patents for the United States. (Rendered by William T. Sisson from *Papers of William Thornton*, ed. C. M. Harris [Charlottesville: University Press of Virginia, 1995], title-page frontispiece.)

back the project a short time.[52] Alone, Fitch was struggling to keep the steamboat project going, but he was very pleased when one day his old friend Voight showed up and agreed to again support him.[53]

Trial 5. A new Thornton condenser, constructed of thicker copper, was installed and passed the initial test, making the boat ready for a trial in September or shortly thereafter. The speed of the craft with the eighteen-inch cylinder, however, did not improve as anticipated, and the disappointed Fitch reported, "yet we did not much exceed our performance the summer before with a 12 inch cylinder. Which all allarmed me beyond measure."[54]

Voight determined the lack of performance to be the fault of the condenser and the air pump (actually a vacuum pump, which draws the steam from the cylinder into the condenser), so he immediately set about designing and building new ones. Fitch, though supportive but unsure of how to solve the problem, "remained a silent spectator."[55] When the new devices were installed, the boiler was fired up in preparation for a trial, but a rainstorm forced a postponement. The fire in the boiler box was extinguished, and Fitch, Voight, and the workers retired for the evening.

A strong wind from the northeast came up sometime during the night and fanned some of the embers, not completely extinguished, setting the boat on fire. It burned to the water line, and when it was discovered the next morning by Fitch, he deliberately sank it to save the machinery. The hull with the equipment was raised and salvaged, but the problem-plagued Steamboat Company would have to begin once more.

The vessel was repaired, and without an explanation from Fitch, it was decided to replace the side paddles with a stern wheel in the rear of the boat.[56] The old machinery, including the latest version of the condenser and air pump, was installed during March 1790. During this period, Fitch was rather badly treated by Dr. Thornton and Mr. Stockton, two members of the company, who accused him of "stupidity, Botching, being a man [that] could not be depended upon, with Drinking, Tipling, and every opprobrious name that could be artefully invented."[57] Success, however, was not far away.

Trial 6. On Monday, April 12, 1790, the boat was taken onto the river, and the engine worked so forcefully that it broke one of the pulleys, which was soon replaced with a much stronger one. Four days later, on April 16, a joyous Fitch reported,

> We got our works compleated, and tried o[u]r Boat again. And altho the Wind blew very fresh at N. E., we reigned the Lord high admirals of the Delaware, and no Boat on the river could hold way with us, but all fell a-stern, altho several Sail Boats which were very light, and [with] heavy sails, that brought their Gunwails well down to the Water came to try us. We also passed many Boats with Oars, which were strong[ly] manned, and [with] no loading, who seemed almost to stand still when we passed them. We also ran Round a Vessel that was beating to windward in about 2 miles, which had about 1½ miles a start of us. And [we] came in without any of our works failing. Which fully convinced us of what we had pursued so long, and with such imbarrassments.[58]

After several more trials, the Steamboat Company was convinced of the boat's worthiness and started to provide demonstrations to prominent individuals in Philadelphia. A justifiably proud and satisfied Fitch wrote, "This had been effected by little Johney Fitch and Harry Voig[h]t, one of the Greatest and most useful arts that was ever introduced into the World. And altho the World nor my country does not thank me for it, yet it gives me a heart felt Pleasure, that neither men or Angles can take from me dureing my existence, if it be to all Eternity. Therefore [I] am sure of going to Heaven when I die, for Heaven is pleased to make me happy [even] if my Country are pleased to see me in Rags penury and distress."[59]

This was Fitch's shining moment of well-deserved glory. He and Voight had done it—created a steamboat to go about seven miles per hour on the Delaware River—and it remained his passionate desire to be recognized and rewarded with adoration by his fellow man.

❧✿❧

Cabins designed by Dr. Thornton were constructed on the deck to provide comfort for the passengers, and with travel time roughly competitive with the stagecoach, the steamboat was put into regular service between Philadelphia and Trenton (see fig. 7.15). Despite advertising and the offer of free food for the trip, an insufficient number of passengers were carried during the summer of 1790 to make the operation profitable. It would be another seventeen years before Americans would embrace steamboat travel on the much slower boat of Robert Fulton (moving at four to five miles per hour), but then the shores of the Hudson River were rocky and not easily traversed by stagecoach.

For poor Johnny Fitch, the time and place were wrong, and he was never able to engage a benevolent benefactor who shared his steamboat dreams. James Watt had found an understanding and fair-minded Matthew Boulton; James Rumsey, Dr. James McMechen and the Rumseian Society; and, later, Robert Fulton would find Robert Livingston. These mergers of creative genius with practical business and political acumen are nearly always required to produce success from new ideas, as the talents to create new products and to convince a sometimes reluctant public are significantly different. Poor Johnny Fitch had to do it all almost single-handedly, and though he finally achieved a technical triumph, commercial success still eluded this driven and determined man.

On August 26, 1791, Fitch suffered his life's greatest disappointment, when the United States government issued on the same day steamboat patents not only to him, but also to James Rumsey,

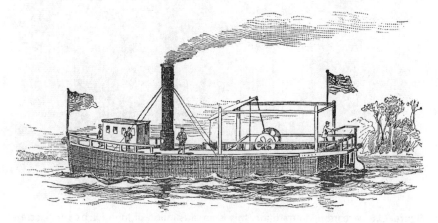

Figure 7.15. John Fitch's and Henry Voight's stern paddle steamboat with a passenger cabin installed, providing service along the Delaware River between Philadelphia and Trenton. (From Alexander Johnston, *History of the United States* [New York: Henry Holt, 1889], p. 157.)

Nathan Read, and John Stevens for similar steamboat designs. He believed that his patent would predate the others, allowing him to claim priority, but this was not to be. Without federal patent protection, his efforts were doomed.

The steamboat company had previously made an agreement with Aaron Vail, the U.S. consul at Lorient, who committed to financing the construction of a steamboat in France. To this end, Vail had obtained copies of Fitch's drawings, and a French patent in Fitch's name was granted on November 20, 1791.[60]

John Fitch, now dejected and distraught over U.S. efforts, sailed in the spring of 1793 for France with dreams of building a successful steamboat. The country, however, was in a state of insurrection, and no interest could be sustained for construction of the engine. Vail sent Fitch to London, but they found that British law did not allow export of a steam engine. Not being able to get back to France because of the war with England, a dejected Fitch found his way back to America, arriving in Boston in the summer of 1794, where he worked as a laborer for some time.

Figure 7.16. According to tradition, this is an illustration of John Fitch's small, screw-propelled steamboat being operated on Collect Pond in New York City about 1796 or 1797. (From Robert H. Thurston, *History of the Growth of the Steam-Engine* [New York: D. Appleton, 1901], p. 240.)

Figure 7.17. Grave of John Fitch in Bardstown, Kentucky, where the beleaguered inventor died. (From Westcott, *Life of John Fitch*, title-page frontispiece.)

A broadside published in 1846 by John Hutchings claimed that he remembered as a boy that Fitch operated a steamboat on Collect Pond in New York City in 1796 or 1797. The account, however, is flawed historically and technically, and since there is no other evidence to substantiate the event, it is not covered further in this account. However, this has not stopped steam navigation writers from reporting the supposed trial, as shown in figure 7.16.

After several years of aimless existence, Fitch removed to Kentucky, where he found that squatters had settled on his land. In exchange for meals and whiskey from a tavern keeper in Bardstown, he offered a portion of his land and hired a lawyer to claim what was rightfully and legally his. During these distressing times he had trouble sleeping, and a local doctor prescribed opium pills, which he provided to Fitch one at a time. The depressed Fitch saved up a quantity of the pills and washed them down with a bottle of

Table 7.1. Summary of the efforts of steamboat pioneers

Steamboat pioneer	Concept date	Steamboat trial date(s)	Notes
Papin	1690	1707	Human, not steam power
Hulls	1736	None	Concept only for steam tug
D'Auxiron	1770	None	In 1774 boat accidentally sunk
De Jouffroy	ca. 1771	1778 Unsuccessful 1783 *Pyroscaphe*	Some success on 1783 trial, but work interrupted by the French Revolution
Miller	ca. 1786	1787 *Edinburgh* 1788 Steamboat	Human, not steam power Engine by Wm. Symington
Dundas	ca. 1798	1802 Tugboat *Charlotte Dundas*	Hull by Alexander Hart with engine by Wm. Symington; towed 140 tons on a canal
Henry	1779	None	Concept only
Rumsey	1783	1786, 1787, 1790, 1792, 1793	Limited success with water jet propulsion
Fitch	1785	1787, 1788, 1789, 1790	Introduced steamboat service on the Delaware in 1790, which was not financially viable

whiskey, ending his life in 1798 as a disappointed and dejected man (see an illustration of his grave in fig. 7.17). With his passing, the United States lost a truly great inventor, who should be better recognized for his creativity, indomitable will, and persevering spirit.

First presented in chapter 4, the table of steamboat pioneers and their contributions has been updated with the efforts of William Henry, James Rumsey, and John Fitch. Every few chapters, the table will be augmented with other contributors, and in chapter 13, where the inventor of the steamboat is identified, it will be presented in its final form.

Notes

1. Fitch, *The Autobiography of John Fitch*, ed. Frank D. Prager (Philadelphia: American Philosophical Society, 1976), p. 25, is the source of this quotation, and much of the material in this chapter is taken from this book. In 1790, when John Fitch was forty-seven years old, he had failed in his personal relationships but was still struggling to introduce steamboat service on the Delaware River; at this time he started writing his steamboat account. Through the encouragement of a friend and adviser, Rev. Nathaniel Irwin, he also added the story of his life up to that time. This embittered genius,

believing his words might injure several of his contemporaries, delivered the manuscript to the Philadelphia Library in 1792 with instructions that it was not to be opened for thirty years. As the document did not undergo contemporaneous review, some aspects of his account have been challenged as being inconsistent with other evidence. Nevertheless, his autobiography remains as the most comprehensive account of his life.

2. Thompson Westcott, *Life of John Fitch, The Inventor of the Steamboat* (Philadelphia: J. B. Lippincott, 1857), p. 27. This well-written account provides a comprehensive summary of Fitch's life taken from a number of sources. Westcott, however, seems intent on portraying John Fitch as the inventor of the steamboat.

3. *Autobiography of John Fitch*, p. 22.

4. Ibid., p. 26n. The book was identified by Prager as either Thomas Salmon, *The Modern Gazetteer* (London, 1746) or his *New Geographical and Historical Grammar* (London, 1749).

5. *Autobiography of John Fitch*, p. 27. In Westcott, *Life of John Fitch*, p. 34, it is claimed that his father taught him surveying.

6. Using the same unit, the rod, for both linear and area measurements is somewhat confusing and was therefore later abandoned. Tapes calibrated in feet and inches have replaced surveyor chains, but old chains may still be seen in museums.

7. *Autobiography of John Fitch*, p. 27n. Prager identifies Roger Wolcott as governor of the province from 1750 to 1754.

8. Ibid., p. 32.

9. Ibid., p. 34.

10. Ibid., p. 47.

11. Brooke Hindle, *Emulation and Invention* (New York: W. W. Norton, 1981), p. 10.

12. Carroll W. Pursell, *Early Stationary Steam Engines in America* (Washington, DC: Smithsonian Institution Press, 1969), pp. 5–6.

13. *Autobiography of John Fitch*, p. 52.

14. Ibid., p. 53.

15. Ibid., p. 111.

16. Ibid., p. 113. The book was Benjamin Martin, *Philosophia Britannica* (London, 1759).

17. Ibid., Prager's note 81.

18. Pursell, *Early Stationary Steam Engines*, p. 18.

19. *Autobiography of John Fitch*, pp. 8, 8n9, 9, 9n10.

20. Ibid., p. 65, Prager footnote. Ewing (1732–1802), theologian and natural philosopher, was provost of the University of Pennsylvania, an

officer in the American Philosophical Society, and a member of Fitch's land-jobbing company.

21. Ibid., p. 153.

22. Ibid., p. 155.

23. Ibid., p. 156.

24. Quoted in ibid., p. 160.

25. Ibid., p. 162.

26. Ibid., p. 164.

27. Ibid., p. 168.

28. From this it can be concluded that neither Fitch nor Voight was aware of the inherent inefficiencies in small models. This follows from the fact that the friction of moving a piston in a cylinder is directly proportional to the diameter of the cylinder, whereas the force moving it is proportional to the square of the diameter. Hence, increasing an increase in diameter of, say, four to one brings an attendant fourfold performance improvement.

29. *Autobiography of John Fitch*, 169, Prager's note.

30. Ibid., p. 117.

31. Ibid.

32. Thomas Boyd, *Poor John Fitch, Inventor of the Steamboat* (New York: G. P. Putnam's Sons, 1935), p. 162; Westcott, *Life of John Fitch*, p. 160.

33. As reported in Westcott, *Life of John Fitch*, p. 162, Fitch sent a letter to Benjamin Franklin on September 4, 1786, offering to sell the model to the American Philosophical Society for one hundred dollars. He candidly described several deficiencies, noting inadequate steam generation, overheating, and sticking; he went on to say, "yet . . . the piston moves with considerable Velocity, when unloaded." This would seem to end speculation that it could have powered a steamboat. Ella May Turner, *James Rumsey, Pioneer in Steam Navigation* (Scottdale, PA: Mennonite Publishing House, 1930), p. 136, and James Flexner, *Steamboats Come True: American Inventors in Action* (1944; reprint, Boston: Little, Brown, 1978), p. 104, also support the position that a trial did not take place.

34. Boyd, *Poor John Fitch*, p. 172.

35. *Autobiography of John Fitch*, p. 172. This is an astute observation by Fitch, that sometimes a defect or deficiency in a design cannot be detected by analysis or observation but only through an actual trial. This aspect of engineering still plagues designers; hence, engineers still learn by doing.

36. Ibid., pp. 174–75.

37. Ibid., p. 178.

38. Quoted in Turner, *James Rumsey*, p. 115.

39. Quoted in *Autobiography of John Fitch*, p. 179.

40. Ibid., p. 183. The boat ordered from Brooke and Wilson specified a 40-foot keel, a length of 45 feet, and an 11-foot beam, but perhaps it was delivered with the dimensions mentioned here.

41. Ibid. Fitch opposed the pipe boiler, which had been used by Rumsey. It was supported by Voight, and through pressure extended by the Steamboat Company; it was installed against his will.

42. John Fitch, *The Original Steam-Boat Supported; or, A Reply to James Rumsey's Pamphlet Shewing the True Priority of John Fitch and the False Datings, &c. of James Rumsey* (Philadelphia: Zachariah Poulson, 1788), certificate no. 2.

43. *Autobiography of John Fitch*, p. 184.

44. Westcott, *Life of John Fitch*, p. 257.

45. As far as Fitch went with the analysis, he was correct, as this was the extent of scientific understanding at that time. As power also has a time component, however, it varies as the cube, not the square, of velocity (i.e., speed).

46. James Rumsey, *A Short Treatise on the Application of Steam, Whereby Is Clearly Shewn from Actual Experiments, That Steam May Be Applied to Propel Boats or Vessels of Any Burthen against Rapid Currents with Great Velocity* (Philadelphia: Joseph James, 1788), certificate no. 14.

47. Joseph Barnes, *Remarks on Mr. John Fitch's Reply to Mr. Rumsey's Pamphlet* (Philadelphia: Joseph James, 1788), next to last paragraph in narrative.

48. Flexner, *Steamboats Come True*, p. 138, notes that Fitch's pamphlet "did not carry much conviction."

49. Westcott, *Life of John Fitch*, p. 237.

50. Ibid., p. 270.

51. Copper is sold by the pound, with eight-pound copper being of the necessary thickness (about 0.17 inches) to weigh eight pounds per square foot.

52. *Autobiography of John Fitch*, p. 187.

53. Part of the controversy with Voight was his extramarital affair with Fitch's landlady, Mrs. Krafft, which resulted in two illegitimate children. See Flexner, *Steamboats Come True*, for an excellent account of this turbulent period.

54. *Autobiography of John Fitch*, p. 189.

55. Ibid.

56. The exact time the side paddles were replaced by stern paddles is not clear and may have been done prior to this point. Westcott, *Life of John Fitch*, says the stern paddles were employed when the 8 × 60 foot hull was built.

57. *Autobiography of John Fitch*, p. 191.

58. Ibid.

59. Ibid., p. 193.

60. Boyd, *Poor John Fitch*, p. 271.

Figure 8.1. Oliver Evans. (Rendered by William T. Sisson from Robert H. Thurston, *History of the Growth of the Steam-Engine* [New York: D. Appleton, 1901], p. 154.)

8.

Oliver Evans
1755–1819

Therefore, he that studies and writes on the improvements of art and sciences labours to benefit generations yet unborn, for it is not probable that his contemporaries will pay any attention to him, especially those of his relations, friends and intimates; therefore improvements progress so slowly.

—Oliver Evans

Oliver Evans was born in 1755 near the colonial village of Newport in New Castle County, Delaware, a short distance southwest of Wilmington.[1] At a very young age he demonstrated a capacity for invention, and by age twenty-two had perfected a method for making fine wire from iron bars. He went on to an extraordinary career of creating useful devices and was America's first systems engineer, physics textbook author, amphibious vehicle inventor, and high-pressure steam engine developer and advocate. He was a visionary powerhouse, an engineer's engineer with a remarkable intellect and thinking that was years ahead of his time.

During intervals in his life he achieved a moderate amount of business success and wealth, but at sixty-four he died an embittered and dejected man, neither appreciated nor revered by his contemporaries. This chapter can in no way do justice to the genius of this creative individual, but will rather simply highlight some of his achievements leading toward and focusing on his steamboat contributions.

The fifth child of Charles and Ann Stalcop Evans, Oliver was born on September 13, 1755, having been preceded by Sarah, Jonathan, John, and Theophilus.[2] Seven more siblings would follow, including two brothers, Joseph and Evan. His younger brothers as well as two of his older brothers would at times participate with him in his many business ventures.

Records of his early life are sketchy. His first biographer, Henry Howe, suggested that when he was a young boy and not allowed to burn candles at night, he studied by the light of wood shavings.[3] Whether or not this is true, however, is not known, but it is apparent that young Oliver had a thirst for mechanical knowledge, seeking out and learning from books and mastering writing in a clear, lucid style. In 1872, George Latimer described the young Evans as having "been born with an inventive mind, which manifested itself early in life, walking with his head downward, his hands behind him, as if in deep study."[4]

At the age of sixteen, in 1771, Oliver was apprenticed to a wheelwright or wagon maker probably in Newport, where he acquired practical knowledge of using iron and wood and became acquainted with blacksmithing. Evans later recalled that in December 1772, he became aware of some boys putting a small amount of water into the barrel of a musket, sealing it with a tight wad, and placing it in a blacksmith's fire. The water vaporized into steam, resulting in a small explosion called a "Christmas cracker." Thinking about this, Oliver was determined to understand how the power of steam could be harnessed to propel a wagon. About this he later wrote, "I labored for some time without success; at length, a book fell into my hands, describing the old [Newcomen] atmospheric engine. I was astonished to observe that they had so far erred as to the use of steam only to form a vacuum, to apply the mere pressure of the atmosphere, instead of applying the elastic power of steam, for original motion; a power I supposed was irresistible."[5] Using what he called the elastic power of high-pressure steam, Oliver reasoned that he should be able to propel a wagon. This suggestion at his young age of eighteen or nineteen met with ridicule from others; hence, he went on to other pursuits, waiting for a more opportune time.

At the age of twenty-two he turned his attention to the manufacture of fine wire, which could be cut and bent to form teeth that could be used in the manufacture of handheld devices called cards. Many of these staplelike pieces of wire were inserted through pricked holes in leather to create the card, which was effectively a very coarse, short-bristled brush. A pair of these could be used together in a process called carding to straighten the fibers of cotton or wool prior to weaving.

Through the help of George Latimer, Esq., a justice of the peace in Newport, he engaged a blacksmith to build a machine that produced high-quality metal staples so well and so fast that the current hand-bending process was immediately abandoned. Oliver, however, was not to profit from this, as his invention was to be taken up and used by others.

When Oliver was twenty-seven he and his younger brother Joseph moved to Maryland's eastern shore, establishing a store in a village called Tuckahoe, on the dividing line between Caroline and Queen Anne counties. This partnership, however, was short-lived, as Oliver, along with brothers John and Theophilus, purchased a portion of the family farm on Red Clay Creek in July 1782 (see fig. 8.2). The tract of land sold by their father contained about four hundred acres and included the remains of an old 1742 mill that would capture Oliver's powerful visual imagination.

Oliver Evans married Sarah Tomlinson on April 22, 1783, at the Old Swede's Episcopal Church in Wilmington. The couple initially returned to the store in Tuckahoe, but in the fall of that year, Oliver and his two older bothers decided that he should return to Newcastle County to oversee the rebuilding of the old flour mill. He wrote, "I then began to study how I could make the mill exceed all others, and having been successful in inventing several useful

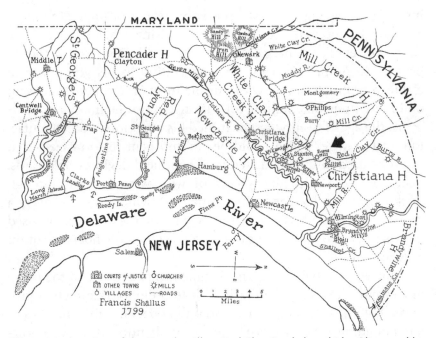

Figure 8.2. Location of the Evans's mill on Red Clay Creek, bought by Oliver and his brothers from their father. This location became the site of America's first automated gristmill. (From Greville Bathe and Dorothy Bathe, *Oliver Evans* [Philadelphia: Historical Society of Pennsylvania, 1935], between pp. 8 and 9.)

improvements, I had hopes of succeeding in this also. I first conceived the great design of applying the power that drives the mill-stone, to perform all those operations which hitherto [were] effected by manual labour."[6]

At this point one can understand the thinking of an individual who in the twentieth century would be called a *systems engineer*. Evans envisioned the milling problem in its totality with continuous flow production and without the intervention of manual labor, which can produce unwanted product variation and introduce undesirable contaminants. These concepts are central to modern manufacturing processes and were introduced by Evans over two hundred years ago. It would, however, take Evans seven or eight years of work to implement his vision completely.

Evans had no plans to change the fundamental way wheat was milled into flour, only to automate each step so as to eliminate manual labor, thereby reducing the cost of operations. Another benefit he later discovered was an improvement in yield and quality. Through the automation process he found that human input, with errors and sometimes poor judgment, was minimized, hence reducing losses and waste as well.

The process began with the wagonload deliveries of bushels of wheat to the mill by the farmers. In mills of the time, the wagon's contents were emptied onto the ground floor of the mill and hoisted to the loft using rope and pulley. To eliminate human labor, Evans installed a water-powered bucket elevator to carry the grain to the mill's third floor. Thereafter the grain descended by gravity to the second floor, where it passed by a fan that exhausted impurities, allowing the cleaned wheat to continue to the millstone on the first floor. There it was ground into meal, which was lifted by another elevator to the third floor for cooling and drying (see fig. 8.3).

One of the mill's most perplexing design problems was to spread the moist and warm meal evenly on the floor and, after a sufficient cooling and drying time, channel it via the bolting hopper to the first floor for packing in barrels. As a boy had in the past done the spreading work, the device invented by Evans to mechanically accomplish this function was called a *hopper boy*. The effort to invent this seemingly impossible device was later described by Evans:

> Here appeared an absolute impossibility—to make a machine that would both spread and gather at the same time, then seemed absurd, and the discovery cost me months of the most intense thinking, for absurdity always presented itself to baffle and deter me; but being highly stimulated by a hope of making the great improvement at first contemplated [i.e., the automated mill], and which I believe would not be completely executed without such a machine, I persevered with a zeal and indefatigability peculiar only to *inventors*, without success, until the idea struck my mind of laying the meal by a machine, first in a large circular ring around

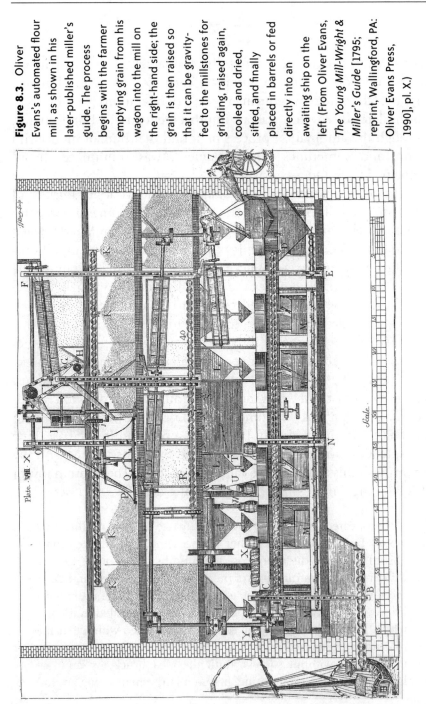

Figure 8.3. Oliver Evans's automated flour mill, as shown in his later-published miller's guide. The process begins with the farmer emptying grain from his wagon into the mill on the right-hand side; the grain is then raised so that it can be gravity-fed to the millstones for grinding, raised again, cooled and dried, sifted, and finally placed in barrels or fed directly into an awaiting ship on the left. (From Oliver Evans, *The Young Mill-Wright & Miller's Guide* [1795; reprint, Wallingford, PA: Oliver Evans Press, 1990], pl. X.)

the bolting hopper, and then gathering it from the ring to the center. I perceived with infinite pleasure, I had discovered the principles on which I could make a machine capable of performing these seeming impossibilities.[7]

The reasoning provided by the creative mind in this paragraph is enlightening, and it says a great deal not only about Evans, but about inventors in general. Persevering "with a zeal and indefatigability" to achieve what on the surface appears impossible requires a measure of faith and tenacity found only in the rarest of individuals. And when the breakthrough finally comes, the "infinite pleasure" from within of having been successful with the celebratory cry "Eureka!" is an immense satisfaction that *only* an inventor can ever experience. Although because of setbacks and disappointments Evans would later in life bitterly announce to his family that he was giving up inventing, this outburst was only a cry of exasperation, because invention ran deep in the blood of this immensely creative individual—he was simply born to invent.

To produce income for the partnership with his brothers, Evans had the mill operating the traditional (old) way by September 5, 1785, which required three men and one half-time boy to produce twenty barrels of flour per day. He immediately set about installing his improvements and selected the fatiguing, manpower-intensive bolting hopper to automate first. In this process, the dried and cooled meal is separated by the use of sieves into what was called superfine, middlings, and carnell.[8] The superfine flour commanded the highest price, and Evans sought to maximize its output and minimize the coarser products. It proved a difficult challenge:

> But although I was so perfect in the principle, yet in the execution
> of this machine I met with so many unforeseen difficulties in
> forming and adjusting it, to produce all the necessary operations
> already described, and also to load to raise and lower itself, in pro-
> portion to the quantity of meal it might have to operate on, I
> found the expenses so far exceed my calculations, that I was much
> discouraged. At length, however, I succeeded in making it a
> machine equal to any other in use for utility, and perfection in its

performance, nor could any have been more difficult to invent and execute.[9]

The bane of the development engineer is underestimating the difficulty of the task, often by overlooking one or more problems that then manifest themselves upon testing. Evans was tenacious, though, and when faced with problems, he persisted until they were under control.

Since costs to automate the bolting hopper had exceeded his predictions, he solicited the Brandywine millers to subscribe to the expenses of additional improvements, and in return they would be allowed to send their millwright to view all the improvements and copy them. As Evans was neither a miller nor millwright, however, they did not believe he could improve on their milling perfection and refused to become involved.

Evans persevered and finally installed all of his improvements, including the hopper boy, for which he had built a model to explain it to the carpenters. He decided the best way to proceed was through the issuing of licenses for his automated mill, so he sought legal protection and appealed to the states: "I petitioned early in the year 1786, and acts were passed early in the year of 1787, in Pennsylvania and Delaware, in favour of my improvements in the manufacture of flour; and in Maryland and New Hampshire for the same, including steam carriages [i.e., automobiles], to be propelled by the powerful steam engine which I had invented [but not yet developed]."[10] Evans had hoped for protection for a period of twenty-five years, but the patents were granted for fourteen- or fifteen-year periods, a length of time he vehemently argued was too short. To promote his mill ideas, he published a broadside in later 1787 (see fig. 8.4) and sent his brother Evan to canvass millers in several states offering a free license to the first who would adopt his principles. Results, however, were initially discouraging.

In May 1789, Oliver Evans ventured into Maryland and called upon the flour mills owned by the Ellicott brothers along the Patapsco River north of Baltimore. Here, more progressive and enlightened owners adopted his elevator and hopper boy and introduced him to advances in milling they had made. Evans was unsuccessful,

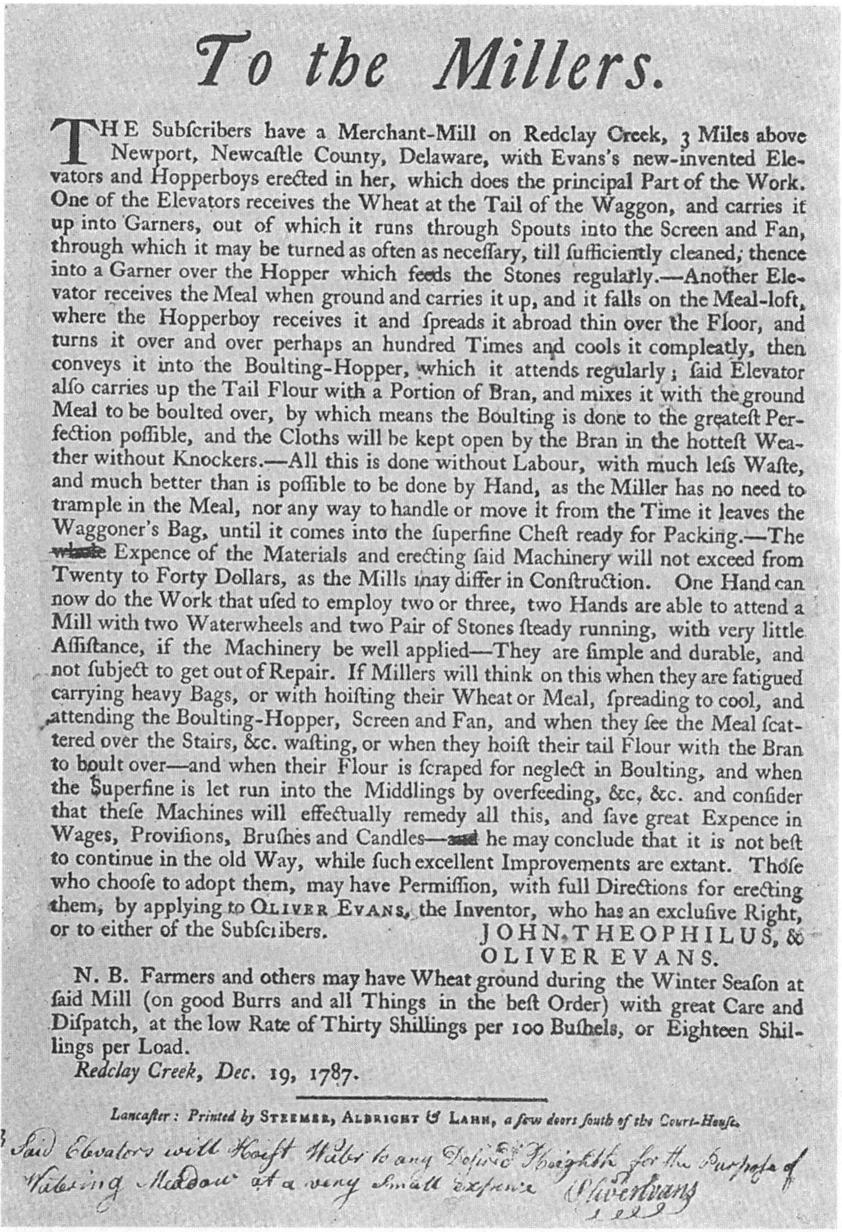

To the Millers.

THE Subscribers have a Merchant-Mill on Redclay Creek, 3 Miles above Newport, Newcastle County, Delaware, with Evans's new-invented Elevators and Hopperboys erected in her, which does the principal Part of the Work. One of the Elevators receives the Wheat at the Tail of the Waggon, and carries it up into Garners, out of which it runs through Spouts into the Screen and Fan, through which it may be turned as often as necessary, till sufficiently cleaned; thence into a Garner over the Hopper which feeds the Stones regularly.—Another Elevator receives the Meal when ground and carries it up, and it falls on the Meal-loft, where the Hopperboy receives it and spreads it abroad thin over the Floor, and turns it over and over perhaps an hundred Times and cools it compleatly, then conveys it into the Boulting-Hopper, which it attends regularly; said Elevator also carries up the Tail Flour with a Portion of Bran, and mixes it with the ground Meal to be boulted over, by which means the Boulting is done to the greatest Perfection possible, and the Cloths will be kept open by the Bran in the hottest Weather without Knockers.—All this is done without Labour, with much less Waste, and much better than is possible to be done by Hand, as the Miller has no need to trample in the Meal, nor any way to handle or move it from the Time it leaves the Waggoner's Bag, until it comes into the superfine Chest ready for Packing.—The whole Expence of the Materials and erecting said Machinery will not exceed from Twenty to Forty Dollars, as the Mills may differ in Construction.   One Hand can now do the Work that used to employ two or three, two Hands are able to attend a Mill with two Waterwheels and two Pair of Stones steady running, with very little Assistance, if the Machinery be well applied—They are simple and durable, and not subject to get out of Repair. If Millers will think on this when they are fatigued carrying heavy Bags, or with hoisting their Wheat or Meal, spreading to cool, and attending the Boulting-Hopper, Screen and Fan, and when they see the Meal scattered over the Stairs, &c. wasting, or when they hoist their tail Flour with the Bran to boult over—and when their Flour is scraped for neglect in Boulting, and when the Superfine is let run into the Middlings by overfeeding, &c. &c. and consider that these Machines will effectually remedy all this, and save great Expence in Wages, Provisions, Brushes and Candles—and he may conclude that it is not best to continue in the old Way, while such excellent Improvements are extant. Those who choose to adopt them, may have Permission, with full Directions for erecting them, by applying to OLIVER EVANS, the Inventor, who has an exclusive Right, or to either of the Subscribers.   JOHN, THEOPHILUS, &
                                                              OLIVER EVANS.

N. B. Farmers and others may have Wheat ground during the Winter Season at said Mill (on good Burrs and all Things in the best Order) with great Care and Dispatch, at the low Rate of Thirty Shillings per 100 Bushels, or Eighteen Shillings per Load.

Redclay Creek, Dec. 19, 1787.

Lancaster: Printed by STEEMER, ALBRIGHT & LAHN, *a few doors south of the Court-House.*

Said Elevators will Hoist Water to any Desired Heighth for the Purpose of Watering Meadow at a very Small expence   Oliver Evans

Figure 8.4. Broadside published by Oliver Evans in 1787 to interest millers in licensing the improvement he had designed and patented. With his improvements he touted, "One Hand can now do the Work that used to employ two or three." (From Bathe and Bathe, *Oliver Evans*, between pp. 20 and 21.)

however, in soliciting a partnership for the development of steam wagons, as the Ellicotts were concerned about the cost and difficulties of such a venture.

By 1790 Evans had made additional improvements to his automated mill, employing different technology such as screw elevators, but he had essentially completed his development. The task then at hand was one of marketing, which probably was a factor in relocating to Wilmington and selling his one-third share in the Red Clay Mill property. He constructed a model of his mill, which he displayed for a while at his home on Market Street and then sent to England to try to stimulate interest there.

He applied for a federal patent, which was granted in 1790, and published a detailed description of his mill with an engraving that appeared in the *Universal Asylum and Columbian Magazine* in January 1791. Based on information in the magazine, a broadside followed, explaining the elevator, hopper boy, and conveyer. He planned to have published additional articles with engraved plates and eventually combine these together into a pamphlet about milling. Unfortunately the magazine ceased publication in 1792 for financial reasons, but the engraving and the January article became the embryo that would later evolve into *The Young Mill-Wright & Miller's Guide.*

During this time, Evans's brother Joseph was traveling through several states to promote the automated mill, and according to the May 1792 edition of the *American Museum,* over a hundred mills had adopted Evans's machinery.[11]

Evans had embarked on a fairly successful marketing campaign and, with the help of his brother, most likely could have sustained the effort into producing a comfortable royalty income. His passion, however, was invention, which has two pillars, one in science and the other in technology. Through intense study, determination, and hard work Evans had acquired a significant knowledge in both of these, and he seemed staunchly determined to leave this knowledge as his legacy to the young men of the day. Perhaps at the time he was opting for posterity rather than profits.

He embarked on writing a pamphlet for millwrights and millers and ended up with a lengthy book approaching five hundred pages, divided into five parts, with twenty-six engraved plates. Part I (160 pages) makes up one-third of the book and deals with first principles. Like any good physics book, it includes detailed chapters on mechanics and hydraulics. Part II (68 pages) provides design guidelines for undershot, tub, and overshot mill wheels; part III (66 pages) contains descriptions of Evans's improvements, including the elevator, conveyer, hopper boy, drill (screw conveyer), and descender (flat belt elevator); part IV (44 pages) was the young miller's guide for operation of a mill; and part V (90 pages), written by Thomas Ellicott, a seasoned thirty-eight-year Bucks County millwright, provided practical advice on the construction of mills.

Writing the book was an intensive, difficult, financially humbling, and physically extracting effort for Evans, who lamented that it forced him to use spectacles and produced many gray hairs on his head. Afterward, upon resumption of an active life, he noted that the gray hairs went away, and he did not require his spectacles again for another ten years. A significant part of the difficulty he faced involved the physics sections on mechanics and hydraulics:

> I wrote to a considerable length before I discovered that the different parts would not harmonize. I suspected the theories, and finally found them to be fundamentally wrong, tending to lead the practical millwright into the most expensive errors. I conceived that I could not do a greater good than to correct those errors, discover true theories, and arrange a system to guide the practitioner. I threw aside all I had written, began anew, and got so deeply engaged in it, that nothing could divert my mind from the abstruse subject (for I found it far more difficult than I apprehended) for three years, until I was reduced to such abject poverty that my wife sold the tow cloth which she had spun with her own hands for clothing for her children, to get bread for them.[12]

Evans was to learn the true theories were indeed difficult to grasp, but once he understood, he was able to explain complicated principles with lucid writing. An example of an important principle for steamboats is found in the chapter on hydraulics in his *Young Mill-Wright & Miller's Guide*, where he states in a footnote, "The effects of striking fluids with equal apertures are as the cubes of their velocities."[13] He goes on to provide proof of this but is focused on the effect imparted to a waterwheel for a gristmill. However, as Evans knew and could have stated at the time, the effect of moving a boat through the water would also vary as the cube of its velocity, as the only difference between the waterwheel and a boat is the point of reference. In 1795, however, he had not yet drawn this conclusion, only reaching it several years later.

On a handwritten sheet of paper dated September 3, 1812, interleaved in a copy of his subsequent steam engine book, Evans states and eloquently proves in about a hundred words "that the power to propel boats is as the cubes of their velocities."[14] The accepted doctrine at the time, developed independently by others, was the power varied as the square of the velocity (see chap. 6), which was wrong but nevertheless persisted until after 1850.

After completing his book, and in the process exhausting all of his financial resources, he was left with no money to have it published. He had obtained a long list of about five hundred subscribers, many of which were prominent—for example, President George Washington, Thomas Jefferson, Edmund Randolph, and Robert Morris. All of these had agreed to purchase the book at an offering price of $1.50, but the size had grown for an initial estimate of 350 pages, including twenty engravings, to about five hundred pages, including front matter, appendix, and twenty-six engravings (see fig. 8.5). He found it necessary to increase the subscribed price to $2.00 and was fortunate to locate John Nicholson, who advanced him, without interest, $1,000 for printing two thousand copies. Some of the books were sold for $3.00, whereas others were given away to encourage millers to adopt his improvements.

The Young Mill-Wright & Miller's Guide survived Evans by forty years: it appeared in no less than fifteen editions from 1795 to 1860, an astounding sixty-five-year lifespan for a technical book. The chapters on mechanics and hydraulics are timeless, as Evans wrote with a singular clarity. This effort was not his final contribution to society, though, as he went on to produce many steam engines, an amphibious vehicle, and another significant book.

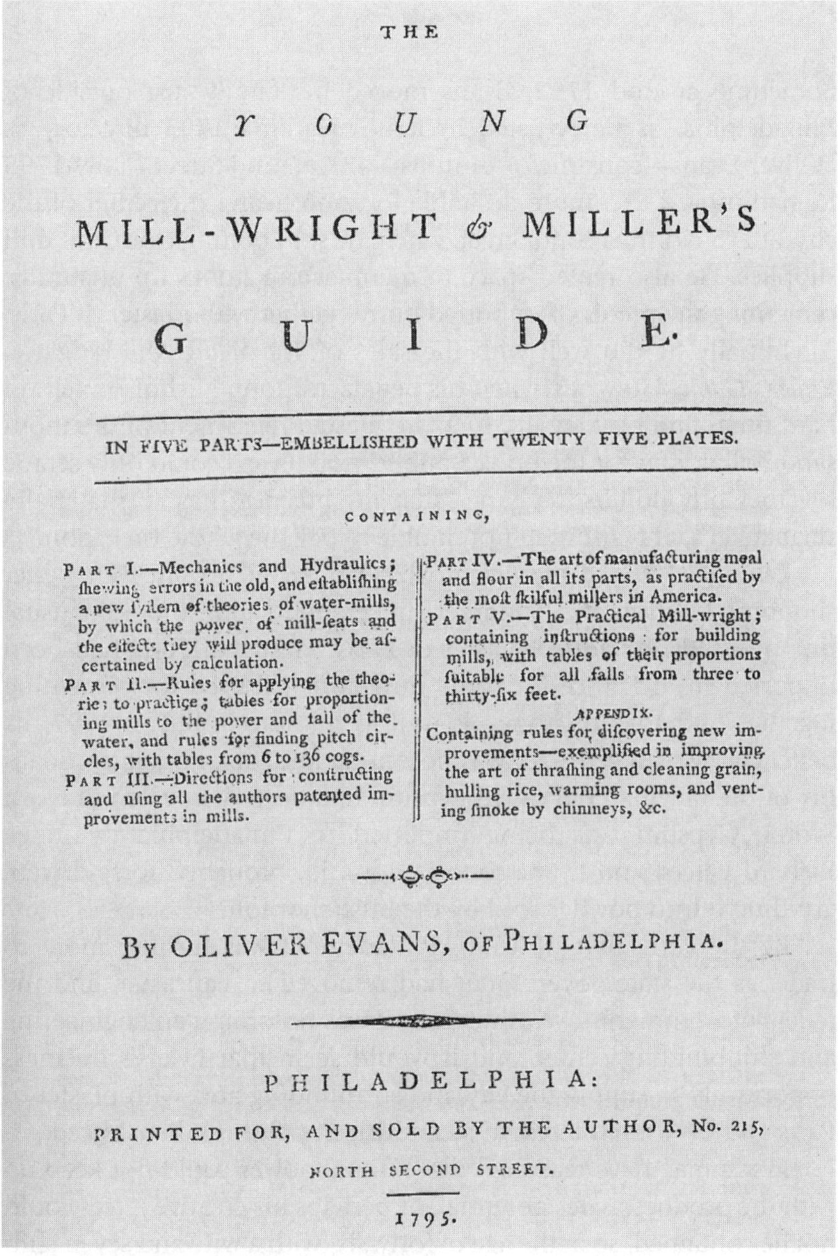

THE

Y O U N G

MILL-WRIGHT & MILLER'S

G U I D E.

IN FIVE PARTS—EMBELLISHED WITH TWENTY FIVE PLATES.

CONTAINING,

PART I.—Mechanics and Hydraulics; shewing errors in the old, and establishing a new system of theories of water-mills, by which the power of mill-seats and the effects they will produce may be ascertained by calculation.

PART II.—Rules for applying the theories to practice; tables for proportioning mills to the power and fall of the water, and rules for finding pitch circles, with tables from 6 to 136 cogs.

PART III.—Directions for constructing and using all the authors patented improvements in mills.

PART IV.—The art of manufacturing meal and flour in all its parts, as practised by the most skilful millers in America.

PART V.—The Practical Mill-wright; containing instructions for building mills, with tables of their proportions suitable for all falls from three to thirty-six feet.

APPENDIX.

Containing rules for discovering new improvements—exemplified in improving the art of thrashing and cleaning grain, hulling rice, warming rooms, and venting smoke by chimneys, &c.

——◈◈◈——

By OLIVER EVANS, OF PHILADELPHIA.

PHILADELPHIA:

PRINTED FOR, AND SOLD BY THE AUTHOR, No. 215,

NORTH SECOND STREET.

1795.

Figure 8.5. Title page of Evans's *Miller's Guide* of 1795. With borrowed money, Evans financed the printing, sold copies, and gave away others to encourage millers to license his gristmill improvements. (From Evans, *The Young Mill-Wright & Miller's Guide*, title page.)

Sometime around 1792, Evans moved his family just outside of Philadelphia, as he is listed in James Hardie's 1793 directory as "Oliver Evans—constructor of mills—437 N. 2nd Street."[15] By 1795 he had moved to a more desirable location nearer the center of the city, at 215 North Second Street, where he sold bolting cloth and mill supplies. He also rented space to manufacture stones for gristmills, cementing the pieces of imported burr together with plaster of Paris.

Initially he did well with the sales of *The Young Mill-Wright & Miller's Guide*. However, when his benefactor John Nicholson fell on hard times and on May 28, 1797, requested repayment of the thousand-dollar loan for the book's publication, Evans could only scrape together fifty dollars, replying "that my business has so materially stagnated that I fear I shall not be able to pay the whole very soon."[16]

Later in 1797, things improved: he moved to an even better shopping location at 275 Market Street, and in addition to the usual line of goods, he added plaster of Paris. Plaster of Paris had been imported from France for some time and was used for cementing together burr for millstones as well as for stucco. But in 1797, its beneficial use for fertilizer was becoming known, and a large quantity of the raw material called gypsum had been discovered in Nova Scotia. Gypsum was being imported to Philadelphia by Judge Richard Peters, and Evans soon had a mill, probably horse-driven, grinding it into powder sold by the bushel or ton.

By the early part of 1800, Philadelphia was turning more to trade, as the state government had removed to Lancaster, and the federal government to Washington. It was becoming an engineering and shipbuilding center, and it would seem that Evans's business opportunity to supply the city and surrounding area with plaster of Paris and his long interest in steam-driven power had coalesced.

Evans may have reasoned that animal power could not keep up with the product's sales potential, or perhaps his creative juices could not be contained, so at the age of forty-six, with a wife and seven children, he hired workmen and started construction of a steam engine. Surprisingly, however, he first attempted a *steam turbine* to be applied to a wagon or boat but quickly realized he should develop his previously patented and more conventional technique, which could also drive mills. He immediately stopped work on the turbine and, two weeks later, resumed constructing a high-pressure steam engine with a cylinder six inches in diameter with an eighteen-inch stroke (see fig. 8.6). The engine developed five horsepower and, coupled to his screw mill (grinder), could produce three hundred bushels, or twelve tons, of plaster of Paris in twenty-four hours.

He had estimated the cost of developing the engine to be $1,000, but expended $3,700, all the money and credit he had available. He later wrote, "I had calculated that if I failed in my experiment, the credit I had acquired would be entirely lost, and without money or credit, at my advanced age, with many heavy encumbrances, my way through life appeared dark and gloomy indeed! But I succeeded perfectly with my little engine, and preserved my credit."[17]

Sometime in 1802, Evans wrote Samuel Jackson in Kentucky that his "little engine" could propel wagons and boats, and the letter found its way into the hands of Captain James McEver. The captain and his partner, M. Lewis Valcourt, were contemplating a steamboat

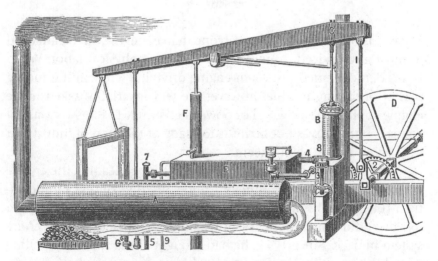

Figure 8.6. Evans's high-pressure, noncondensing steam engine with an overhead grasshopper beam. The boiler and smokestack are in the foreground, and the small steam cylinder is labeled B. (From Thurston, *History of the Growth of the Steam-Engine*, p. 156, which also provides an explanation of its operation.)

to operate on the Mississippi River between Natchez and New Orleans and thus needed a steam engine. Based on the letter, Valcourt ventured to Philadelphia, where he met with Evans and saw his engine grinding plaster. Duly impressed, a larger engine with a nine-inch cylinder and thirty-six-inch stroke was ordered.

While Captain McEver was superintending the building of the eighty-foot vessel with an eighteen-foot beam in New Orleans, Evans designed and constructed the steam engine in Philadelphia. By the end of 1802, the completed engine and boiler had arrived in New Orleans, where Evans's workmen installed the equipment into the hull of the boat. Everything was in place for a trial in early 1803, but then heavy rains caused a severe freshet, floating the hapless vessel half a mile inland. Not being able to salvage the boat, the steam engine and boiler were taken off and installed in a sawmill owned by William Donaldson. After operating reliably for over twelve months, Evans's equipment drew praise from the engine man, who said "that nothing relating to the engine had broke, or went out of order so as to stop the mill for one hour."[18]

Today it is obvious to nearly everyone that an engine can be attached to most any device to minimize or eliminate physical labor. When Evans demonstrated his steam engine driving a screw mill grinding plaster or sawing marble, however, he was invariably asked if the engine could also saw wood or power a gristmill. The answer was "of course," but this was not obvious to many of the men of industry at the beginning of the nineteenth century.

To initially sell his engine, he had to sell the idea that the engine could provide a valuable function, and since the concept of integration (buying subsystems from several sources and putting them together) did not exist, he was forced into manufacturing the entire system himself. Since 1777, he had thought about steam propelling a land vehicle, so in September 1804 he proposed to the Lancaster Turnpike Company to use steam wagons to replace the horses and wagons for transporting goods between Lancaster and Philadelphia. Evans developed and included a fairly detailed cost comparison, showing his steam wagons to be cheaper. In the closing paragraph of his letter, he stated he had no doubt that engines would propel boats against the current of the Mississippi and wagons on turnpike roads with great profits.

Despite his optimism and cost analysis showing a benefit to the company, his proposal, while carefully considered, was rejected for being too radical. It should be noted, as pointed out by Bathe and Bathe, that his estimates for the horses were quite detailed, including expenses for driver's whips and feed troughs, but for his steam wagon he did not include costs for repair tools and spare parts.[19]

In search of other opportunities, Evans approached the Board of Health in Philadelphia with a proposal to construct a dredging machine to remove debris from the city's docks. After some initial negotiations, on April 4, 1805, the Board advanced Evans five hundred dollars to commence construction of the dredge he was to name the *Orukter Amphibolos* (see fig. 8.7). Buckets on a chain pow-

Figure 8.7. The *Orukter Amphibolos* of 1805, designed by Oliver Evans to remove debris from the docks of Philadelphia, is shown with wheels attached. It was driven from the place of manufacture down Market Street to the Schuylkill River and then steamed to the Delaware River. Note that the drive belts would not allow the amphibious vehicle to be steered, so the illustration is not accurate, and the operators not to scale. (From Thurston, *History of the Growth of the Steam-Engine*, p. 157.)

ered by his five-horsepower steam engine accomplished the digging. It was capable of removing nine hundred cubic feet of mud each hour, which was then loaded onto a barge and disposed of at another location.

By the end of June 1805, the twelve-by-thirty-foot, flat-bottomed dredge boat was almost complete at his shop about one mile from the water. To transport it, Evans requested and received permission from the Board of Health to attach wheels and suitable power linkages to drive it to the Schuylkill River. By the second week in July, after an abortive start, he was ready, and the following notice appeared in *Riff's Philadelphia Gazette* on July 13, 1805, and two days later in the *Aurora*:

TO THE PUBLIC

In my first attempt to move the ORUKTER AMPIIBOLOS, or AMPHIBIOUS DIGGER to the water by the power of steam, the wheels and axletrees proved insufficient to bear so great a burthen, and having previously obtained the permission of the Board of Health, (for whom this machine is constructed) to gratify the citizens of Philadelphia by the sight of this mechanical curiosity on the supposition that it may lead to useful improvements.

The workmen who had constructed it, voluntarily offered their labour to make without wages, other wheels and axletrees of sufficient strength, and to receive as their reward one half of the sum that may be received from a generous public for the sight thereof, the other half to be at the disposal of the inventor, who pledges himself that it shall be applied to defray the expense of other new and useful inventions.

The above machine is now to be seen moving round the Center Square at the expense of the workman, who expect 25 cents from every generous person who may come to see its operation; but all are invited to come and view it as well those who cannot as those who can conveniently spare the money.

OLIVER EVANS.[20]

Although the vehicle only went a probable three or four miles per hour, it did amble down Market Street to Center Square (see fig.

Figure 8.8. Center Square in Philadelphia, where Oliver Evans, perhaps America's first automobile driver, provided demonstrations to an amazed public. (From Bathe and Bathe, *Oliver Evans*, between pp. 108 and 109.)

8.8), driving around the city's water works for several days and piquing the interest of the curious. Finally, it was driven down to the river at low tide, had a paddle wheel attached to the stern, had the wheels unbolted, and when the tide came in floated off the under-carriage and steamed off.

As the five-horsepower engine churned, the paddle wheel and a long oar serving as a rudder, the world's first amphibious vehicle made its way down the Schuylkill River to the Delaware and on to the Philadelphia waterfront. Proceeding at four miles per hour with the current, *Orukter* went sixteen miles beyond Philadelphia to Dunk's Ferry (now Beverly, New Jersey) and returned to the city.

To construct a compact engine that would more easily adapt to a boat such as his amphibious dredge, sometime in the winter of 1803 Evans created a new configuration, with a vertical steam cylinder connected to a half-length walking beam. The beam employed a new kind of linkage, which because of its peculiar motion was called a "grasshopper." Although William Freemantle in England had patented the device in 1803, the arrangement was later used on many locomotives and became known in Europe and worldwide as the Evans straight-line linkage. Its construction and operation is deceptively simple and is shown in figure 8.9.

The piston rod, 3, travels straight up and down parallel to the guidepost, 1, provided the linkages AB and BC are equal. Linkage A5 is connected at the lower end to a fixed point. The operation is not immediately obvious (at least to the author), but by referring to the diagram in figure 8.10, the motion can be easily understood. One wonders how Evans himself, a highly visual thinker, mentally pictured the operation.

Imagine two fixed surfaces, one perpendicular to the other and intersecting at point C. Rod AD (twice the length of BC) is positioned as shown, with the center connected by a pin to one end of rod BC, which in turn is pinned at C. It is fairly obvious that the tip of the rod at D can be moved precisely to point C, but what about other positions—will it always glide along the vertical surface? A little trigonometry provides the answer.

Consider any arbitrary angle of *a* degrees, opposite point A. As AB = BC then the angle opposite C must also be *a* degrees, with the apex angle being 180 − *a*. The complementary angle at B would be 2*a*, leaving angles opposite D and C at 90 − *a*. Therefore this arrange-

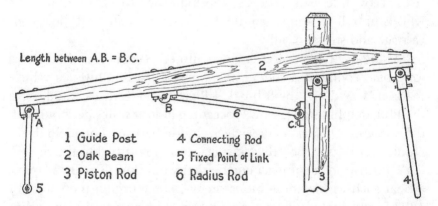

Figure 8.9. Grasshopper beam and spider parallel motion used by Oliver Evans, as shown on the illustration of the *Orukter Amphibolos*. Although not invented by him, his use of this technique in many steam engines resulted in its being called the Evans straight-line linkage. (From Bathe and Bathe, *Oliver Evans*, p. 113.)

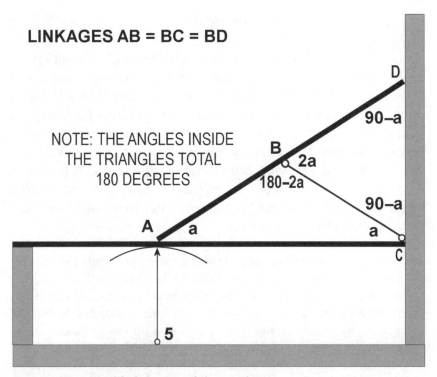

LINKAGES AB = BC = BD

NOTE: THE ANGLES INSIDE
THE TRIANGLES TOTAL
180 DEGREES

Figure 8.10. A simplified diagram of the grasshopper operation. By examination a reader can understand how the two triangles BAC and BCD work in a complementary fashion. (Diagram by Jack L. Shagena.)

ment produced exact horizontal and vertical motion at the tips of rod AD. For this analysis, BD was chosen to equal AB, which *must* equal BC, but this restraint on BD is not necessary in practice.

Evans, rather than constructing a slide for the horizontal surface, chose the simpler linkage A5, which over a narrow travel arc approximates motion in a straight line. The longer the link, the closer the motion will be to perfect, but in practice the length need only be about that of links AB or BC. While this grasshopper beam arrangement worked satisfactorily for the *Orukter Amphibolos*, the height posed a problem for later steamboats, and the cylinder would again resume a horizontal position. High-pressure steam engines, with smaller pistons, did not pose the same leakage problems as did the seals in the large-piston, low-pressure cylinders.

Around the time of the development of his first high-pressure steam engine in 1801, Evans had probably started thinking about a new book that would parallel his *Young Mill-Wright and Miller's Guide*. By 1804, the book was in progess, and Evans published a three-page pamphlet on steam engines, which he circulated among the members of Congress. His intention, no doubt, was to promote a favorable decision on a pending bill to extend for seven years his patent on mills, which was about to expire. In early 1805, he testified before Congress to further his case and later wrote of himself and the incident in the third person:

> During these flattering appearances he issued proposals for publishing by subscription a new work to be entitled "The Young Steam Engineer's Guide," that he might be collecting subscriptions [from members of Congress] while waiting the passage of the bill. But on the third reading a most energetic opposition arose on grounds unexpected; arguments were used which is not unnecessary to state; the friends of the bill were not prepared to answer them, and it was completely lost. His plans were thus proved abortive, all his fair prospects are blasted, and he must suppress a strong propensity for making new and useful inventions.[21]

About this same time a New York senator, Dr. Mitchell, sent Evans's steam pamphlet to *The Medical Repository and Review of American Publications*. At this juncture, Col. John Stevens, who for some time had been curious about Evans's steam engine work, read it. On at least two previous occasions, Stevens had sent Dr. Coxe of Philadelphia to obtain answers from Evans about his engine. As Evans was led to believe that Stevens bore him no ill will, the questions posed by Coxe were answered honestly and in a straightforward manner, but now Stevens was to go on to attack Evans's credibility.

In a letter to the *Repository* printed in the same issue Evans's pamphlet appeared, Stevens claimed that his ideas were not new and suggested he "was in pursuit of an ignis fatuus," a foolish quest of a goal. A series of letters ensued, and finally, on April 13, an incensed

Evans wrote a lengthy counterattack to Stevens, where he validated his steam engine claims by pointing out that his engine had been in operation for three years and that "she will grind 400 bushels of plaster in twenty four hours, or saw 200 feet of marble stone." Once making his point that his steam engine unquestionably worked, he reiterated the seven accusations Stevens had made against him and posed a question: "Now, sir, what benefits do you expect to arise from our having laid me under the necessity not only of defending my character, but my interest? Shall we criminate and recriminate each other in public, until we give good people cause to pronounce us fools? I wish to employ my time to a more useful purpose."[22] Stevens would reply, but this seemed to be the end of the battle of letters, as Evans preferred to pursue his business interests instead.

Because of his setback with Congress in getting his automated mill patent extended and the unsettling controversy with Stevens, his planned book was never completely finished. In 1805, he decided to publish the part he had already done and add the series of letters between himself and Stevens with one last postscript. Thus his effort to provide the world with a comprehensive treatise on steam was aborted, a decision reflected in the revised title: *The Abortion of the Young Steam Engineer's Guide* (see fig. 8.11).

Though much smaller than his companion mill-wright guide, the 130-page book with four illustrations (see example in fig. 8.12) provided a good explanation of high-pressure steam engines and the emphasis Evans placed on boiler safety. Unfortunately, his knowledge and understanding about steam thermodynamics was incomplete, leading to some misstatements about engine performance and fuel consumption. That subject would not be completely understood for another fifty years; however, his fundamental belief in the benefits of high-pressure steam proved to be correct.

In the book he also included information on a couple of inventions by his brother Evan: one was a straw-cutting machine, and another a flour press for closing barrels that employed a toggle-joint principle. The steam guide was reprinted several times but was not nearly as successful as the mill-wright guide, which was reprinted fifteen times spanning a period of sixty-five years.

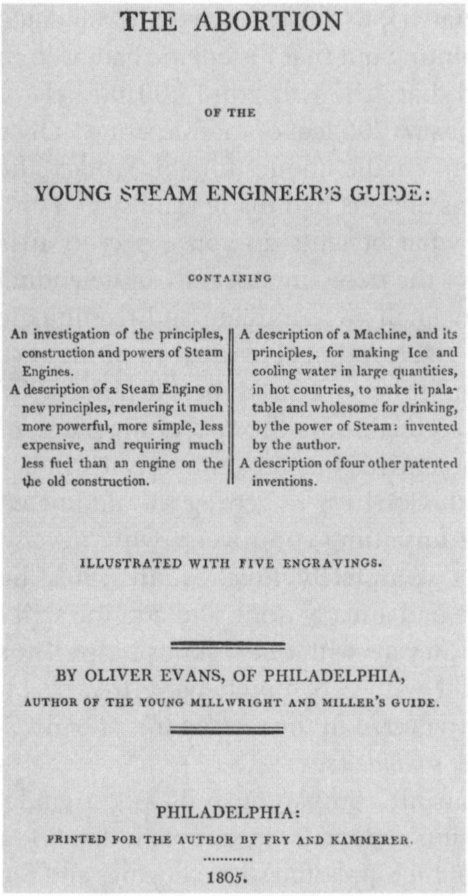

THE ABORTION

OF THE

YOUNG STEAM ENGINEER'S GUIDE:

CONTAINING

| An investigation of the principles, construction and powers of Steam Engines. A description of a Steam Engine on new principles, rendering it much more powerful, more simple, less expensive, and requiring much less fuel than an engine on the old construction. | A description of a Machine, and its principles, for making Ice and cooling water in large quantities, in hot countries, to make it palatable and wholesome for drinking, by the power of Steam: invented by the author. A description of four other patented inventions. |

ILLUSTRATED WITH FIVE ENGRAVINGS.

BY OLIVER EVANS, OF PHILADELPHIA,
AUTHOR OF THE YOUNG MILLWRIGHT AND MILLER'S GUIDE.

PHILADELPHIA:
PRINTED FOR THE AUTHOR BY FRY AND KAMMERER.
....
1805.

Figure 8.11. Because of many disappointments in his personal life before getting his steam guide published, Evans described the book as an abortion of his planned efforts. (From Oliver Evans, *The Abortion of the Young Steam Engineer's Guide* [1805; reprint, Wallingford, PA: Oliver Evans Press, 1990], title page.)

From the time Evans first started pondering the benefits of elastic steam until he produced his first steam engine, he never doubted that his high-pressure technique would find widespread application, especially in transportation. In his pamphlet *The Principles of Steam Engines*, presented to Congress in 1804, he writes,

I have an engine in operation in the most simple form without a condenser, which is capable of performing three times the work with equal fuel, compared to the English engine:[23] and succeeds according to the theory, working with steam, generally equal in power from 50 to 100 lbs. to the inch; doubling the fuel appears to produce about 16 times the power and effect. Its great power and simple structure, fits it for propelling boats up the Mississippi, and carriages on turnpike roads; two of the most difficult applications; therefore, will apply to all others as a powerful agent.[24]

In early 1807, Evans provided William Donaldson with a steam engine for a new sawmill set up in the Louisiana Territory.[25] This was

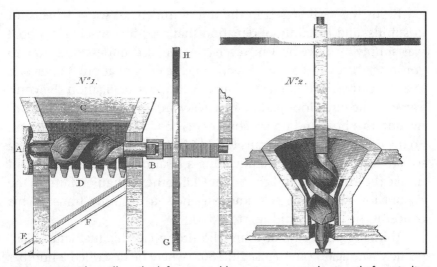

Figure 8.12. The mill on the left, powered by water, was used primarily for grinding plaster and corn, and according to Evans it worked exceedingly well. The hand-operated mill on the right was used to break charcoal for a steel furnace. Evans invented these mills during the winter of 1795–96, and by the time they were published in his steam guide in 1805, he noted, "[The plaster mill] is now getting pretty general use, and there has at least a dozen of inventors started up already, all claiming the invention; so easy it is to invent a machine already in use." (From Evans, *The Abortion of the Young Steam Engineer's Guide*, image between pp. 70 and 71; quote p. 72.)

Donaldson's second Evans engine: in 1803, he had acquired the engine sent to New Orleans for the ill-fated steamboat project of Capt. James McEver and Louis Valcourt. Pleased with the superlative performance of Evans's engines, this fact was introduced by Donaldson to Maj. Richard Claiborne, who had a keen interest in operating a steamboat on the Mississippi River.

So on July 12, 1807, Claiborne wrote Evans about his plan to propel a boat by means of duck-foot paddles and invited Evans to partner in providing the steam engine, confidently predicting, "The attempt cannot fail!" Evans must have been intrigued with the thought of his engine chugging a steamboat up and down the mighty Mississippi, something that he had predicted, but he was dismayed at the prospect of using the highly impractical and ungainly duck-foot paddles as the means of propulsion.

Evans was well versed in the use of undershot water wheels for gristmills, and certainly understood that a paddle wheel on a boat was simply a mill wheel in reverse. As he also understood how to construct sturdy water wheels with an efficiency of about 33 percent, Evans clearly saw them as the answer to water navigation. By comparison, the duck-foot paddle was awkward and fragile, and its connection to a boat and power source was complicated and inefficient. So the practical-minded Evans found himself in a quandary, wanting to have one of his engines powering a boat on the Mississippi, but unlike the exuberant and confidant Claiborne, he had no faith in the propulsion technique. Not knowing how to tactfully transmit his concerns, he initially did not respond.

Thinking perhaps that Evans did not receive his first letter, Claiborne wrote again on August 24, 1807, where he described a number of ongoing activities in the Louisiana Territory and repeated his boat offer. He also enclosed a copy of his previous letter. Still not getting a response, the persistent entrepreneur sent another letter on September 10, in which he extols the virtues of his duck-foot propulsion and ends on a patriotic note: "If it should be my good fortune to throw in a mite for the benefit of my country, I shall be happy,—and if it should be my good fortune, or the good fortune of any person to produce a useful application for Mr. Evans's steam engine as applied to navigation, it will be a national benefit to the U.S., as well as the world."[26]

This altruistic philosophy must have reverberated in Evans's mind, as it was the personal driving force in life that motivated him. But first and foremost, he was a no-nonsense, barefoot engineer who did not want to become involved in a project that he believed could not succeed, and Claiborne just seemed too adamant about trying out his impractical duck-foot propulsion.

The ink on Claiborne's previous letter was hardly dry when he again took his pen in hand, writing on September 14 mentioning that he had no money to pursue the boat project but that he could only offer ideas. Nevertheless, still hoping to entice Evans's interest in the significant navigation opportunities in the territory, he added a postscript: "P.S. I recommend you to procure one of Safongs maps of the Territory of Orleans, which was lately executed and published in Philadelphia, and it will give you the Geography of the Country. You will there see the advantages of Lake navigation."[27]

On March 6, 1808, Evans finally responded to Claiborne, but the letter was delayed in the mail several months, finally reaching him on July 7, 1808. Two days later, a disappointed Claiborne wrote a closing response, and it is evident that Evans had informed him of the impracticability of duck-foot propulsion and declined his offer to be his partner in a steamboat project. Of this turn of events, Bathe and Bathe write, "It is to be regretted that the correspondence between Evans and Major Claiborne closed with this letter, for if Claiborne's mechanical ideas were impractical, he himself was a progressive thinker and an able man."[28] Reading Claiborne's letters and sensing his innate desire to benefit mankind, it is indeed unfortunate that he and Evans could not have come to some arrangement. Evans, the engineer forever with both feet on the ground, had great difficulty dealing with ideas that did not fit his orderly and logical mind, and while he seemed to be a companionate man, he did not have the business finesse to regularly build bridges between opportunity and engineering. He suffered from the lack of an understanding business partner, such as Watt had found with Boulton, and Fulton with Livingston. To be such an individual would have required the intellect and talents of an extraordinary man, as Evans often had his mind made up and technically was nearly always right.

By 1811, Evans was producing high-pressure steam engines that were receiving the praises of users throughout the United States. In addition to powering grist- and sawmills, he sought out and found steamboat opportunities, since one of his early dreams had been steam navigation. In his "Account of Steam Engines of Oliver Evans" from February 22, 1812, ten were identified in use and ten in construction. Two under construction being built were for steamboats.[29] One was an engine rated at twenty horsepower and destined for a passage boat on the Middlesex River and Canal. In a test in 1812, the vessel achieved a speed of nearly eight miles per hour.

Another, forty-horsepower engine appears to have been planned for a boat built by the Pittsburgh Engine Company, founded by Evans, his son George, and Luther Stephens, who had invented an improved rotary steam valve adopted by Evans. This hundred-pas-senger boat would be 121 feet long with a beam of seventeen feet and was planned to operate between Natchez and New Orleans. That year he wrote to both Robert Fulton and John Stevens about his improved rotary valve, high-pressure Columbian Steam Engine, touting its superior performance and suggesting that it be ordered for boats they were considering. Fulton expressed some interest and requested that Evans propose an engine for a boat he was planning. Evans set about doing the calculations and sent them to Fulton, hoping to receive an engine order. However, in a written reply to Evans dated March 28, 1812, Fulton arrogantly attacked his calculations:

> Your boat is 120 feet long 15 feet Beam to draw two feet of water, to drive her 10 miles an hour in still water you propose a 40 horse power, you will I know be astonished when I tell you that 160 horse power is required to drive her 10 miles per hour, which engine could not put in the Boat therefore it is impracticable, I have always told you that the discovery of the science of steam boats was the cause of my success, and now I have proved it by practice, the sci-ence is not known even by those who copy me, I tell you as a friend the power required for 10 miles an hour, you will excuse my giving the reason until I beat all who are pirating me and until law has confirmed my rights I will then show you the secret.[30]

There must have been subsequent discussions between the two inventors, with claims and counterclaims, as the size of the boat being argued was almost identical to the one planned by the Pitts-burgh Engine Company. Evans was to write a note on one of the blank interleaved pages in his 1805 *Steam Guide* stating, "Robert Fulton has bet with me a beaver hatt that my Boat building for the Mississippi will not run 10 miles per hour and a suit of clothes that my Boat will not run nine miles per hour. I take him up."[31]

Based on the size of the boat and draught, the displacement would have been about 112 tons. Thus the ratio of motive power to ten tons of displacement figures to be 3.57. From appendix B, figure B.1, this ratio would produce a steamboat that could achieve about twelve miles per hour. Evans, not Fulton, was correct.

Eight years earlier, Evans had requested that Congress extend his mill patent, which was to expire in January 1805. This attempt was unsuccessful, but the ever-persistent Evans petitioned Congress again in 1807, this time prevailing. On January 21, 1808, Congress passed "An Act for the Relief of Oliver Evans," assuring him of fourteen more years of royalty income. Not being satisfied with the thirty to forty dollars he was obtaining for a pair of grinding stones, a few years later he arbitrarily increased the fees to one hundred dollars, understandably raising the ire of the millers (although Thomas Jefferson paid; see fig. 8.13). Quite simply, Evans was an engineer, not a marketeer.

Many of the millers, some of whom had taken over from their fathers, refused to pay, prompting the litigious Evans to bring suit. In one action, *Evans v. Robinson*, brought against a Maryland miller, Evans was awarded a judgment of $1,850. The fact that

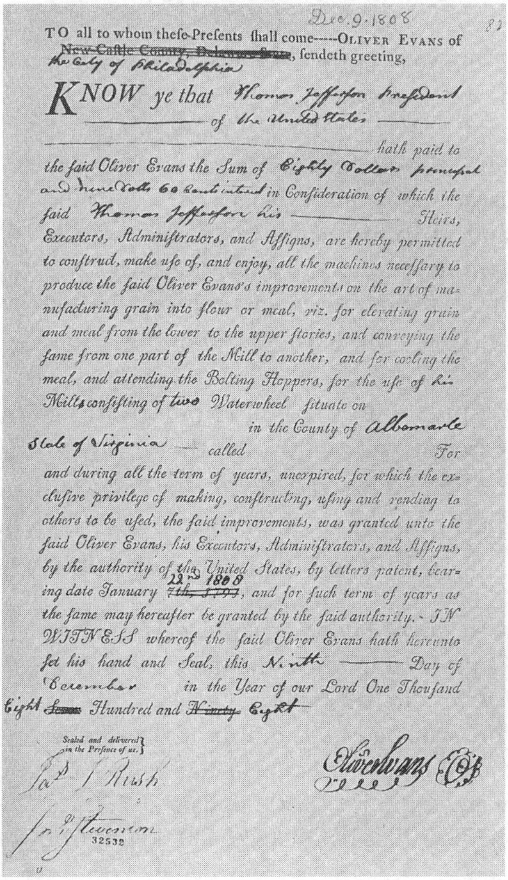
Dec. 9. 1808

TO all to whom these Presents shall come-----OLIVER EVANS of ~~New-Castle County, Delaware State~~ the City of Philadelphia, sendeth greeting,

KNOW ye that Thomas Jefferson President of the United States hath paid to the said Oliver Evans the Sum of Eighty Dollars principal and nine Dolls 60 Cents interest in Consideration of which the said Thomas Jefferson his Heirs, Executors, Administrators, and Assigns, are hereby permitted to construct, make use of, and enjoy, all the machines necessary to produce the said Oliver Evans's improvements on the art of manufacturing grain into flour or meal, viz. for elevating grain and meal from the lower to the upper stories, and conveying the same from one part of the Mill to another, and for cooling the meal, and attending the Bolting Hoppers, for the use of his Mills consisting of two Waterwheel situate on in the County of Albemarle State of Virginia called For and during all the term of years, unexpired, for which the exclusive privilege of making, constructing, using and vending to others to be used, the said improvements, was granted unto the said Oliver Evans, his Executors, Administrators, and Assigns, by the authority of the United States, by letters patent, bearing date January 22nd 1808 ~~7th, 1791~~, and for such term of years as the same may hereafter be granted by the said authority. IN WITNESS whereof the said Oliver Evans hath hereunto set his hand and Seal, this Ninth Day of December in the Year of our Lord One Thousand Eight ~~Seven~~ Hundred and ~~Ninety~~ Eight

Sealed and delivered in the Presence of us.

Oliver Evans

32532

Figure 8.13. Milling License acquired by Thomas Jefferson in 1808. It was prepared on a form that was printed in 1791 and signed by Evans on the bottom right. When President Jefferson signed the legislation extending Evans's automated mill patent, Evans's agents in Virginia immediately sent the president a bill for $80.00 plus interest, for a total of $89.60. Jefferson, who was unaware that his mill builder had infringed on Evans's patent, immediately sent the payment. (From Bathe and Bathe, *Oliver Evans*, between pp. 160 and 161.)

Evans, a Philadelphian, won in a Maryland court with a Maryland jury alarmed the other Baltimore millers, some of whom begrudgingly paid.[32] Others, however, in 1813 petitioned Congress for relief, which resulted in a committee report with the lengthy but descriptive title of, "A Petition of Sundry Inhabitants of Baltimore County in the State of Maryland Complaining against the Oppressive Exercise of, &c. of the Patent Rights Granted to Oliver Evans for the Invention of Certain Mill Machinery." This action threatened Evans's income and for a while consumed his thinking, as is reflected in a letter he wrote to his son George in Pittsburgh on March 10, 1813:

> By letters that I have received and conversations with members of Congress, I understand that Congress has been highly irritated against me by the unrighteous misrepresentations of interested millers.
>
> There is danger that truth may not prevail against them. The event is uncertain, and my whole thoughts, time, and money, will necessarily be engaged in my defense until the case is determined. I must therefore abandon every other enterprise. . . . You will please stop the work of the steamboat immediately; settle with and pay off the carpenters and all others engaged . . . and make it known that if any others will take it up and go with it, that I will finish the engine and apply all my improvements and warrent [sic] it to succeed, if it be built as I have directed on the following terms.

While Evans was preoccupied with the matter before Congress, he still found the time to carefully think through the terms for the steamboat's engine. In what would be called, in the twentieth or twenty-first century, a cost-performance trade-off matrix, Evans went on in the letter to suggest, for the time, a truly innovative approach:

> If she runs loaded with the weight of a hundred passengers and baggage, at the rate of 12 miles per hour through still water, the price of the steam engine set up with all shafts, wheels, etc. and all my improvements set in motion, shall be 14,000 dollars. If 11 miles per hour, 12,000 dollars, and if 10 miles per hour, 10,000 dollars, if 9 miles per hour, 9,000 dollars, if 8 miles, 7,500 dollars, if 7 miles, 6,000 dollars, if 6 miles, 4,500 dollars if 5 miles, 3,000 dollars. The

purchasers may choose the velocity that I may proportion the engine to produce but the velocity [selected] must finally determine the price. The boat is laid for 121 feet keel and 17 feet beam.[33]

Evans did not encumber the potential purchaser with the details of the engine, boat design, or any other particulars but boiled the decision down to performance versus cost, something easily understood by an entrepreneur.

Although his steamboat business approach was forward-thinking and eminently logical, smoldering underneath was a hidden bitterness towards the claims being made by the millers. In his 1813 pamphlet *Patent Right Oppression Exposed; or Knavery Detected*, written under a pseudonym, Evans reflected his animosity toward the millers with a statement about his unfinished steamboat: "it now stands, showing its ribs like a picture of death, a monument of the triumph of ignorance, malice, and persecution."[34]

On March 10, 1813, a group of four individuals publicly announced the formation of the independent Pittsburgh Steam Engine Company. Two of the principals included Evans's son George and rotary valve inventor Luther Stephens. About this time, the trusted Luther Stephens traveled to Philadelphia and proposed to Evans that this company finish the steamboat, with Evans bearing one-quarter of the cost. On May 19, 1813, Evans wrote his son George, accepting the proposal, but lamenting, "It appears to me that my head has got weak I cannot stand intense study so well my memory seems less retentive than formerly and my mind requires more rest I have almost determined for relaxation to take my wife and son Cadwalader and travel to Pittsburgh and home again if health permits."[35] Despite this confession of reduction in his mental capacity, Evans went on in the letter to discuss how the exhausted steam from his engine could be passed through light copper pipes to heat a factory. He also provided calculations showing that the proposal would indeed work.

Evans was successful against the millers, as Congress did not pass an act providing them relief, and his mill patent prevailed. In 1815, he was able to extend his steam engine patent seven years in addition to the previously granted fourteen years. As the original patent was granted February 14, 1804, this carried his protection to the year 1825.

In the *National Intelligencer* of March 18, 1815, Evans announced, "I have resumed the construction of a Steam-Boat at Pittsburgh on my original plan."[36] It would take, however, until the later part of 1816 before the steamboat would be completed, and when launched in December, she was named the *Oliver Evans*. Shortly thereafter, the steamboat was purchased by a group of investors from Pittsburgh, renamed the *Constitution*, and put into service on the Ohio and Mississippi Rivers. The boat made several successful trips, but in April 1817, on the Mississippi near St. Francisville, north of Baton Rouge, her boiler exploded, killing eleven men. After being towed to New Orleans, she was sold, the boiler was rebuilt, and she resumed service on the Mississippi River as a freight boat.

Sometime in 1815, Evans produced a forty-five-horsepower high-pressure steam engine for a new steamboat named the *Aetna*, which was 114 feet long with a beam of eighteen feet. The vessel was completed early the next year, and an advertisement appearing in the *Aurora* on May 4, 1816, stated that the boat "WILL start from the lower side of Market street wharf, Tomorrow Morning at 8 o'clock precisely, for Wilmington and return the same evening" (see fig. 8.14).[37] Drawing about 3½ feet of water, unofficial estimates of her speed were about eight to nine miles per hour.

In 1817, another high-pressure engine, probably of the same design as the forty-five-horsepower engine for

Figure 8.14. Using a forty-five-horsepower high-pressure engine built by Oliver Evans, the steamboat *Aetna* began operation on the Delaware River in 1816. (From Bathe and Bathe, *Oliver Evans*, between pp. 234 and 235.)

the *Aetna*, was built for a sister steamboat called the *Pennsylvania* (see fig. 8.15). She joined the *Aetna* in July 1818, plying the waters of the Delaware River, leaving from the same dock but on a time-staggered schedule. This appears to be the last steam engine for a boat built while Evans was still living.

On April 15, 1819, Oliver Evans died in New York after a month-long illness, having reached the age of sixty-four. His 1795 publication, *The Young Mill-Wright and Miller's Guide*, would outlive him by forty years, and his contributions to high-pressure steam engines would usher in a new era in steamboat navigation and rail transportation.

Oliver Evans was a highly original thinker who could envision a completed product or system, analyze how to achieve a practical implementation, and finally reduce his thoughts to practice. His thinking was years ahead of his time, arousing suspicion in his contemporaries, and because of his self-assured manner and litigious propensities, he offended many of the important business and political leaders of the day. As a result, he was not accepted as an equal into the business community and never found a significant amount of ready capital to develop his ideas. He principally had to struggle along with his own resources, but managed, in the end, to be financially successful. His ideas, though strange-sounding and advanced for the time, were so profound that they simply could not be ignored.

His 1980 biographer, Eugene Ferguson, wrote of him with a deep understanding, and in his closing paragraph observed, "Whether we praise or damn Oliver Evans for the legacy he left to the world, it is clear that his life

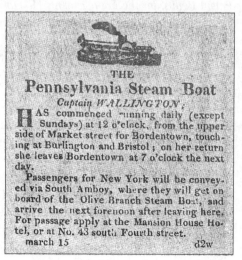

THE
Pennsylvania Steam Boat
Captain WALLINGTON;

HAS commenced running daily (except Sundays) at 12 o'clock, from the upper side of Market street for Bordentown, touching at Burlington and Bristol; on her return she leaves Bordentown at 7 o'clock the next day.

Passengers for New York will be conveyed via South Amboy, where they will get on board of the Olive Branch Steam Boat, and arrive the next forenoon after leaving here. For passage apply at the Mansion House Hotel, or at No. 43 south Fourth street.

march 15   d2w

Figure 8.15. The steamboat *Pennsylvania*, with another high-pressure steam engine built by Oliver Evans, was in operation by March 1819. (From Bathe and Bathe, *Oliver Evans*, between pp. 234 and 235.)

made a difference, not only while he was alive but for as long as our technological society persists. Many a dreamer and visionary can safely be dismissed as a person who, in [Hezekiah] Nile's words, 'would never be worth any thing because he was always spending his time on some *contrivance* of another.' Others, like Oliver Evans, must be taken seriously."[38]

Notes

1. Greville Bathe and Dorothy Bathe, *Oliver Evans; A Chronicle of Early American Engineering* (Philadelphia: Historical Society of Pennsylvania, 1935), p. 1. This excellent account is the primary source of information for this chapter. The epigraph is from Oliver Evans's philosophy, written on a blank page in a copy of his book *The Abortion of the Young Steam Engineer's Guide* that was willed to his son and published in Bathe and Bathe, *Oliver Evans*, p. iv.

2. It is perhaps interesting to note that a dozen years earlier, both Rumsey and Fitch were also fifth-born children in their families.

3. Henry Howe, *Memoirs of the Most Eminent American Mechanics* (New York: J. C. Derby, 1856), p. 68.

4. George A. Latimer, *A Sketch of the Life of Oliver Evans, a Remarkable Mechanic and Inventor* (Wilmington, DE: John C. Harkness, 1872), p. 4.

5. Quoted in Bathe and Bathe, *Oliver Evans*, p. 4.

6. Oliver Evans, *To His Counsel, Who Are Engaged in Defense of His Patent Rights, for the Improvements He Has Invented* (1817), Milton S. Eisenhower Library, The Johns Hopkins University, microfilm S40778, p. 5.

7. Ibid., pp. 6–7.

8. Oliver Evans, *The Young Mill-Wright & Miller's Guide* (1795; reprint, Wallingford, PA: Oliver Evans Press, 1990), part 4, p. 169.

9. Evans, *To His Counsel*, pp. 8–9.

10. Ibid., p. 10.

11. Eugene S. Ferguson, *Oliver Evans, Inventive Genius of the American Industrial Revolution* (Greenville, DE: Hagley Museum, 1980), p. 29.

12. Evans, *To His Counsel*, p. 16.

13. Evans, *The Young Mill-Wright & Miller's Guide*, part 1, p. 83n. Using the word "effects" to describe the power delivered to the waterwheel is technically imprecise, but it did convey to the reader the proper concept.

14. Quoted in Ferguson, *Oliver Evans*, p. 54.

15. Quoted in Bathe and Bathe, *Oliver Evans*, p. 33.

16. Ibid., p. 60.

17. Ibid., p. 68.

18. Oliver Evans, "Account of the Steam Engines of Oliver Evans," *Archives of Useful Knowledge* 2 (1812): 367.

19. Bathe and Bathe, *Oliver Evans*, p. 99.

20. Quoted in ibid., p. 109.

21. Oliver Evans, *The Abortion of the Young Steam Engineer's Guide* (1805; reprint, Wallingford, PA: Oliver Evans Press, 1990), pp. iv–v.

22. Ibid., p. 123.

23. Increasing the steam pressure in an engine results in operating with higher efficiency relative to fuel consumed. To raise water at 212 degrees into steam requires a baseline of energy, known as the latent heat of fusion, with additional heat energy beyond this providing an increase of steam pressure. With increasing steam pressure, there is a corresponding increase in engine power at a relatively small increase in fuel, as the baseline energy is the same in both cases.

24. Oliver Evans, *The Principles of Steam Engines* (1804), quoted in Bathe and Bathe, *Oliver Evans*, p. 301.

25. Bathe and Bathe, *Oliver Evans*, p. 132.

26. Ibid., p. 137.

27. Ibid., p. 138.

28. Ibid., p. 147.

29. Oliver Evans, "Account of the Steam Engines of Oliver Evans," pp. 363–66.

30. Quoted in Bathe and Bathe, *Oliver Evans*, p. 186.

31. Ibid., pp. 186–87. Evans should have easily won the bet.

32. Ibid., p. 189.

33. Quoted in ibid., p. 191.

34. Quoted in ibid., p. 192.

35. Quoted in ibid., p. 193.

36. Oliver Evans, "To the People of the United States," *National Intelligencer*, March 18, 1815, quoted in Bathe and Bathe, *Oliver Evans*, p. 222.

37. Quoted in Bathe and Bathe, *Oliver Evans*, pl. XLII, between pp. 234 and 235.

38. Ferguson, *Oliver Evans*, p. 64. The Nile quote is from Hezekiah Nile, addenda to *Weekly Register* 3 (1812–13): 1.

Figure 9.1. Nathan Read as a member of Congress around 1800. (Rendered by William T. Sisson from an illustration drawn and engraved by Charles Saint-Memin, Philadelphia.)

9.

Nathan Read
1759–1849

To the Honorable Congress of the United States: The petition of Nathan Read . . . respectfully showeth . . . that he has also invented a portable steam-engine, which may be constructed with less expense, is much lighter, occupies less space, and requires far less fuel, than any other within his knowledge. Your petitioner has likewise discovered an improved method of applying the power of steam to the purposes of navigation.

—Nathan Read

Nathan Read was born in Warren, Massachusetts, on July 2, 1759, the son of Maj. Reubin Read and his wife, the former Tamsin Meacham.[1] He entered Cambridge College in 1777 and graduated in four years. After studying medicine for a short time, he opened an apothecary in Salem, Massachusetts, where he found time to engage in mechanical pursuits.

Learning of the steamboat work of James Rumsey and John Fitch around 1788, Read turned his attention to steam transportation, inventing a portable multitubular boiler and an improved steam cylinder. He applied these devices to designs for propelling a steamboat and steam carriage and in 1790 petitioned Congress for a patent.

In 1791 he was elected to the American Academy of Arts and Sciences and in 1795 moved to a farm in Danvers. A year later he and his associates established an iron manufacturing company, putting to use a nail-making machine of his invention, which he patented in 1798. Starting in the fall of 1800, he served in Congress for two terms and in 1802 was appointed as a justice for the Court of Common Pleas for Essex County. In 1807 he moved to a farm in Belfast, Maine, where he lived in good health until his death on January 20, 1849.

Read's ancestors immigrated to America from England, where they had lived in Newcastle-upon-Tyne, later settling in the county of Kent. They arrived in America about 1632 and settled in the vicinity of Boston, living there for many years. Read's grandfather acquired a large tract of land in Warren, Massachusetts, sixty-five miles west of Boston, where he operated a farm and raised his family. His only son, Reubin Read, was born and resided on an adjacent homestead, later marrying Tamsin Meacham. She was a first cousin to the celebrated Gen. Nathaniel Greene of Rhode Island, who served under George Washington during the Revolutionary War. Reubin, also an officer in the war, attained the rank of major.

Nathan Read was the third of eight children born to this union, and little is know about his early life, as he did not write his memoirs but did prepare an abbreviated autobiography. Growing up on a farm, it is likely that he was acquainted with blacksmithing, milling, and mechanics, and no doubt these early experiences and images left an imprint, as later he would be attracted to invention and would become an owner of the Salem Iron Factory. His parents encouraged study in the ministry, as they probably perceived that the young lad was very intelligent. Such was also the case for Oliver Evans, born four years earlier; at that time, the most intellectually capable young men were bent toward the church, an important institution for providing guidance for its congregation.

When Nathan was seventeen, he started his study for college and one year later, in 1777, entered Cambridge College. Following his parents' desire, he studied Hebrew, learning so well that upon the death of his instructor, Professor Sewall, he instructed the Hebrew class himself until a successor was appointed.

Read graduated from Cambridge College in 1781 and was scheduled to deliver the valedictory address. He had distinguished himself as a scholar, "but in consequence of hard times the students were dismissed from college much sooner than usual; and, of course, the address was not delivered."[2] For two years he taught school in Beverly and Salem, Massachusetts, then went to Harvard as a tutor in 1783. He was to remain there in that capacity until 1787.

Deciding to study medicine, he proceeded to work with Dr. Edward A. Holyoke of Salem. This relationship lasted for about eighteen months, whereupon Read, still looking for his life's work, left the practice and opened an apothecary on Main Street in Salem. It was during this period that his inventive and inquisitive mind turned to mechanical pursuits that would challenge his intellect for much of his life.

In 1788, upon learning of the steamboat work of Rumsey and Fitch, Read became interested in the power of steam and its potential for moving boats and carriages. Most likely his attention was drawn to these individuals by the publicity surrounding their sometimes adversarial claims to the invention of the steamboat. The controversy has already been discussed (see chapters 6 and 7), with claims and counterclaims being lodged by each combatant.

Read intuitively understood the requirement for powerful yet lightweight machinery for transportation and applied his substantial intellectual ability to improve upon existing designs of the steam boiler and steam engine. His biographer asserts, "He believed, moreover, that the modes of propulsion used by Rumsey and Fitch—setting-poles, oars, paddles, or the ejection of water from the stern of the boat—were not only awkward in their operation but unreliable."[3]

Having knowledge of Rumsey's boiler invention and that a tube boiler in the form of a tight spiral had been reported in the *Philosophical Transactions* upwards of twenty years earlier,[4] Read set his mind to work on how to create a smaller and lighter-weight steam generator that he would call portable. Reasoning that if a single tube could produce a given amount of steam in a fixed volume, then through the use of multiple tubes, even more steam could be produced (see fig. 9.2). Hence, the multitubular approach he envisioned would be more compact, lighter, and better suited to transportation requirements. His logic was correct.

During the early part of 1788, the same year that Read initiated his steam work, Rumsey went to England to prepare and obtain a patent on his tube boiler. In his patent application he notes, "As my boilers, steam vessels, or stills may be infinitely varied in their forms and application, it is impossible to give Drawings of them all, but those that follow will sufficiently explain the nature of the principles

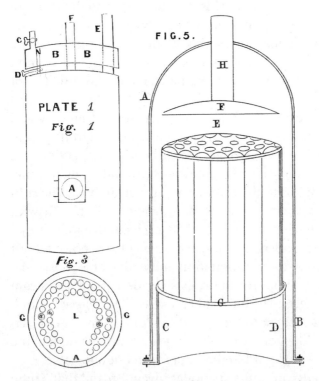

Figure 9.2. Nathan Read's boiler (left), with many water tubes in parallel, along with a visually similar configuration (right) patented by James Rumsey about the same time. (Respectively from David Read, *Nathan Read, His Invention of the Multi-tubular Boiler* . . . [New York: Hurd and Houghton, 1870], between pp. 50 and 51; and James Rumsey, *Boilers for Steam Engines, for Distillation, &c.*, British Patent No. 1673, December 6, 1788 [London: George Eyre and Wm. Spottiswoode, 1854], fig. 5.)

I have laid down."[5] Rumsey illustrated a number of water tube and fire tube configurations, one of which coincidentally bears a remarkable similarity to the boiler of Read's. (Both are shown side by side in the above figure.)

A close inspection, however, reveals that the designs are different, as Read used the tubes for water, while Rumsey allowed the heat to pass though the tubes with the water in the spaces between. Therefore Read's is a water tube boiler, whereas Rumsey's is a fire tube. Although this is a significant distinction, it is not of profound importance, as the objective of a boiler is to expose heat to water, and the final configuration is dictated by technological and safety considerations.

In a January 8, 1791, letter to Thomas Jefferson, one of the commissioners of patents, Read avows, "I do not know that any other person but myself hath ever constructed a tubular boiler,"[6] and this is no doubt true. He sincerely believed that using multiple tubes instead of a single tube was a new invention. In fact, neither Read nor Rumsey actually invented the tube boiler: in 1766, William Blakely had patented a primitive water tube boiler in England, but Rumsey had provided a thorough explanation of the concept and is generally credited with the invention.[7]

Read's multitube boiler can be thought of as simply placing in parallel a number of single tubes used by Rumsey on his steamboat. Evaluated today, the improvement would be considered an innovation (see chap. 2) obvious to anyone working with such devices, but at the time was considered a new device and hence an invention under then existing law.

Another Read invention was a cylinder that admitted steam to both ends of a piston (see fig. 9.3). He also proposed using high-pressure steam to extract more power from a construction of a given size. The idea is sound but was not new, as Boulton and Watt had a so-called double-acting steam engine in operation as early as 1784.[8] Their particular configuration, however, because of its size and weight, was entirely unsuitable for installation on a steamboat or carriage, which was Read's vision.

In his patent description of the steam cylinder, Read addresses the four valve cocks, *a*, *b*, *c*, and *d*, which regulate the flow of steam

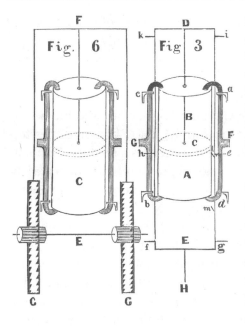

Figure 9.3. Read's improved steam cylinder, shown with valves for directing steam into either end, causing the piston to move back and forth. On the left he shows the ratchet system for turning an axle. (From D. Read, *Nathan Read*, between pp. 78 and 79.)

to one or the other either ends of the piston. He points out, "Sliding plates or regulators, like those made use of for other engines for letting in and shutting off steam from the cylinder, may be substituted for the cocks."[9]

However, it was the development of the sliding valve that delayed the introduction of the double-acting steam engine by Boulton and Watt, as the concept had been envisioned by James Watt many years earlier, but technology had to catch up.

As in the case of the boiler, Nathan Read had conceptually developed a worthwhile idea but had not worked out the important design details, so it remained a concept, not a practical reality. It does not appear that Read ever produced working devices of either the boiler or the steam cylinder, although models were made for the patent office.

Read next applied the portable boiler and the improved steam cylinder to the propulsion of a carriage and steamboat. The carriage is shown in figure 9.4, where a single boiler, C, supplies steam to two cylinders that independently, through ratchet and pinions, drive the vehicle's front wheels. A steering wheel, L, provided for turning, but it would be necessary to rotate the driving mechanisms as well, and how this can be done from the drawing is not at all clear.

As a result, the drawing is merely conceptual and not useful for the practical embodiment of a workable idea. For example, the pipes from the steam cylinder could not be fixed unless all units rotated together

Figure 9.4. Read's conceptual steam carriage. One of his improved steam cylinders is used to drive each of the carriage's larger front wheels. (From D. Read, *Nathan Read*, between pp. 78 and 79.)

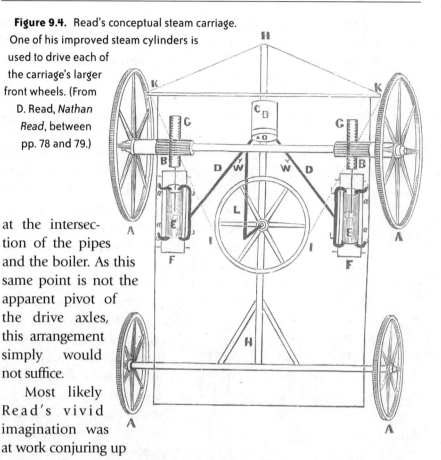

at the intersection of the pipes and the boiler. As this same point is not the apparent pivot of the drive axles, this arrangement simply would not suffice.

Most likely Read's vivid imagination was at work conjuring up a potential application of his new boiler and steam cylinder, and the details of his carriage implementation did not at the time seem too important. He would also apply the same components—or, as we know them today, subsystems—to a boat. To provide some evidence that his steamboat proposal would indeed work during the summer of 1789, Read undertook an experiment.

Trial. In 1789, Read attached paddle wheels on both ends of an axle and mounted the device across the gunwales of a small boat. Because a crank turned the wheel, most likely the axle was bent into a double U shape to allow both hands to provide the power. Using his own fractionally generated horsepower, Read propelled himself across an arm of the sea called Porter's River that separates Danvers from Beverly. Dr. John Prince of Salem witnessed the event.[10]

A drawing of his proposed steamboat is shown in figure 9.5, and this illustration, like the steam carriage, is conceptual, with much of the apparatus required to make it a practical entity omitted. A boiler, B, provides steam alternately to either side of the cylinder, C. The piston therein moves back and forth and is connected to gear racks on both sides. One rack will cause the pinion to rotate the axle on the upstroke and the other on the downstroke. This is similar to the forward motion of the carriage, but in the latter case two different steam cylinders were employed. The axle connected to the paddle wheel propels the boat forward. Initially Read claimed the paddle wheels as an invention but later learned that the application of such wheels to move a boat were not new and withdrew that part of his patent specification and claim.

Paddle wheels were quite old, simply a complementary adaptation of the waterwheel that dated back hundreds of years (see fig. 9.6). Read initially believed that their application to a boat for steam navigation was new, and in America this appears to have been true. Fitch had employed a stern wheel for one of his steamboats, and in 1797, Samuel Morey would reinvent side paddle wheels. In other parts of the world, however, they had already been tried, and

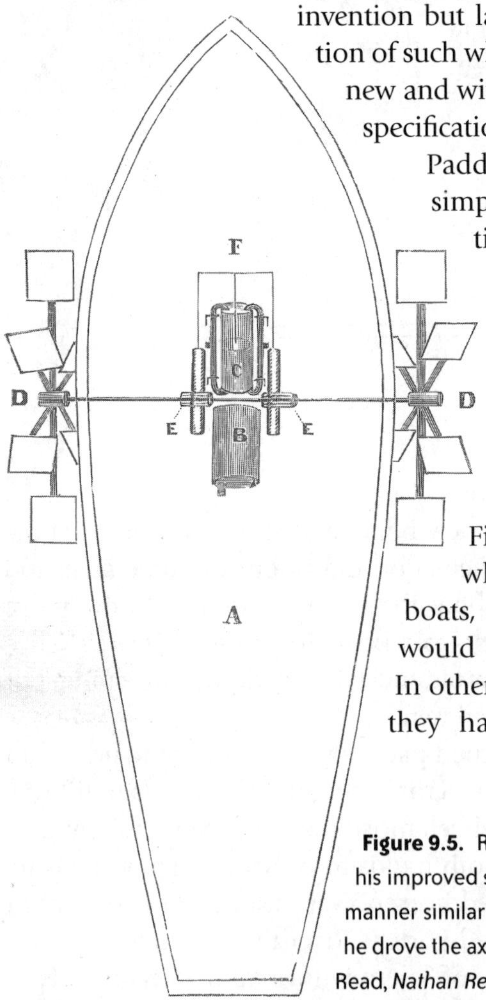

Figure 9.5. Read's conceptual approach of using his improved steam cylinders to propel a boat. In a manner similar to what was employed on the carriage, he drove the axle with ratchets and pinions. (From D. Read, *Nathan Read*, between pp. 102 and 103.)

Figure 9.6. The first known paddle wheels to propel a boat are shown in this illustration. Oxen provided the motive power, and an unmanned oar served as the rudder. (From S. C. Gilfillan, *Inventing the Ship* [Chicago: Follett, 1935], p. 72. Gilfillan dates the image to be from ca. 527 CE but states that it was first known only in 1643.)

David Read notes, "in looking over the old volumes of the 'Transactions of the Royal Society,' [Nathan Read] chanced to notice an article, relating to an experiment a long time previous, in France, in which it was related, that paddle-wheels and oars both had been tried, to see if they would not control the action of a ship of war in a calm."[11]

Now realizing that paddle wheels were not new to boats, and not realizing that this would likely make no difference in his American patent application, Read abandoned them, substituting a chain pump, which he called a "rowing machine." This device can be thought of as an elastic paddle wheel stretched between two sprockets attached to the side of a boat. As it is turned, the paddle boards, sometimes called float boards, move through the water, imparting forward motion. On the sprocket at the stern of the boat, the boards return to the bow in the air upside down, then proceed to the bow sprocket and the water once again.

No doubt Read was unaware that such a propulsion technique had been already envisioned, constructed, and tried by John Fitch and Henry Voight in the middle of 1786. Their trial was unsuccessful, but the concept was workable if properly applied. The construction and maintenance of such a device is more complicated than the well-understood paddle wheel, which either on the stern or on both sides of a vessel became the principal means of propulsion until the screw propeller was developed several decades later.

Read submitted his inventions to the American Academy of Arts and Sciences, and on January 15, 1790, a subcommittee reviewed and endorsed his improvements, stating that they "justly entitle him to the patronage of the Government of the United States."[12] Five days later a group of eleven distinguished men of Boston added their endorsement, attesting they believed his ideas to be real improvements and original inventions.

Read delivered a letter dated February 8, 1790, to Congress, then sitting in New York City, in which he states,

> To the Honorable Congress of the United States: The petition of Nathan Read of Salem, in Massachusetts, respectfully showeth; . . . that he has also invented a portable steam-engine, which may be constructed with less expense, is much lighter, occupies less space, and requires less fuel, than any other within his knowledge. Your petitioner has likewise discovered an improved method of applying the power of steam to the purpose of navigation; and has formed a plan to facilitate land carriage by the same agent. The machinery for communication motion to boats, vessels, land carriages, etc., is very simple, and takes up but little room.[13]

On the same day, the following is taken from the *Journal of the House of Representatives*:

> A petition of Nathan Read, of Salem, in the State of Massachusetts, was presented to the House and Read, praying the aid of Congress, and an exclusive privilege for construction sundry machines and

engines, which he has invented for improving the art of distillation, for facilitating the operation of mills and other water-works,
and for promoting the purpose of navigation and land carriages.

Also, a petition of John Stevens, junior, praying that an exclusive privilege may be granted him, for an improvement on the
steam engine, which he has invented, by a new mode of generating
steam.[14]

Read was present for the reading before Congress and was
offended by the skeptical smirks of its members when the mention
was made of steam land carriages. Taken aback by the reaction, he
would later remove all references to this invention. It is also likely
that John Stevens, of Hoboken, New Jersey, was also present during
this meeting of Congress, as he was in New York City at the time.
Read and Stevens were later introduced and discussed steam navigation. Read explained the principles of his multitubular boiler and its
application to steam navigation, supposedly unaware of Stevens's
interest in steamboats.

Congress was in the process of examining a patent bill to promote the useful arts, which had been introduced the previous year
and would be passed on April 10, 1790. The petitions by Read,
Stevens, and others were therefore not acted upon, and it would
become necessary for inventors to apply for patents directly with the
Board of Commissioners, consisting of the secretary of state, secretary of war, and attorney general.

On April 16 and 23, 1790, Read applied to the commissioners
with a request for a patent on his steam cylinder followed by a
request for patents on his steam carriage, steam boat, and paddle
wheels. There were three patents issued in 1790,[15] but the commissioners were faced with pending legislation (which eventually
failed) that would have required the recipients of federal patents to
give up any rights granted to them by a state. Therefore, on January
25, 1791, Henry Remsen, secretary for the Board, wrote Read that a
decision had been made not to proceed with patents for the new
application of steam until Congress passed a supplemental bill. He
noted that it would therefore be unnecessary for applicants to
appear as planned, on the first Monday in February.

On January 1, 1791, prior to receiving Remsen's letter, Read had written Thomas Jefferson, the effective head of the patent board, to withdraw his two previous petitions, of April 16 and 23, 1790, and substitute a single petition. Missing from the new petition was the steam carriage, and a chain wheel was substituted for the paddle wheels. Remsen would write back, pointing out the omission and offering Read the opportunity to resubmit the steam carriage, but shaken by the reaction of Congress, Read declined to do so.

The Board met on April 4, 22, and 23, and Remsen wrote Read on July 1, 1791, stating,

> Sir:—I received your letter of the 18th of May last a few days since. The Commissioners, at their meeting in April, agreed to grant patents to *all of the claimants* of steam-patents, so far as they had applied steam to useful purposes, *without taking it upon themselves to ascertain whether those claimants were really the inventors, as they severally alleged in their petitions.* Accordingly John Fitch for applying steam to navigation; James Rumsey for generating steam, applying it to navigation, and to raise water; yourself, [and] John Stevens for generating steam, applying it to raise water, to work a bellows, and to propel a vessel; and Engleback Cruise [i.e., Engle-hart Cruse] to apply steam to raise water, are all to have patents.[16] (emphasis added)

Patents to each of the five inventors were subsequently issued, all with the date of August 26, 1791, providing conflicting rights as outlined in Remsen's letter to Read. After introducing the claims of each patent as a prologue to a lengthy argument, it is the contention of biographer David Read that "It will thus be seen that the patents of Rumsey, Fitch, and Stevens, clash in several particulars; but that neither of them interferes with the patent of Read."[17]

Certainly for the generation of steam, there most assuredly did exist a conflict with Read and the others. Because the patent board was unable to sort out the priorities of the boilers as well as for the steamboats, this did not bode well for the new patent law. Without a clear priority and protection, the patent act did not accomplish the objectives of the framers of the Constitution. It would be, moreover, another forty-five years before a useful and workable patent law would be passed by Congress and implemented.

There is no evidence that Read ever attempted to construct a boiler, steam cylinder, steam carriage, or steamboat (but see a conjectural depiction of his steamboat design in fig. 9.7). It may have been the case that the patent issued to him was not to his liking, but more realistically Read appears to have been a good big-picture thinker who was not enthusiastic about the practical implementation.

On October 20, 1790, at the age of thirty-one, Read married Elizabeth Jeffrey, daughter of William Jeffrey, Esq., clerk of the court in Essex County. Five years later, in April, they moved to a farm in Danvers, and the following year he and some associates began the Salem Iron Factory. One of the company's products was nails made on a machine invented by Read, which was put into operation in 1797 and patented by him on January 8, 1798.

In October 1800, he was appointed a member of Congress and was duly elected on November 5, 1800, for a two-year term. In February 1802, Governor Strong appointed him a special justice of the Court of Common Pleas for Essex County. Five years later, in November 1807, he and his family moved to a larger farm in Belfast, Maine, where he became chief justice of the court in Hancock County.

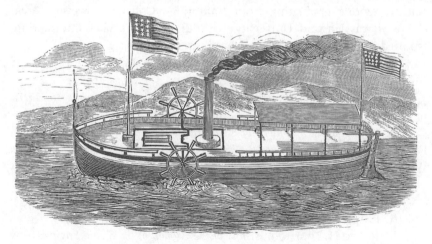

Figure 9.7. Conjectural drawing of Read's steamboat, showing the paddle wheels and pair of American flags, one with twelve stars and the other with sixteen. Note that the tiller is unattended. (From D. Read, *Nathan Read*, between pp. 156 and 157.)

On May 21, 1815, Read was elected an honorary member of the New England Linnaean Society. He lived until February 20, 1849, planning to write his memoirs but instead preparing a brief autobiography.

Read had a great visual imagination and was an original mechanical thinker. However, the assertion made by his biographer and "friend and nephew" that Nathan Read had "just claims for the essential part he took in both the invention of the steamboat and locomotive engine"[18] does not appear to be valid.

Notes

1. Read, "Autobiography of Hon. Nathan Read," *New-England Historical and Genealogical Register,* October 1896; and "The Read Families of Western (Warren) and Sudbury, Mass.," *New-England Historical and Genealogical Register,* July 1896. These articles, along with David Read, *Nathan Read: His Invention of the Multi-tubular Boiler . . .* (New York: Hurd and Houghton, 1870), are the principal sources for information in this chapter. (The epigraph for this chapter is quoted in D. Read, *Nathan Read,* pp. 99–100.) In his autobiography Nathan Read identifies his mother as Tamsin Meacham whereas David Read has her as Tamison Eastman. Both claim she was a cousin of Revolutionary War general Nathaniel Greene. Nathan reports graduating from Cambridge College; however, David claims it was Harvard University. In both cases, the later information, reported in Read's autobiography, has been used.

2. "Autobiography of Hon. Nathan Read," p. 435.

3. D. Read, *Nathan Read,* pp. 46–47.

4. Letter from Nathan Read to Thomas Jefferson, January 8, 1791, quoted in D. Read, *Nathan Read,* p. 52.

5. James Rumsey, *Boilers for Steam Engines, for Distillation, &c.,* British Patent No. 1673, December 6, 1788 (London: George Edward Eyre and William Spottiswoode, 1854), p. 3.

6. Quoted in D. Read, *Nathan Read,* pp. 52–53.

7. *Steam: Its Generation and Use,* 38th ed. (New York: Babcock & Wilcox, 1972), p. 4.

8. Robert H. Thurston, *History of the Growth of the Steam Engine* (New York: D. Appleton, 1901), p. 119.

9. Quoted in D. Read, *Nathan Read,* p. 80.

10. Ibid., pp. 93, 164, 182.

11. Ibid., pp. 103–104.

12. Quoted in ibid., p. 98.

13. Quoted in ibid., pp. 99–100.

14. *Journal of the House of Representatives of the United States, 1789–1793*, Monday, February 8, 1790, American Memory, http://memory.loc.gov/, p. 154 (accessed February 20, 2004).

15. Brooke Hindle and Steven Lubar, *Engines of Change: The American Industrial Revolution, 1790–1860* (Washington, DC: Smithsonian Institution Press, 1986), p. 79.

16. Quoted in D. Read, *Nathan Read*, p. 115.

17. Ibid., p. 118.

18. Ibid., p. v.

Figure 10.1. Samuel Morey. (Rendered by William T. Sisson from an unrefined portrait found in Frederick H. Getman, "Samuel Morey, a Pioneer of Science in America," *Osiris* 1 [January 1936]: pl. 2.)

10.

Samuel Morey
1762–1843

As nearly as I can recollect, it was as early as 1790 that I turned my attention to improving the steam-engine and in applying it to the purpose of propelling boats. . . . [In 1796 in New York] I invited the attention of Chancellor Livingston thereto, and he, with . . . Mr. [John] Stevens and others, went with me in the boat from the ferry as far as Greenwich and back, and they expressed very great satisfaction at her performance and with the engine. . . . He offered me at that time for what I had done seven thousand dollars . . . but I did not deem this sufficient, and no bargain was made.
—Samuel Morey, letter to William A. Duer, October 31, 1818

Samuel Morey grew up and lived most of his adult life in Orford, New Hampshire, retiring to Fairlee, Vermont, late in his career. At an early age, he exhibited a significant capacity to understand scientific and mathematical principles. When he was twenty-eight, he started thinking about using steam to power a boat, and several years later, in 1793, he secretly ran his first steamboat on the Connecticut River. Over the next several years, he made a series of steam navigation improvements, achieving five miles per hour in 1796 and discovering the advantages of two side paddle wheels in 1797. Over his remarkable inventive career, he patented some twenty devices, many of them relating to steam, including several steam engines.

This creative inventor was descended from George Morey, one of the first settlers of Bristol, Rhode Island. Samuel's grandfather later found his way to Lebanon, Connecticut, about thirty miles southeast of Hartford, where his father was born on May 27, 1735. There he met Martha Palmer, also of Lebanon, and they were married on July 14, 1757.

The couple moved to Hebron, Connecticut, a small town west of Lebanon, where Samuel was born on October 23, 1762.[1] He was their second child, preceded by a brother, Israel (named after his father), and would be the only one of seven children to achieve distinction.[2] By trade his father was a blacksmith,[3] but it is likely that his parents also eked out a meager living tilling the fertile land and trading crops for imported groceries such as sugar, rum, dry goods, and pewter tableware in the town of Lebanon, ten miles from Hebron.[4] The family's future, however, would be in the northern, timbered lands of New Hampshire and Vermont, where the senior Israel and his son Samuel would distinguish themselves as pioneers.

About 160 miles north along the Connecticut River was an untamed frontier in New Hampshire, where land was selling for one dollar per acre. The town of Orford there had received a charter from King George III in September 1761, and less than two years later, in March 1762, the governing body of sixty-five men decided that "the first six settlers [could] have fifty acres each . . . [and] the first six women who settled here shall have one cow each."[5] In addition, these early pioneering families were to enjoy water rights along the Connecticut River for the establishment of grist- and sawmills.

After the fall harvest in mid-October 1765, Orford's first pioneer, John Mann, set out from Hebron with his wife on horseback, reaching the wilderness in eight days. Several months later, in the dead of the winter of January 1766, came Israel Morey and his family. They left Hebron with four-year-old Samuel, his older brother, Israel, and a six-month-old sibling on an ox-drawn sled (see fig. 10.2), trudging through the snow, heading for new opportunity. The road to Charlestown, in the southern part of New Hampshire, was passable; however, north beyond this point, the crude road continued only as a trail too narrow for the sled. For the last sixty miles, Israel chose to proceed up the frozen Connecticut River, the reason January was chosen for the trip.[6]

Arriving safely in Orford, Israel Morey set about building a log cabin for his family from local timber. A historian later described the significant construction task that Morey faced, noting "that the pine trees were, on the average, two hundred feet high and of enormous size."[7]

Other pioneers soon arrived, and four years later, in 1770, Israel Morey and four other men were appointed to lay out and survey the roads and streets of the village. Morey started the town's first industry, constructing mills on Jacob's Brook, a tributary of the Con-

Figure 10.2. A two-wheeled ox cart that was the counterpart of the sled used by Israel Morey to move his family to Orford. This type of cart was used for nearly two centuries, and the one shown here depicts a scene near Yorktown, Virginia, about 1850. (From James Smillie, *Ladies National Magazine* [Philadelphia: C. J. Peterson, 1844].)

necticut River. By 1772, he had two mills operating, one grinding grain and one sawing wood. The settlers soon recognized him as a capable leader, and he gained their respect as a result, opening the town's first blacksmith's shop and first store. He kept a tavern in his home serving food and drink and obtained a charter to operate a ferry across the Connecticut River.[8]

Prior to America's war for independence, Morey commanded the Orford Military Company, which drilled on the village green, achieving a fair amount of discipline. In May 1776, he was appointed a member of the Committee of Safety, and in August he was promoted to the rank of colonel of the Twelfth Regiment on Foot. He would serve nearly twenty years in the American cause, retiring in 1794 with the rank of brigadier general.[9]

As Morey had risen from the rank of private to general, he was obviously possessed with qualities that inspired others to follow his lead. The general was described as a "placid, easy gentleman with benignant countenance who was wont on summer days to ride horse-back dressed in light colored garments, much in Quaker style, with a cloak thrown over one arm, the very personification of quiet enjoyment."[10] No doubt young Samuel inherited much of his father's intelligence, sense of timing, and industriousness.

Little is known about Samuel's early education, other than that he exhibited a thirst for knowledge and easily assimilated information of a mechanical nature. It is likely that his father taught him the principles of surveying, and he learned about levers, pulleys, ropes, and boats from the ferry across the Connecticut River. Israel's mills along Jacob's Brook used falling water on overshot wheels for grinding grain and sawing wood, and observation of these provided Samuel an early education in hydraulics, water power, and gearing. He gained much knowledge about metallurgy and the fabrication of metal parts used in the mill and elsewhere by watching and asking questions about his father's blacksmith trade. This work with fiery sparks flying was no doubt a source of fascination for young Samuel's inventive mind.

Morey's formal education was limited to that provided by schools in the area of Orford, but his inquisitive intellect gleaned knowledge from any source available, most likely a small number of technical books. Over the years his genius was to be applied to mechanics, as he was able to envision the creation of many useful devices long before technology rendered them practical.

Shortly after the close of the Revolutionary War, many Americans sensed unbounded opportunities for the fledgling republic, leading to a period of industrial development of labor-saving devices. The country was predominantly an agricultural nation, with large families providing the source of labor, but continual expansion to the west drained the available labor pool. In addition, British imports had been cut off during the war, and the need for manufactured goods would encourage Americans to eventually become self-sufficient.

This need for self-sufficiency was addressed particularly in the New England states, where a shorter agricultural season yielded time during the winter months for other pursuits. New Englanders responded with so-called Yankee notions, producing in cottage industries useful devices that were sold from the back of a horse and wagon in the south during the warmer months. Such was the economic environment into which the creative mind of Samuel Morey was thrust during his most productive years of mechanical improvements and inventions.

As Frederick Getman points out in his well-written 1936 paper about Samuel Morey, "A typical New England mechanic of this period has been described as 'a Yankee of Yankees' by birth, and of a temperament thoughtful to dreaminess. His natural bent is strongly toward mechanical pursuits and he finds his way early in life into a workshop. To 'get up' a machine is his one aim and ambition. If he succeeds, supported by patents, he may revolutionize an industry, forcing opponents who produce the old way altogether out of the market, while benefiting the consumer and making his own fortune at the same time."[11] Apparently such was the young Samuel Morey, who with his father's support built a workshop near the family home, where he spent many hours pursuing one device or another. The cold New England winters required a large amount of wood to be burned to keep warm as well as to cook, and at some point Samuel contemplated harnessing the power of boiling water and steam. As with all inventive individuals, he was a highly visual thinker, and his creative cortex started envisioning how steam energy could be put to useful work. At the age of thirty-one, he obtained his first patent for a steam spit (see fig. 10.3), which used steam to drive a small motor rotating a metal rod for roasting meat over an open fire. It is said that he could sell the devices as fast as he could build them.

Roads in New England, as well as throughout the new nation, were little more than poorly marked trails, making water transportation along navigable streams and canals the preferred method of travel. Whereas it was fairly easy to go downstream in flat-bottomed boats, proceeding in the other direction was very difficult and afforded a real opportunity for steam navigation. It is likely that Morey had some knowledge of the attempts of James Rumsey and John Fitch in the late 1780s; however, it is doubtful that he directly benefited from their pioneering work. He set his genius to solve the steam navigation

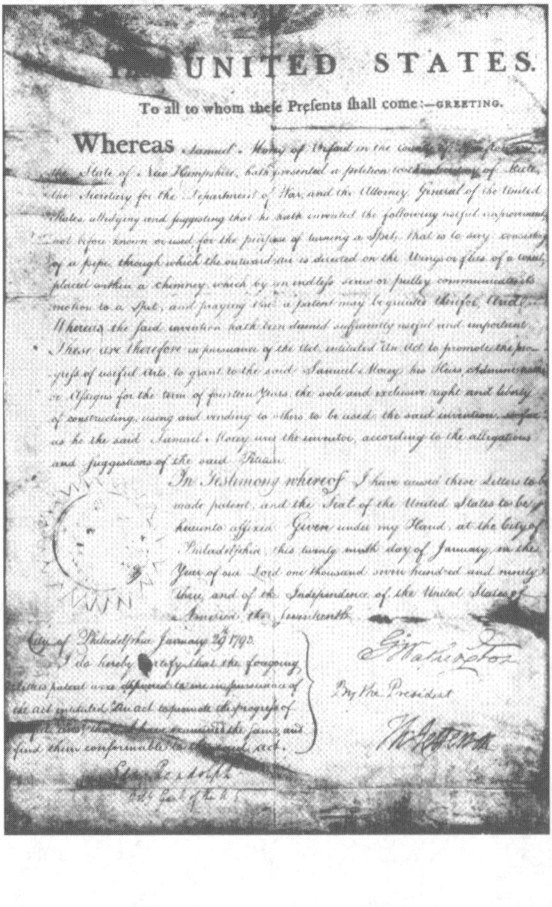

[illegible] UNITED STATES.

To all to whom theſe Preſents ſhall come:—GREETING.

Whereas Samuel Morey of Orford in the County of [illegible] the State of New Hampshire, hath presented a petition to the [illegible] of State, the Secretary for the Department of War, and the Attorney General of the United States, alledging and suggesting that he hath invented the following useful improvement, not before known or used, for the purpose of turning a Spit, that is to say: consisting of a pipe, through which the outward air is directed on the wings or flies of a wheel placed within a chimney, which by an endless screw or pulley communicates motion to a Spit, and praying that a patent may be granted therefor. And Whereas the said invention hath been deemed sufficiently useful and important: These are therefore in pursuance of the Act entituled "An Act to promote the progress of useful Arts," to grant to the said Samuel Morey, his Heirs, Administrators or Assigns, for the term of fourteen Years, the sole and exclusive right and liberty of constructing, using and vending to others to be used, the said invention, so far as he the said Samuel Morey was the inventor, according to the allegations and suggestions of the said Petition.

In Testimony whereof I have caused these Letters to be made patent, and the Seal of the United States to be hereunto affixed. Given under my Hand, at the City of Philadelphia, this twenty ninth day of January, in the Year of our Lord one thousand seven hundred and ninety three, and of the Independence of the United States of America the Seventeenth.

G° Washington

By the President

Th: Jefferson

City of Philadelphia January 29th 1793.

I do hereby certify that the foregoing Letters patent were delivered to me in pursuance of the act entituled "An act to promote the progress of useful Arts," that I have examined the same, and find them conformable to the said act.

Edm: Randolph
Att. Genl. of the U. S.

Figure 10.3. Samuel Morey's first patent, signed by President George Washington, for his steam-operated turning spit of 1793. (From the Baker Library, Dartmouth College, Hanover, New Hampshire, and reproduced from Alice Doan Hodgson, *Samuel Morey: Inventor Extraordinary of Orford, New Hampshire* [Orford, NH: Historical Fact Publications, 1961], p. 4.)

problem, and many years later, in 1818, he wrote a fortunately still-preserved letter to William A. Duer, where he recalled some of his early steamboat work:

> Sir—In answer to your enquiries relative to my experiments with steam-boats many years ago at New-York, previous to the construction of them by the late Chancellor Livingston and Mr. Fulton. I will state the simple facts as briefly as possible, and as nearly as I can at this time recollect.
>
> As nearly as I can recollect, it was as early as 1790 that I turned my attention to improving the steam-engine and in applying it to the purposes of propelling boats—I began my experiments in this vicinity on [the] Connecticut River.[12]

Morey does not elaborate in his letter to Duer about his initial Connecticut River trial, but other accounts provide some information.

Trial 1. One spring morning in 1793, Capt. Samuel Morey, most likely assisted by one other individual, launched a small, twenty-two-foot craft fitted with a steam engine and extended-bow paddle wheel into the Connecticut River (see fig. 10.4). Morey chose a Sunday morning for his experiment, reasoning that should the trial not be successful, most people would be attending church—hence, the ridicule would likely be mitigated. His concerns, however, were

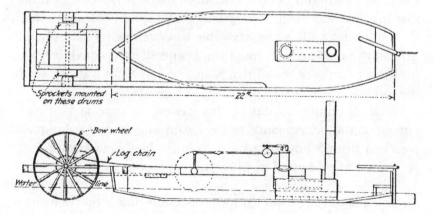

Figure 10.4. Conjectural drawing of Morey's first steamboat, with a bow wheel drawn from contemporary descriptions. (From Guy Hubbard, "Development of Machine Tools in New England," *American Machinist* 59 [August 16, 1923]: 241.)

unfounded, as the tiny vessel filled with equipment made four miles per hour as he steamed from Orford upstream a distance of four miles to Fairlee, Vermont, on the other side of the river, and back to Orford. It was a very successful trial.[13]

The confidence gathered through this trial and others that followed prompted him to obtain a patent for a steam engine and the use of a paddle wheel to propel a boat. The document bearing the signature of George Washington is dated March 25, 1795, and a copy is preserved at the New Hampshire Historical Society at Concord.[14]

Morey constructed a model of his steamboat, possibly for later submission to the Patent Office, and ventured to New York in search of more advanced metal fabrication and shipbuilding artisans. Like any other inventor, he would eventually require capital to bring his steamboat to practical reality, and through showing his model to several individuals, he learned of Chancellor Robert R. Livingston's interest in steam navigation.

Trial 2. Before Morey could invite the Chancellor for a demonstration, however, he needed to construct a larger steamboat. He writes, "I went to New-York and built a boat, and during three successive summers [1793–95] tried many experiments in modifying the engine and in propelling. Sickness in my family calling me home, I had the boat brought to Hartford, as a more convenient place, and there run her in presence of many persons."[15] It is likely that his boat was taken to Hartford under its own steam power, as the Connecticut River was navigable to Warehouse Point, some ten miles north of the city.[16] Since the ice generally prevented river traffic for three months of the winter, Morey had to await warmer weather the following year before returning to New York.

Trial 3. Sometime during this period, it appears that Morey enhanced the performance of his steam engine and subsequently switched from a bow to a stern paddle wheel before returning to New York, as evidenced by his written account:

> The next season [1796] having made sundry improvements in the engine, I went again to New-York and applied the power to a wheel in the stern, by which the boat was impelled at the rate of about five miles per hour. I invited the attention of Chancellor Livingston

thereto [see figure 10.5], and he, with Judge Livingston, Mr. Edward Livingston, Mr. Stevens and others, went with me in the boat from the ferry as far as Greenwich and back, and they expressed very great satisfaction at her performance, and with the engine.

Chancellor Livingston requested me to continue my endeavors to devise a better mode of propelling, and I continued my experiments through that summer, encouraged by his promise, which were to give me a considerable sum, provided I succeeded in making a boat run eight miles per hour. He offered me at that time for what I had done seven thousand dollars for the patent right on the North [Hudson] River and to Amboy; but I did not deem this sufficient, and no bargain was made. I never received any thing from him.[17]

Trial 4. Encouraged by the Chancellor's enthusiasm and his interest in a boat capable of achieving eight miles per hour, Morey continued his experiments. He already had in his fertile mind a number of improvements that could increase the speed of the steamboat, and hoping to obtain funding in excess of that already offered, he was spurred on. Some accounts have reported a figure offered by Livingston to be as high as one hundred thousand dollars, but such an amount seems quite unlikely in light of written evidence that only seven thousand dollars was offered for a boat with a respectable but slightly slower speed, of about five miles per hour.[18] In Morey's letter to William Duer, he continues:

Being desirous of devising a more effectual mode of propelling satisfactory to Chancellor Livingston and others, I continued my exertions; and as it had been sickly in New-York, I went to Bordentown [New Jersey] on the Delaware in June 1797, and there constructed a steam-boat, and then devised the plan of propelling by means of two wheels, one on each side. The shaft ran across the boat with a crank in the middle worked from the beam of the engine, with a shackle bar (commonly so called) which mode is in principle the same as that now used in the large steam-boats. I found my two wheels answered the purpose very well, and better than any other mode that I had tried; and the boat was openly exhibited at Philadelphia. For that time I considered every obstacle removed,

Figure 10.5. Robert R. Livingston, left, as Chancellor of New York State, is shown administering the oath of office to George Washington, the country's new president. The chancellor, who would later become Robert Fulton's steamboat partner, was given a demonstration of Morey's 1796 steamboat and encouraged him to construct a faster model. (Print from a painting by Alonzo Chappel [New York: Johnson, Fry, 1870].)

and no difficulty remaining or impediment existing to the construction of steam-boats on a large scale, and I took out patents for my improvements.[19]

He goes on later in the letter to state that after Livingston and Fulton started operating their steamboats, he told both of them that the application of two wheels was his invention and therefore patentable, but he never pursued the discussion beyond that point.

Although no mention is made in Morey's writings about showing the new design to the Chancellor, it appears that at some point Livingston became aware of Morey's success, perhaps as a result of the Chancellor's visit to Orford or Morey's subsequent visit to his Clermont estate. In 1801, Thomas Jefferson appointed Livingston minister to France, where he met with Joel Barlow and Robert Fulton, who was staying with Joel and his wife, Ruth. As a result of a conversation in Paris between Barlow and Livingston regarding Fulton's proposed use of side paddle wheels for a steamboat, Barlow wrote Fulton, who was away at the time, "You converted him as to the preference of wheel above all other modes, but he says they cannot be patented in America because a man (I forget his name) has proposed the same thing there. That unnamed man was identified in a footnote to the letter as 'Capt. Samuel Morey.'"[20]

Because of the success of the steamboat with paddle wheels, Morey attracted the attention of other investors:

> The notoriety of these successful experiments enables me to make very advantageous arrangements with Dr. [Burgess] Allison and others to carry steam-boats into effectual operation; but a series of misfortunes to him and others concurring soon after, deprived them of the means of prosecuting this design, defeated their purpose, and disappointed my expectations. But I did not wholly relinquish the pursuit, from time to time devising improvements in the engine. I recollect to have had repeated conversations with Chancellor Livingston and Mr. Fulton on these subjects. The Chancellor once visited me at this place, and at his request and expense I went to see him at Clairmont [sic].[21]

Morey, though having successfully demonstrated side paddle wheels, was not able to find financial backing to make the necessary improvements to place the boat in passenger and freight service. Although no mention is made in the October 1818 letter to Duer, it appears that Morey tried over the next two years to develop his steamboat further on his own, incurring a large amount of debt.

When he returned to Orford from New York in March 1800, he found his entire family—his wife, Hanna; their one-year-old daughter, Almira (see fig. 10.6); and an older adopted daughter—all gravely ill. As he and Hanna had previously lost two infant sons, he was desperately concerned about their welfare, but further compounding his woes was debt. Vermont and New York creditors were suing him for $585, and he avoided travel to the two states for fear of arrest and imprisonment. He wrote to a New York creditor, "I could hardly find it possible to write you sooner after my return, on account of the unhappy situation of my family. I now have the satisfaction to see then much better, a girl we brot up and who has been raving distracted

Figure 10.6. Silhouette of Samuel Morey and his beloved daughter, Almira, who almost died when she was one year old, made in Hartford, Connecticut, in 1824. (From Hodgson, *Samuel Morey, Inventor Extraordinary*, p. 2.)

has in a great measure recovered in her reason again, Mrs. Morey is a-gaining cleaverly and our little child I hope is getting well again."[22]

To raise the capital for the steamboat's development, Morey had mortgaged some of his patent rights on his steam engine, and the notes were coming due. Furthermore, it is likely that he had also incurred some debt with fabricators of components for his steam engine and boat, but the prospects of a commercially successful design from which he could recoup the investment were not in sight.

His principal assets, the patents on the steam spit and the steam engine as well as his boat, were at risk, so arrangements were made to transfer these rights to his New York creditors, a most unfortunate turn of events, as this effectively ended his steamboat pursuits. However, with his family's health improving, Morey still had his most valuable asset, his creative fertile mind, to concentrate on invention.

Toward the end of the year, on November 17, 1800, he patented a new steam engine to raise water and sold the rights for its use in Massachusetts and Maine to Mr. Giles Richards and Dr. Benjamin Haskell of Boston for the almost unbelievable amount of twenty thousand dollars.[23] It is most likely, however, that this sum was to be paid in installments over a fairly long period of time as the steam engines were built. It is also equally likely that the full amount was never received.

By June 15, 1803, Morey patented yet another steam engine, which also reportedly brought a large sum of money, but no mention is made of this engine or the previous one being used in a boat; applications were to stationary tasks such as sawing timber, grinding coffee, and making staples and boards for carding at the factory of Glen Richards & Company in Boston.

Based on early 1800 tax assessment records, the value of Morey's home was steadily increasing, indicating he was making significant improvements in its size and appearance (see fig. 10.7). In 1803 the assessment increased from $200 to $300, in 1804 to $350, and in 1805 to $400. To finance these improvements, he must have been fairly successful with steam engine business arrangements and lumbering, while his steamboat work remained neglected for many years.

After John Fitch's death in 1798, Chancellor Livingston was successful through his political influence in having Fitch's exclusive

Figure 10.7. Samuel Morey House. The sign reads, "Ell built 1773 by Orford's first minister, Obadiah Noble. Front built 1804 by Samuel Morey." (Photograph by Jack L. Shagena, about 1999.)

right to navigate the waters of New York transferred to him via legislation passed by the state's General Assembly. It was this act and three extensions thereof that enabled the Fulton-Livingston monopoly in the waters of New York State for many years.

In 1801, shortly after securing the monopoly, Livingston carried his abiding interest in steam navigation to Paris as the U.S. minister there, eventually partnering with Robert Fulton. The two entered into an agreement that ultimately resulted in the commercially successful *North River Steam Boat* (*Clermont*) in 1807. They were able to defend their New York monopoly until it was finally declared unconstitutional by the U.S. Supreme Court in 1824. The power of interstate commerce had been given by the Constitution to the federal government, and the New York State monopoly usurped that power.

There is much similarity between Fulton's design and that of Morey's ten years earlier, leading Morey to note, "I have often made passages in the steam-boats, and do not see in their construction any new principle."[24] Morey strongly believed that Fulton getting the credit for the invention of the steamboat was an injustice. Years later, when someone would refer to Fulton's steamboat, he would bitterly retort, "Curse his stomach, he stole my patent."[25]

Morey's inventive career was hardly over with the demonstration of his steamboat in 1797, as during the period from 1799 to 1815 he was granted eight patents, primarily associated with mechanical power. His patent of July 14, 1815, for the revolving steam engine finally afforded him a degree of recognition as being an inventor. The rights were sold to John L. Sullivan, who built the engines for his steam-powered tugboats, and it was the combative Sullivan, as well as others, who carried the battle to Fulton and Livingston about the unfairness of the monopoly they enjoyed on the waters of New York State.

Several years later, Sullivan, who had been impressed with Morey's steam engine, wrote a paper for the *American Journal of Science and Arts*, edited by Prof. Benjamin Silliman, titled, "On the Revolving Steam-Engine, Recently Invented by Samuel Morey," and published in 1818 (see figs. 10.8 and 10.9). In this article he explained the operation of the engine with the use of four engraved plates.

In an 1820 issue of Silliman's journal, Morey, who had by this

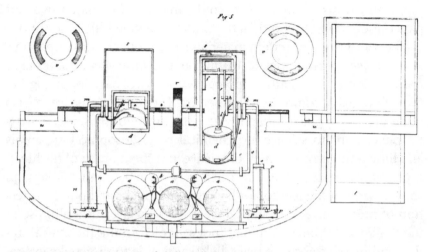

Figure 10.8. Cross-section of Sullivan's steam-powered tow-tug, showing Morey's revolving steam engine with two steam cylinders driving the shaft and flywheel across the top of the gunnels. The boiler is in the hull, and one of the paddle wheels is shown on the right. (From John Sullivan, "On the Revolving Steam Engine Recently Invented by Samuel Morey," *American Journal of Science and Arts* 1 [1818], pl. 1.)

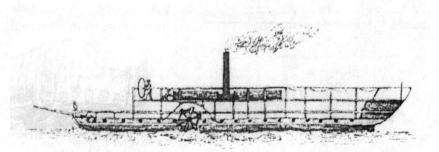

Figure 10.9. A crude sketch of John L. Sullivan's steam-powered tug, using Morey's revolving steam engine. (From John Sullivan, "Mr. Sullivan on the Revolving Engine: In Reply to Mr. Doolittle," *American Journal of Science and Arts* 2, no. 2 [November 1820]: fig. 6.)

point turned his attention to scientific investigation, had a two-part paper titled, "On Heat and Light," which contained his observations on the combustion of certain materials such as oil, tar, and turpentine. Silliman (see portrait, fig. 10.10) introduced Morey in a footnote as follows: "I presume that no apology will be necessary for giving Mr. Morey's valuable communications entire. They are the practical results of an ingenious practical man who as he ingenuously states, 'having no pretensions to science, no chemical or philosophical apparatus and little or no access to men of science, has spent much of his life in experiments.' Such results are often very valuable, and perhaps, in some cases, not the less so, for having been sought without the direction of preconceived, theoretical views."[26] This gives us a glimpse of the thinking of Morey. Although not schooled in the theoretical understanding of science, he nevertheless was able to undertake experiments to garner necessary information in order to develop a practical product such as the steamboat. Morey continued his scientific investigation, publishing two articles in Silliman's journal in 1821. One was his observations of heated rosin, titled, "Bubbles Blown in Melted Rosin," and the other about a mineral found in the Orford area, titled, "On Fetid Crystallized Limestone." With these efforts, Morey demonstrated that he was not only an inventor but a scientist as well.

Though not recognized during his lifetime, the last two Morey patents were precursors of the internal

Fig 10.10. Benjamin Silliman, editor of the *American Journal of Science and Arts*, published articles written by Morey, providing him a literary outlet in a respected technical journal. (Rendered by William T. Sisson from Getman, "Samuel Morey," pl. 2.)

combustion engine, which powers most of the world's automobiles. On April 1, 1826, he obtained a patent signed by J. Q. Adams for a "Gas or Vapor Engine," and on November 19, 1833, one for "Mode of Decomposing and Recomposing Water in Combustion with Spirits of Turpentine." This discovery would provide the source of information for an 1826 paper titled, "An Account of a New Explosive Engine, Generating a Power That May Be Substituted for That of the Steam Engine." The paper was published in Silliman's *Journal of Science and Arts* as well as in the *Journal of the Franklin Institute*.

Trial 5. Toward the later part of Morey's life, he and his wife retired to Bonny Vale, a residence on Fairlee Pond, now Lake Morey, across the Connecticut River in Vermont. There he built a boat and equipped it with an engine believed to be of the vapor type. He launched the craft on the lake and operated it for a year or so. Getman writes about the boat: "It mysteriously disappeared and no trace of it has been found despite numerous attempts to determine its fate. Some contend that it was sunk by the malicious hand of an enemy, while others are inclined to believe that Morey himself scuttled the craft during a period of melancholy brought on by prolonged brooding of the injustice done to him early in his career by Fulton and Livingston."[27] On October 27, 1874, a group from the New Hampshire Antiquarian Society, led by G. A. Curtice and Edward W. Howe, went to Fairlee Pond and interviewed individuals as to the location of the sunken boat. According to some, it had been seen, under favorable conditions of lighting, about twenty-five feet from the shore. Supposedly, Morey's enemies, sometime between 1800 and 1825, loaded the boat with rocks to sink it. The investigation team dragged the area of the suspected location but was unable to locate any evidence of its remains.

Morey died on April 17, 1843, at the age of eighty-one and was buried in the cemetery in the village of Orford. This remarkable man—inventor, scientist, author, and steamboat pioneer—was years ahead of his contemporaries in his technological vision, and he deserves to be better recognized for his many contributions to America's early technological triumphs.

Based on the table first presented in chapter 4, the table of steamboat pioneers and their contributions has been updated with the

Table 10.1. Summary of the efforts of steamboat pioneers through Morey

Steamboat pioneer	Concept date	Steamboat trial date(s)	Notes
Papin	1690	1707	Human, not steam power
Hulls	1736	None	Concept only for steam tug
D'Auxiron	1770	None	In 1774 boat accidentally sunk
De Jouffroy	ca. 1771	1778 Unsuccessful 1783 *Pyroscaphe*	Some success on 1783 trial, but work interrupted by the French Revolution
Miller	ca. 1786	1787 *Edinburgh* 1788 Steamboat	Human, not steam power Engine by Wm. Symington
Dundas	ca. 1798	1802 Tugboat *Charlotte Dundas*	Hull by Alexander Hart with engine by Wm. Symington; towed 140 tons on a canal
Henry	1779	None	Concept only
Rumsey	1783	1786, 1787, 1790, 1792, 1793	Limited success with water jet propulsion
Fitch	1785	1787, 1788, 1789, 1790	Introduced steamboat service on the Delaware in 1790, which was not financially viable
Evans	1785–86	1805 Amphibious 1812 Middlesex 1816 *Aetna* 1819 *Philadelphia*	About 1787 proposed boat with paddle wheels to Fitch and yielded priority to him
Read	1788	1789	Human, not steam power
Morey	1790	1793, 1794, 1795, 1796, 1797	Demonstration for Livingston in 1796 at 5 mph; side paddle wheels in 1797

efforts of Oliver Evans, Nathan Read, and Samuel Morey. Every few chapters, the table will be augmented with other contributors, until chapter 13, where the inventor of the steamboat is identified.

Notes

1. George Calvin Carter, *Samuel Morey, the Edison of His Day* (Concord, NH: Rumford Press, 1945), is a general overview of Morey's life; however, Carter did not use notes to cite sources. In this chapter on Morey, older references have generally been cited, some of which Carter mentions.

2. Frederick H. Getman, "Samuel Morey, a Pioneer of Science in America," *Osiris* 1 (January 1936): 283. This article, pp. 278–302, provides an excellent account of Morey's life.

3. Alice Doan Hodgson, *Thanks to the Past: The Story of Orford, New Hampshire* (Orford, NH: Historical Fact Publications, 1965), p. 12. The

book contains much information about Orford's early settlers, including Israel and Samuel Morey.

4. Margaret E. Martin, "Merchants and Trade of the Connecticut River Valley, 1750–1820," *Smith College Studies in History* 24, nos. 1–4 (October 1938–July 1939): 4–5.

5. Getman, "Samuel Morey," p. 283.

6. Hodgson, *Thanks to the Past*, p. 12.

7. Getman, "Samuel Morey," p. 282.

8. William A. Mowry, *Who Invented the Steamboat?* (Contoocook, NH: Antiquarian Society, 1874), p. 22.

9. Hodgson, *Thanks to the Past*, pp. 360, 377.

10. Getman, "Samuel Morey," p. 283.

11. Ibid., p. 284.

12. Morey, letter to William A. Duer, October 31, 1818, p. 2.

13. Guy Hubbard, "Development of Machine Tools in New England," *American Machinist* 59 (August 16, 1923): 241.

14. Carter, *Samuel Morey*, p. 34.

15. Morey, letter to William Duer, October 31, 1818, p. 2.

16. Martin, "Merchants and Trade of the Connecticut River Valley," p. 7.

17. Morey, letter to William Duer, October 31, 1818, p. 2.

18. *National Cyclopaedia of American Biography*, ca. 53 vols. (New York: James T. White, 1901), 11:168.

19. Morey, letter to William Duer, October 31, 1818, p. 3.

20. H. W. Dickinson, *Robert Fulton, Engineer and Artist: His Life and Work* (London: John Lane, 1913), p. 146.

21. Morey, letter to William Duer, October 31, 1818, p. 3. Dr. Allison's first name is supplied in David Read, *Nathan Read, His Invention of the Multi-tubular Boiler . . .* (New York: Hurd and Houghton, 1870), p. 129n.

22. Quoted in Hodgson, *Thanks to the Past*, pp. 14–15.

23. Ibid., p. 15.

24. Morey, letter to William Duer, October 31, 1818, p. 4.

25. Getman, "Samuel Morey," p. 287. Hodgson, *Thanks to the Past*, p. 17, reports the quote, "Blast his belly! He stole my patent!"

26. Benjamin Silliman, note to Samuel Morey, "On Heat and Light," *American Journal of Science and Arts* 2 (1818): 118n.

27. Getman, "Samuel Morey," p. 300.

Figure 11.1. Robert Fulton. (Rendered by William T. Sisson from Robert H. Thurston, *History of the Growth of the Steam-Engine* [New York: D. Appleton and Company, 1901], p. 251.)

11.

Robert Fulton
1765–1815

Men of true genius glow with lib'ral spirit,
And bind a garland round the bust of merit;
While blockheads, void of wisdom's grateful light,
Bury distinction in eternal night.

—Robert Fulton, quoting Morehead,
in *Treatise on the Improvement of Canal Navigation* (1796)

Fame and fortune, or rather the reverse order, was Robert Fulton's lifelong pursuit. Of humble Pennsylvania parentage, this brilliant and industrious late-eighteenth-century entrepreneur—or "projector," as they were then called—was determined from an early age to excel. During his life he would turn his attention from one activity to another, anticipating that each presented an opportunity for making money while at the same time affording adoring recognition from the public. Fulton would pursue, among other activities, miniature painting; historical and portrait painting; canal and inclined plane design; and invention and development of a marble cutting device, earth excavator, and a rope-making machine.

To raise money to support his various activities, he borrowed from associates, sold interest in his inventions, designed and constructed a panorama in Paris, and built and hustled a submarine and torpedo system sequentially to the French, Dutch, English, and Americans. Along the way he wrote and illustrated a number of treatises, convincing proposals, and corresponded with some of the most important people of the day. After meeting Robert R. Livingston in Paris in early 1802, when he was thirty-six years old, Fulton would finally give the steamboat serious attention. Even then, during its development he juggled other activities with the English and Americans on machines of war.

In 1803, he had his first steamboat success on the Seine in Paris. Four years later, with an English steam engine, he ran the *North River Steam Boat* (*Clermont*) on the Hudson River, resulting in the firm establishment of the beginning of steam navigation in America. To place in perspective all of his many and sometimes abortive efforts before the steamboat, his life up to that time will be reviewed.[1]

Robert Fulton Sr., of Scotch-Irish descent, came to America as a young man, arriving at the port city of Philadelphia. By 1735, he had migrated west to the small but fast-growing inland town of Lancaster, Pennsylvania, where he became active in the community. He was a charter member of the volunteer fire company, a founding member of the Juliana Library, and a founder and dedicated member of the First Presbyterian Church.

The elder Fulton established himself as a tailor, and after acquiring some wealth married Mary Smith in February 1759, who was a sister of Colonel Smith of nearby Chester County. In August of the same year, the couple moved into a modest brick house on Center Square, now known as Penn Square, in Lancaster. Close by was the home of patriot and inventor William Henry, who since 1750 had been producing rifles that would later be used in winning America's freedom from Great Britain.

During the next 5½ years, their union produced three daughters: Elizabeth, Isabella (called Bell), and Mary, who went by the nickname of Polly to distinguish her from her mother. On February 8, 1765, Fulton purchased, at a foreclosure sale, a 394-acre farm twenty-five miles south of Lancaster, close to the Maryland-Pennsylvania border. It was here in a 2½-story stone farmhouse in the township of Little Britain that Robert Fulton Jr. was born, on November 14, 1765 (see fig. 11.2). The house still stands and is identified as his birthplace by an official State of Pennsylvania historical marker.

For Robert Sr., the transition to farming was not successful. By January 1772, after several unsuccessful crops, he was forced to sell the farm. It was disposed of in the same manner as he had bought it, at a sheriff's sale, and even household articles were sold, leaving the family with just their personal clothing. Mary and Robert moved with their family, now including a second son, back to Lancaster. He resumed his tailoring business but died two years later, leaving the family with little support.

These were difficult times, most likely a significant factor in young Robert's determination for financial well-being throughout

Figure 11.2. Today the birthplace of Robert Fulton, depicted here, is identified by a Pennsylvania roadside historical marker that recognizes Fulton as an inventor, but it does not credit him with inventing the steamboat. (From Thomas W. Knox, *The Life of Robert Fulton and a History of Steam Navigation* [New York, G. P. Putnam's Sons, 1886], p. xiv.)

his life. One of the assets left by his father's death was an apprentice, Samuel Chapman, an eleven-year-old boy who had come to the family as a pauper. The custom at the time allowed for a family to care for a young lad who was indentured to work for the master, usually until he reached the age of majority. As Mary found it very difficult to care for her own family, she asked the overseers of the poor to assume responsibility for the boy. This concept of apprenticeship would, however, be later used by Mary to provide for young Robert's artistic training.

Up until the age of eight, Robert's mother taught the lad, but then he went to a Quaker school under the tutelage of a strict schoolmaster, Caleb Johnson. From the limited knowledge we have, it seems that Robert was full of ideas from an early age and not disposed to learning from "dusty books." He acquired the moniker "Quicksilver Bob" for his ability to mentally calculate the distance performance of guns, coupled with his love of playing with the liquid metal mercury, or quicksilver as it often called.[2] He is reported to have produced lead pencils (see fig. 11.3) and made improvements to rifles, but specifics are hazy, and he never wrote much about his early life, as he nearly always looked forward, seldom backward. It is worthwhile to note that this forward-looking characteristic is found in many inventors, as the successes and failures of yesterday provide boundaries for today's thinking, whereas the mind strives to create a new device for tomorrow.

The town of Lancaster during Robert's formative years was an exciting place, becoming the temporary capital of America during the Revolutionary War. For a short while, it was filled with leaders of the day, as well as prisoners of war who were on their honor not to leave but were otherwise free to roam the village. The imprints of these people and the paintings of Benjamin West, which the young Fulton saw in the home of William Henry, influenced this bright young man.[3] Robert developed a talent for art and he painted signs for local businesses and sketched the events of the day.

Exactly how he acquired his skills as an artist is not clear, but as a result of his talent for drawing and his manual dexterity, around the age of seventeen he was apprenticed to a jeweler in Philadelphia

Figure 11.3. According to tradition, the young Fulton was late to school one day. He explained his tardiness on being delayed by completing work on a lead pencil, which he is showing to the schoolmaster. (From Frank P. Bachman, *Great Inventors and Their Inventions* [New York: American Book, 1918], p. 33.)

to learn the trade. His master, Jeremiah Andrews, a jeweler from London, had found his way to Philadelphia via New York and advertised an assortment of shoe buckles, lockets, brooches, and hair work, a kind of ornamentation that was popular at the time. Unlike John Fitch, who found his masters in the clock trade, Benjamin Cheney and later his brother Timothy Cheney, to be oppressive, Fulton thrived in his apprenticeship.

"[H]air worked in the neatest manner," as advertised by Andrews,[4] was a way of honoring the dead by integrating the departed's hair into a locket or other complex design. The jewelry often included a tiny painting, which Fulton learned to do with great skill. From this he branched out into other small paintings and was listed in White's *Directory of the City of Philadelphia* in 1785 as being located at the corner of Second and Walnut streets, with the profession of miniature painter. This was also the location of Andrews's jewelry shop.

Around the end of 1785, Fulton developed some congestion in his chest and sought the advice of a doctor. Various remedies were prescribed for what was probably pneumonia or tuberculosis, but he continued to spit up blood. On the advice of friends he ventured to

Bath, Virginia (now Berkeley Springs, West Virginia), to partake of the famed curative powers of the warm waters. He arrived in the town in early May 1786.

This was the home of James Rumsey, who was working on his steamboat at the time, and it has been suggested that perhaps Fulton met Rumsey and they discussed steam navigation. This, however, is unlikely, as from June 1785 to July 1786, Rumsey was working for George Washington to make the Potomac River navigable and was away from Bath most of the time. Fulton may have heard about steamboat work from other sources, as in mid-1793 Rumsey had confided his ideas to John Wilson of Philadelphia, who had been visiting Bath, and to his mill partner George Bedinger in March 1784.

Most likely, though, Fulton was first interested in improving his health and secondly had his future focused on painting, following in the footsteps of Pennsylvanian Benjamin West, who had achieved fame in London. As the resort town was a gathering place of the influential people of that time, it is likely he did some painting there and was encouraged to continue his studies in Europe and England.

His health improved, and he soon returned to Philadelphia in early June. On June 6, 1786, he advertised, "Robert Fulton, miniature painter and hair worker, is removed from the northeast corner of Walnut and Second Streets to the west side of Front Street, one door above Pine Street, Philadelphia."[5] Front Street was one block off the Delaware River, where John Fitch was busily working trying to perfect his steamboat (see map in fig. 11.4). It is likely that Fulton saw or at least heard of Fitch and his efforts, but that interested him little, as he was then pursuing a painting career. By the following summer, he had borrowed forty guineas and sailed to London with a letter of introduction to Benjamin West.[6]

What Fulton had planned, a short visit to acquire additional artistic skills, became unpredictably extended, when on April 14, 1789, he wrote his mother lamenting, "Painting Requires more studdy than I at first imagened in Consequence of which I shall be obliged to Stay here some time longer than I expected."[7] Up to this point in his life he had always been the quick master of new challenges, and this revelation must have been quite an awakening for

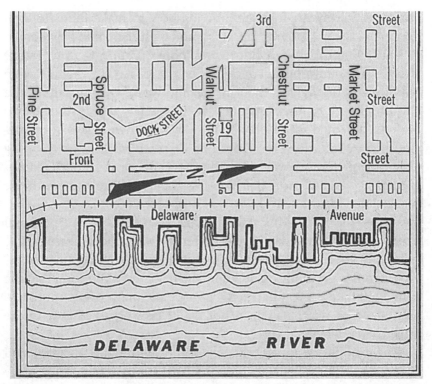

Figure 11.4. Partial map of Philadelphia, showing the area where Fulton was apprenticed (Second and Walnut streets), his miniature painting location (Front Street one door above Pine Street), and the docks on the Delaware River where John Fitch was working on his steamboat. (Adapted by Jack L. Shagena from a map in *Philadelphia* [Philadelphia: William Penn Association, 1937], p. 398. Copyright 1937 Commonwealth of Pennsylvania, Pennsylvania Historical and Museum Commission.)

this financially struggling young man of twenty-three years in a strange country.

In a manner akin to his early childhood in Lancaster, these years in London must have shaped his character, forcing his genius in new directions. He changed his address often, searching for more affordable accommodations or new friends that could support him financially or from whom he could borrow. Even in early 1792, after he had achieved a modicum of success in the art field, he wrote his mother a letter lamenting his difficult times and alluding to money owed: "Many, many a Silant Solitary hour have I spent in the most unnerved Studdy Anxiously pondering how to make funds to Support me til the fruits of my labour should be sufficient to pay them."[8]

In 1790, probably by borrowing money, Fulton went to France to study art, returning to London after a brief stay. By the following year, he was finally successful in having two of his portraits and a pair of subject pictures exhibited at the Society of Artists and two portraits accepted for display by the Royal Academy. Two years previously, in 1789, however, Fulton had written his mother asserting that "my pictures have been admitted this year into the Royal Academy,"[9] an assertion that was less than truthful. Even with the current success at the Academy, he exaggerated the number of paintings to eight, honing his skills for his later avocation of proposal writing. At this early stage in his career, he was quickly learning the technique of how to pitch a product by "stretching the truth." Although it is generally assumed by the reader of a proposal that the writer will take a few liberties to enhance the merit of his approach, outright lying never has been nor could ever be acceptable in the professional business world.

The exhibitions apparently drew some attention to his work, and Fulton received an invitation to visit Lord Courtenay at his country home in Devonshire, called Powerhorn. He painted a portrait of Courtenay and wrote home saying that he was beginning to earn a little money and that he hoped to be free of debt in six months. The painting pleased Lord Courtenay, with the result that Fulton was introduced to his friends, the Duke of Bridgewater and Lord Stanhope. Stanhope would play a significant role later in his life.

Fulton remained at Powerhorn for 1½ years, and in a January 20, 1792, letter to his sister Mary mentioned that he had been "doing some business for Lord Courtenay" but was silent as to its nature. It was well known that Courtenay was a homosexual, but his lifestyle at that time did not prevent him from maintaining relationships with influential people: "He decks himself out like a doll and paints his face like a p[rostitute]," Philip wrote in her Fulton biography, adding, "Even had he so desired, it is difficult to believe that Fulton could have remained detached from the life-style of Courtenay and his fiends, on whom he was financially dependent."[10] Unlike Henry,

Rumsey, Evans, and Morey, who married and stayed married, and John Fitch, who married, fathered two children, then deserted his wife and later proclaimed he had no interest in sex, Fulton appears to have been interested in both sexes.

In February 1793, James Rumsey's water jet steamboat, the *Columbia Maid*, had been completed in London and was tested on the Thames, achieving a speed of four knots. Rumsey had died suddenly the preceding December, so the trial was conducted by one of his investors, Daniel Parker. It is not known whether Fulton was present, but he did know Rumsey and was aware of the effort, as he recorded in his notebook under an entry titled "Messrs. Parker and Rumsies experiment for moving boats." After analyzing the experiment, Fulton concluded, "It therefore appears that the Engine was not loaded to its full power, that the water was lifted four times too high and that the tube by which the water escaped was more than five times too small."[11]

In his 1908 biography, *Robert Fulton, Engineer and Artist*, engineer H. W. Dickinson points out the fundamental error in Fulton's thinking, as the would-be inventor was not yet grounded in engineering principles:

> This shows that Fulton did not realise where the cause of Rumsey's failure lay. If water, contained in the boat itself, is forced out through an orifice at the stern at twice the speed at which the boat moves, the efficiency may be as much as 75 per cent; that is, looked at in another way, the efficiency would be greater than that of any other forms of propulsion. The water has, however, to be taken in at the bow and come to rest relative to the boat; this means a loss of energy which greatly reduces efficiency, and this was Rumsey's case.[12]

In addition, it is also clear that Fulton's concern for raising the water to four times the necessary height was in and of itself not a serious problem, as this simply resulted in storage of potential energy that would be recovered almost 100 percent upon letting the water subside. The conclusion that the water jet output escaped through an orifice too small was also incorrect, as a small jet was indeed necessary to impart a velocity to the water to achieve forward thrust.

At this point in Fulton's technical development, he had not grasp the fundamentals of hydraulics, but this did not stop this genius with an ego of gigantic portions from abandoning his artistic career and taking up canal engineering. The reasons he gave up painting have been the subject of much speculation. A few reasons seem clear.

Fulton held Benjamin West in high esteem (see fig. 11.5 for a portrait of West), and it is most likely the great master provided him

Figure 11.5. Benjamin West, president of the Royal Academy. West provided encouragement to Fulton during his early years in England. Fulton would later correspond with the famous painter and acquire some of his works, taking them back to America. (From a print based on a portrait by Sir Thomas Lawrence, 1894.)

with the same advice he offered all aspiring artists: "Give your heart and soul wholly to art, turn aside neither to the right or left, but consider the hour lost in which a line has not been drawn, nor a masterpiece studied."[13] For Fulton this was tough advice, as his head was teeming with new ideas, and the persistent and dedicated attention to detail was not part of his makeup. He was outgoing and persuasive; he loved to hold forth on political, social, and technical issues; and his ideas and thoughts were engaging to others. He learned principally through verbal intercourse; introverted and introspective concentration did not suit his demeanor.

Painting requires a high degree of attention, usually by a single individual, to minute detail. Critics of art will scrutinize an effort requiring tens or hundreds of hours of work on a relatively small canvas in a matter of seconds or minutes. The artist knows, for example, in the case of a portrait that special attention must be given to eyes and mouth; otherwise, regardless how well the rest of the work is executed, the effort will be for naught. Fulton knew this and also knew of his weakness of not being able to focus intensely on detail for long periods of time, but he did not realize the parallel between art and engineering.

Had Fulton, through his genius, quickly developed a talent approaching the high order of West, no doubt he would have continued to paint. But this level was not easily, if ever, within his personal reach, and his riches and recognition would have to be achieved by other means. Morgan points out that he had acquired or was acquiring the status of "gentleman," and a gentleman did not work for others. He was also temperamentally unsuited to take orders, and these reasons "largely account[ed] for his learning engineering by observing rather than working in it."[14] The transition from art to engineering would not, however, be without difficulty.

Another factor in changing careers was certainly opportunity. As Flexner noted during this time, "England had gone canal mad . . . [and] some ventures were making profits as high as 1,000 per cent."[15] Fulton saw the potential for jumping on the canal boat and riding it to fortune. Over the next several years, he would examine a number of canal projects and publish an impressively written and illustrated canal treatise. He obviously hoped through the publication to establish his credentials, which would lead to profitable contracts. His plan, however, was to fail.

On September 30, 1793, he wrote to Lord Stanhope, who was planning a canal connecting the English Channel with Bristol. Fulton promoted the elimination of canal locks by pulling boats up inclined planes from one level to another, and also mentioned in the letter he had some thought on steamboats. In Stanhope's reply he pointed out that inclined planes had been proposed sixteen years earlier (in fact they were ancient), but as the Earl himself was working on a steamboat, he would be interested in learning more about Fulton's ideas: "I have received yours of the 30th of September, in which you propose to communicate to me the principles of an invention which you say you have discovered, respecting the moving of ships by the means of steam. It is a subject on which I have made important discoveries. I shall be glad to receive the communication which you intend."[16] Buoyed by Stanhope's attention, Fulton responded on November 4, 1793, noting that he had been interested in steamboats since June 1793 and had experimented with various means of propulsion, concluding a paddle wheel was best (see sketches in fig. 11.6).

He also correctly suggested, "With regards to the formation [shape] of ships moved by steam I have been of the opinion that they should be long, narrow and flat on the bottom, with a broad keel, as a flat vessel will not occupy so much space in the water; it consequently has not so much resistance."[17] In this letter it is interesting that Fulton had proceeded along the line of sound development: envisioning a new idea; design, construction, and testing of a model; and finally, using feedback to correct deficiencies and retest.

Stanhope ignored Fulton's propulsion advice, and the copper-sheathed *Ambi-Navigator Kent* was built using a twelve-horsepower Boulton and Watt steam engine. With its duck-foot propulsion, a speed of only three miles per hour on the Thames was achieved, and the idea was abandoned.

Exactly one year after his letter to Stanhope, on November 4, 1794, Fulton was still thinking about a steamboat as he wrote a

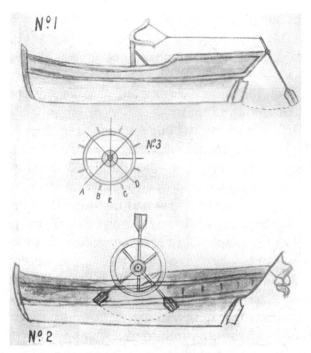

Figure 11.6. Steamboat sketches made by Robert Fulton in 1793 showing an oar at the stern to propel a boat forward, as well as the use of paddle wheels. (From Alice Crary Sutcliffe, *Robert Fulton and the* Clermont [New York: Century, 1909], p. 131.)

letter to Boulton and Watt about an engine:

I shall esteem it a favour to be informed of the Expense of a Steam Engine with a Rotative movement of the purchase of 3 or 4 horses, which is designed to be placed in a Boat.[18] You Will be so good as to mention what sized boat it would occupy, as I wish to have it in as little space as Possible, and what you consive will be the Expense when finished Compleat in the Boat. Whether you have one ready of the dimentions specified or how soon one might be finished. With [what] Weight of Coals which it will consume in 12 hours, and what Quantity of purchase you allow to each horse, as I am anxious to supply some Engines of the above dimensions as soon as Possible. Your Emediate Answer will much oblige.[19]

The letter was incredibly naive: not only did it ask the world's foremost manufacturer of steam motive power to provide price and delivery of an engine, but it also asked them to specify the boat. Furthermore, they were to provide the engine installation costs and furnish the performance of coal consumption once installed. Obviously, if this were a trivial task, for which there was identified a financial payback, Boulton and Watt would have simply moved in that direction without any prompting from Fulton. There is no evidence to suggest that Fulton received a reply.

Over the next two years, Fulton continued to observe and study canals, inclined planes, aqueducts, locks, and canal boats. His active mind envisioned one new idea after the other, and at some point he found time to read Adam Smith's *The Wealth of Nations*, grasping the fundamental truth that a country's wealth resulted from turning labor into capital. So it is not surprising that his book, titled *A Treatise on the Improvement of Canal Navigation*, when published in March 1796, was a mixture of economics, philosophy, and technical considerations (see title page in fig. 11.7). He had discovered, he believed, a basic truth: that to increase the production of every square mile of land it was essential to interlace the countryside with small canals that would allow an easy access to markets for all producers. Furthermore, if partners traded, prejudices would be eroded, and they would over time become economically interdependent, transferring from one to another mechanical improvement, which would provide additional enhancements in productivity.

On July 30, 1795, prior to the publication of his treatise, Fulton printed in the *London Morning Star* newspaper a signed article announcing his forthcoming book and inviting a number of engineers by name to provide comments on the merits of small canals, inclined planes, and wheeled boats. Without any details of his forthcoming approach, and without a reputation as a canal builder, those mentioned may have been flattered but disinclined to comment. When the treatise was published in March 1796, he signed it, "By R. Fulton, Civil Engineer," attempting to give his ideas more standing than they would have received by being identified as an artist. About this Dickinson writes, "Although the title at that time was assumed rather loosely—civil engineering not being a definite profession— yet Fulton had no right to use it, for he had not been in practice, nor had he assisted in carrying out an engineering works."[20]

Fulton attributed his interest in canals as having stemmed from a 1793 paper written by the Earl of Stanhope in which the "many difficulties" in canal construction are manifested and for which he perceived to have some solutions. In the front matter of the treatise,

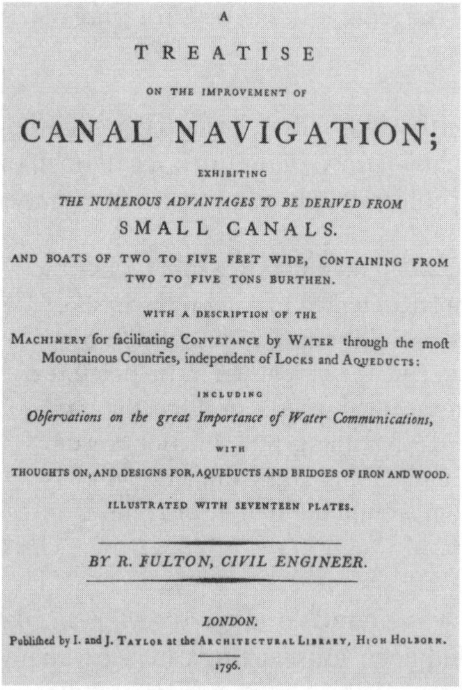

A

TREATISE

ON THE IMPROVEMENT OF

CANAL NAVIGATION;

EXHIBITING

THE NUMEROUS ADVANTAGES TO BE DERIVED FROM

SMALL CANALS.

AND BOATS OF TWO TO FIVE FEET WIDE, CONTAINING FROM
TWO TO FIVE TONS BURTHEN.

WITH A DESCRIPTION OF THE

MACHINERY for facilitating CONVEYANCE by WATER through the moſt
Mountainous Countries, independent of LOCKS and AQUEDUCTS:

INCLUDING

Obſervations on the great Importance of Water Communications,

WITH

THOUGHTS ON, AND DESIGNS FOR, AQUEDUCTS AND BRIDGES OF IRON AND WOOD.

ILLUSTRATED WITH SEVENTEEN PLATES.

BY R. FULTON, CIVIL ENGINEER.

LONDON.

Publiſhed by I. and J. TAYLOR at the ARCHITECTURAL LIBRARY, HIGH HOLBORN.

1796.

Figure 11.7. Title page from Fulton's treatise on canal navigation, where the artist-turned-engineer expounded his technical approach in convincing words and illustrations but without the benefit of practical training or knowledge. (From Robert Fulton, *A Treatise on the Improvement of Canal Navigation* [London: I. and J. Taylor, 1796].)

Fulton addressed the several canal engineers mentioned in the *London Morning Star* article from which he had sought comment. Here he would add the names of Telford, Cockshot, Chapman, and Benet, calling on them "to deliberately weigh the following pages on small canals, and favour me with your opinion, or transmit it to the public, in order that they may be put in the possession of the arguments for and against the system."[21] Of these William Chapman, a civil engineer, would respond one year later with a carefully thought-out rebuttal.

In the treatise's preface, Fulton stated why he believed new ideas lie dormant, emphasizing a lack of boldness on the part of his fellow men, a propensity this man with a substantial ego did not share:

The fear of meeting the opposition of envy, or the illiberality of ignorance, is, no doubt, the frequent cause of preventing many ingenious men ushering opinions into the world. Hence, for the want of energy, the young idea is shackeled [*sic*] with timidity, and a useful thought is buried in the impenetrable gloom of eternal oblivion.

But if we consider for a moment, how much men are the sons of habit, we shall find, that almost the whole operations of society are the produce of accident, and a combination of events, rendered familiar by custom, and interwoven into the senses by time; insomuch, that it is mere chance if the ideas are awakened to a sense of particular errors.

It is at this point in the preface that the self-anointed civil engineer, who had never built a single canal, inclined plane, lock, canal boat, or accomplished any civil project professes with unmitigated arrogance,

> But in such case it is fortunate, when they arise in a mind active to investigate, and which feels only contented to rest on the basis of reason; for without this, man must ever remain in a fixed point, and improvements will be at an end: the adventurer must therefore arm himself with fortitude to meet the attacks of illiberality and prejudice, determined to yield to nothing but superior reason; resting assured, that every virtuous mind will commend an exertion to remove the rubbish from around the Temple of Truth, even should the undertaking fail.[22]

This is quite a statement and was bound to elicit knowing smirks from the civil engineers who had been constructing canals for many years, solving difficult problems along the way using inclined planes, locks, and aqueducts, some of the same ideas this novice was proposing. It is not surprising, though, that many others would be impressed by Fulton's reasoning, as he was persuasive and abounded with ideas and apparent solutions, even though most of his approaches were not founded on field-verified technical expertise.

In the preface Fulton enunciates his definition of invention, which would later be repeated when he was attacked for not being truly original: "The component parts of all new machines may be said to be old; but it is that nice discriminating judgement, which discovers that a particular arrangement will produce a new and desired effect, that stamps the merit. And this may perhaps, with propriety, be called either invention or improvement." Not missing an opportunity to toot his horn, he adds, "which certainly exhibits that the artist has that penetration which is usually dignified the term Genius." He succinctly, insightfully, and correctly sums up the essence of the process of invention by saying, "Therefore the mechanic should sit down among the levers, screws, wedges, wheels,

&c. like a poet among the letters of the alphabet, considering them as the exhibition of his thoughts; in which a new arrangement transmits a new idea to the world."

He concludes his preface with a quote from Morehead, perhaps intended to taunt his would-be detractors:

> Men of true genius glow with lib'ral spirit,
> And bind a garland round the bust of merit;
> While blockheads, void of wisdom's grateful light,
> Bury distinction in eternal night.

Following a rambling introduction, Fulton includes chapters on the origin of canals and their benefits to society (quoting from Adam Smith), and he launches into a description of how to form canals, traverse hills with inclined planes, and cross valleys with aqueducts. His carefully couched arguments, profusely supplemented with well-executed and authoritative-looking illustrations (see fig. 11.8), created an impressive-looking treatise. So convincing was his presentation that Lord Stanhope wrote, "Your Book abt Canals has set me you see on fire; particularly the Part about America and your note about the enormous expense of Horses. So I hope at that at last I shall burn to some purpose; provided you keep on blowing on the fire, as you have done."[23] The *Monthly Review*, a popular periodical, devoted five pages to the treatise in early 1797, writing favorably about his theory: "whether or not it may conduce to his own emolument, there can be no doubt of its beneficial tendency with respect to the public."[24]

To provide canals throughout a wide geographic region, Fulton reasoned that it was necessary to reduce the cost per mile of canal construction. Since this cost is related to the size of canals, he intuitively understood that small canals would be cheaper. Therefore, using small canals with small barges (i.e., with a four-ton capacity), more miles of canals could be built, thereby crisscrossing a greater geographic area with inexpensive transportation. This would stimulate production and trade, making the country and its inhabitants richer. The idea was sound as far as it went.

While espousing a global view of society, Fulton failed to under-

Figure 11.8. One of the very impressive illustrations from Fulton's canal treatise. Here water powers an overshot wheel, which through a system of pulleys pulls a barge up an inclined plane. The operator holding the horse is monitoring the barge's ascent. As it would have been unlikely to have a sufficient supply of water on the top of the hill, the approach was impractical—but sure to impress the nontechnical reviewer. (From Fulton, *Treatise on the Improvement of Canal Navigation*, pl. 4.)

stand how a business society interacted and the paramount importance of sound business decisions with regard to investment in transportation infrastructure. Small canals were simply not cost-effective from a profit standpoint, and herein lay a fundamental flaw in his logic. One year later, in 1797, William Chapman was to take him on with his *Observations on the Various Systems of Canal Navigation*. In a reserved, gentlemanlike manner, he pointed out the flaws in Fulton's thinking. Regarding the use of inclined planes as opposed to locks, he notes, "Those who would adopt any favorite system on practical subjects, without the aid of experience to guide them, are liable to be carried away by the warmth of their imagination; and are led to apprehend they have attained a something of universal application."[25]

To understand how a canal is constructed on level ground, it is instructive to refer to the cross-section shown in figure 11.9. For any arbitrary but given depth, here shown as one unit, it is necessary to excavate the sides on a gradual slope. At the time in England, with the soils encountered, for every one unit of depth, the horizontal was extended 1.5 units.[26] Hence, the triangular areas shown shaded must be excavated regardless of the width of the channel, shown in this case as equal to the depth, or one unit.

In addition, the small humps on either side of the canal, called berms, are used to keep surface water from running into the canal and eroding the canal banks. The towpath shown on the left also has a berm on the outside, and the berms as well as the towpath must be constructed regardless of the canal's channel width.

The canal's transportation capacity, usually specified as tons of goods moved, is directly related to the size of the channel. For a channel one unit deep and one unit wide, the capacity is normalized to a factor of one unit, where the cost of excavating the channel would be one unit squared. Likewise, the cost to excavate the sloping sides would be 1.5 units squared, and the cost of the berms and towpath is estimated to be about 0.5 units squared, for a total fixed or

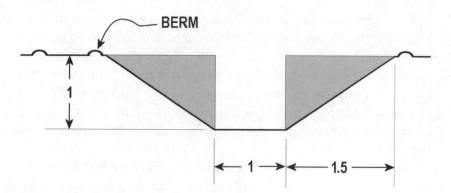

Figure 11.9. Cross-section of a small canal with towpath and berms. The shaded areas represent soil that must be excavated regardless of the canal's width. (Diagram by Jack L. Shagena.)

overhead cost of 2.0 units squared. Therefore the fixed cost for a small canal compared to the cost to provide canal capacity is 2 to 1.

Contrast this with the canal of five units in width and the same depth. The tonnage capacity is now 5, but with the same overhead cost the ratio of fixed cost to capacity cost now becomes 2 to 5, a much more favorable number. Correspondingly, for a width of ten, the ratio becomes even better, at 2 to 10. It should be noted these are approximate and do not include the additional cost of land acquisition and bridging, which is somewhat mitigated by amortizing the project management over a larger base. It should also be noted that the maintenance cost for both small and large canals is primarily associated with the banks and towpath, hence tends to favor large canals.

Two additional, more subtle factors also tend to work against small canals. If a canal were to be constructed only one unit wide and one unit in depth, ignoring vessel-passing considerations, a barge would have to be as deep as it is wide to efficiently utilize the channel's capacity. Such designs are, however, inherently unstable and would easily capsize if not properly loaded with heavy cargo at the bottom.[27] More reasonable designs require a width-draught ratio approximating 3 to 1.

A second factor is related to the day-to-day operational cost and speed through which barges would pass. Typically, any barge would require at least one individual on board to operate the tiller, and a second to drive the horses or mules. Allowing for the two operators and any given number of burden animals, larger canals with larger barges allow for spreading the labor and animal cost over more cargo, hence reducing cost per ton-mile. Working against this, however, is the speed at which the horses or mules can pull the barges. For a given number of animals, larger and heavier barges will move at a slower speed, hence covering less distance in a day. Whereas labor costs are likely to remain constant, animal costs increase for larger barges.

Regardless of operational costs, which likely are borne by independent operators, an investor builds a canal for a financial return, and the payback is directly related to the goods conveyed. Higher-capacity canals, not small canals, provide for maximization of profits. This was a concept that went unnoticed by Fulton and provided a fatal flaw in his logic.

Another suggestion proffered by Fulton was to provide wheels on each barge for the purpose of traversing inclined planes (see fig. 11.10). This was eminently impractical, because such wheels would spend most of their life under water being subjected to corrosive forces as well as collecting mud and debris. This would have had the result that the barges would have become stuck on the inclined planes, disrupting canal traffic. It was far better to place well-maintained wheels under each barge before the operator traversed the ridge with the vessel.

Despite receiving initial favorable comments on his treatise, Fulton was unable to secure a job building a canal. Most likely, even to the most naive financier of a canal project, his ideas were deemed impractical; certainly they would have appeared so to any competent

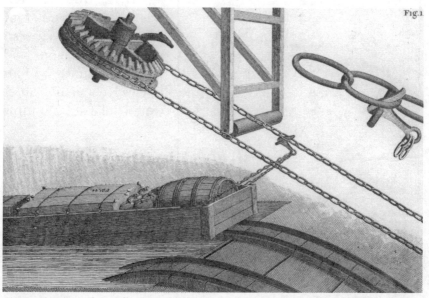

Figure 11.10. In this well-executed illustration from his canal treatise, Fulton shows a barge being moved on an inclined plane. Although the details of the operation are not clear, it can be inferred that the rollers on the bottom of the barge ride on the rails of the inclined plane. As the rollers spend most of their time in the water, they are subject to fouling; hence, this approach portends many problems. (From Fulton, *Treatise on the Improvement of Canal Navigation*, pl. 5.)

practiced civil engineer of the day. Therefore, without practical experience, investors would be wary of betting their money on an unknown who had only performed on paper, and work was not forthcoming as he had hoped, placing him in a difficult financial position.

Fulton had received a patent on May 8, 1794, for an inclined plane, titled, *A Machine or Engine for Conveying Boats and Vessels and Their Cargoes to and from the Different Levels in and upon Canals, without the Assistance of Locks or Other Means Now Known and Used for That Purpose.* Commenting on the invention, Dickinson notes, "It is difficult for a trained mind to see in this specification anything more than a crude idea, ill digested; better methods, worked out in a more practical manner, were already in use."[28]

During the period Fulton was working on his treatise, he invented an earth excavation device he believed was suitable for

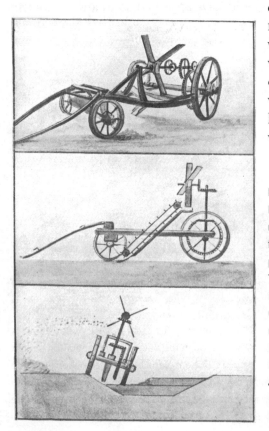

canal construction (see fig. 11.11). The excavator was mounted on four wheels, similar to an open-frame wagon, and was to be pulled by four horses. A small cutter would take up a furrow

Figure 11.11. Three views of Fulton's earth excavating machine, pulled by four horses, from an original sketch in the possession of Lord Stanhope. The design is only conjectural, as important elements such as regulating the depth of the cut are missing, and the design appears impractical. (From H. W. Dickinson, *Robert Fulton, Engineer and Artist: His Life and Works* [London: John Lane, 1913], between pp. 36 and 37.)

about four inches wide, at which point the dug earth would be pulled upward on a conveyor belt with paddles, where it would be propelled to the side by a spinning blade. The four-propeller blade, similar to those on a ceiling fan but much sturdier and turned sideways, was powered by gearing from the real axle and designed to throw the earth away from the excavated canal ditch. As Fulton was not sure he could raise the earth high enough to clear a partially dug-out channel, he also proposed an alternate arrangement that would raise the earth to a higher elevation through the use of a small belt or chain elevator.

Fulton obtained money for the development of his machine from Robert Owen and for some time thereafter kept him informed on improvements and progress. It turned out, however, to be unworkable, as a cursory examination of his drawings by a trained eye would indicate. Dickinson observes that the "digging machine was a crude and impractical apparatus,"[29] and Fulton, finding this out, made no mention of it in his treatise.

One of the devices Fulton invented during the period he was working on the treatise did work out. He designed a "mill for sawing marble or other stone."[30] In 1794, he sent a model of the mill to the Society for the Encouragement of the Arts in London, who awarded him their silver medal for his ingenuity.

As Fulton did not find a ready market for his small canals in England, he sent a copy to George Washington, hoping to interest the president. When an answer finally came in February 1797, it was a polite reply thanking him for the book and saying that the press of business had kept him from examining it. He immediately wrote again to the president, touting the system's benefits, but to no avail.

In 1796, the financially struggling Fulton was fortunate to meet Rev. Edmund Cartwright, a poet and inventor who befriended so-called projectors and individuals with interesting ideas. He stayed with Cartwright for a while, and it is believed that he learned from him about a method of making rope that he later improved upon and patented in France.

He worked hard to interest others in his system of small canals, making a proposal to his friend Benjamin West and his patron

William Beckford, but no financial help was forthcoming because of difficult personal economic circumstances. In mid-April 1797, Fulton's fortunes suddenly improved, though, as he was able to sell a one-fourth interest in his American canal prospects to speculator John Barker Church, brother-in-law of Alexander Hamilton.[31] Church agreed to pay Fulton a total of fifteen hundred pounds, of which he was paid five hundred pounds on May 17, 1797. He was to receive another five hundred in six months, with the final payment to be made in the United States upon the arrival of Fulton in June 1798. It would seem that he had finally succeeded, and a joyous Fulton went off to France to take out a patent and find other opportunities there before going back home. He wrote Robert Owen, saying, "*Thank heavens* (some men would say *please the pigs*) I have succeeded,"[32] and sent him some of the money he owed for the digging machine, promising the rest at a later date.

Around June 1797, Fulton applied for a passport, then sailed across the English Channel to Calais with the promise from the French authorities that his papers would be waiting upon his arrival. It was, however, about three weeks later that his papers arrived, whereupon he headed to Paris, a city he reported to be alive and gay, seemingly unaffected by the Napoleonic Wars. He probably carried a letter of introduction from his friend Benjamin West to an American, Joel Barlow (see fig. 11.12), who was then living in Paris. In any case, he first met Barlow's wife, Ruth, then Joel, and they formed an endearing personal relationship that was to continue for the rest of their lives.

The childless Barlows were older than Fulton, Joel by eleven years and Ruth by nine, and they were to live together for seven years, providing the younger man and surrogate son his first stable home environment since his early Pennsylvania childhood. The warm, encouraging, and mentoring relationship they enjoyed is described by a quotation that is part of a eulogy written by Cadwallader D. Colden shortly after Fulton's death. The author of this sincere and personal message is most certainly Ruth herself as she was then living in New York, a short distance from Colden:

> In the year seventeen hundred and ninety-seven, [Fulton] took his lodgings in Paris, at an hotel in which Mr. Joel Barlow, our cele-

Figure 11.12. Joel Barlow, from a painting by Robert Fulton and engraving by Durand. Joel, his wife Ruth, and Fulton met in Paris soon after the inventor arrived there in 1797 and remained good friends for the rest of their lives. (From Charles Burr Todd, *Life and Letters of Joel Barlow* [New York: G. P. Putnam's Sons, 1886], title-page frontispiece.)

brated countryman, and his lady had their residence. "Here," to use the warm language of one who participated in the sentiments expressed, "commenced that strong affection, that devoted attachment, that real friendship, which subsisted in a most extraordinary degree between Mr. Barlow and Mr. Fulton during their lives. Soon after Mr. Fulton's arrival in Paris, Mr. Barlow removed to his own hotel [home], and invited Mr. Fulton to reside with him. Mr. Fulton lived seven years in Mr. Barlow's family, during which time he learnt the French, and something of the Italian and German languages. He also studied the high mathematics, physics, chymistry, and perspective, and acquired that science, which when untied with his uncommon natural genius, gave him a great superiority over many of those who with some talents, but without any sort of science, have pretended to be his rivals. Mr. Fulton, during his residence with Mr. Barlow, projected the first panorama that was exhibited in Paris. This was a novelty which attracted many spectators, and afforded a handsome emolument."[33]

One of Fulton's first efforts after arriving in Paris was to prepare the drawings and specifications for a French patent on his inclined plane. This was accomplished in the fall and winter of 1787, and the following February 14 he was granted a fifteen-year patent (number 289) containing sixteen pages of description and no less than fifty-six figures engraved on four plates with his name.[34]

During the time Fulton was working on the canal patent application, his imagination was stimulated by discussions with Barlow about the construction of an underwater ship that could covertly attack and destroy an enemy's vessel. It was the submarine, and this instrument of war would completely captivate his imagination and become his favorite project for the rest of his life.

The submarine was not a new device, as sixteenth-century Englishman William Bourne had built a type of submergible craft; his efforts were followed by a Dutchman, Cornelius van Drebble, who made a number of improvements. There were others who made experiments as well, but it was an American, David Bushnell, who while attending Yale College as a senior in 1775 developed the world's first practical submarine. His six-foot, single-man craft, called the *American Turtle* because it resembled two turtle shells attached together, was propelled by hand-cranked screw propellers providing both lateral and vertical movement (see fig. 11.13). The submarine was submerged by allowing water to enter an onboard tank and was raised by expelling the water with foot pumps. A hand-operated rudder, located behind the lateral propeller, controlled the sub's direction.

Bushnell's concept was to carry an explosive charge under an enemy's anchored vessel. A spike or screw with an eye would be secured to the hull, and the line towing the mine, filled with gunpowder, threaded through it. As the submarine moved away, a timing device was activated, and the charge would be pulled directly against the ship's hull. The operator would obviously distance him-

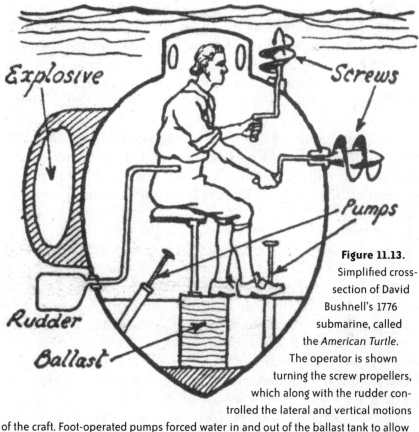

Explosive

Screws

Pumps

Rudder

Ballast

Figure 11.13.
Simplified cross-
section of David
Bushnell's 1776
submarine, called
the *American Turtle*.
The operator is shown
turning the screw propellers,
which along with the rudder controlled the lateral and vertical motions
of the craft. Foot-operated pumps forced water in and out of the ballast tank to allow
the submarine to submerge and rise. (From William McDowell, *The Shape of Ships*
[New York: Roy Publishers, n.d., ca. 1948], p. 155.)

self from the vessel as quickly as possible. There were three attempts
to blow up a British ship in such a way during the Revolutionary
War, but none was successful. But since the British knew about the
submarine, which was an "open secret," it did succeed as a weapon
of terror that produced some confusion.

Joel Barlow was a freshman at Yale the same year that Bushnell
was a senior, and with a total student body of less than 150, it is
almost certain that Joel knew of the submarine work. Barlow graduated in 1778, the same year "Bushnell submitted the detailed

descriptions of his experiments, his apparatus, and his attempts against the British fleet as the thesis for his Master's Degree."[35] Therefore it is probable that Barlow had a copy of the thesis, or became familiar with it, and imparted that knowledge to Fulton. There is a high degree of commonality between Fulton's *Nautilus* and Bushnell's *American Turtle*, but Fulton never credited him as providing either the idea or any design details. But then, of course, Fulton "never admitted that anyone influenced him" in any of his work.[36] Although he was an intelligent and extremely resourceful entrepreneur, Fulton was not a highly original thinker, and all through his life he would borrow from others' ideas in which he saw merit.

On December 13, 1797, Fulton sent the French government a well-written and startling document carefully detailing the terms under which he proposed to use his *Nautilus* submarine to annihilate the British fleet, thereby, he claimed, achieving liberty of the seas. Since he had established an investment company to pay for the development of the submarine, no immediate funding was requested; he only asked to be rewarded an amount of money for each gun sunk on a destroyed British vessel. Other terms in his proposal dealt with contingencies such as remuneration, should peace be declared; French actions to save his life, should he be captured; and an ineffectual and useless clause that France should not use the *Nautilus* against the United States. While promoting a stealth means of destroying the navies of the world, he failed to recognize it could have crippled the fledgling and almost nonexistent U.S. Navy and could have likewise destroyed the merchant fleet. Though ruthless in its pursuit of fortune and fame, Philip noted quite correctly, "Fulton's proposal was a masterpiece of entrepreneurship."[37]

Blinded by what he perceived as the technical brilliance of his submarine, he completely failed to think through an effective weapons delivery approach, as well as also the inevitable countermeasures that would be employed to defeat his attacks. In fact, his system would never achieve its objective. But the concept nevertheless intrigued the French, whose navy was inferior to that of the British, and with some modifications, including cutting the reward per gun in half, the French authorities sent a counterproposal to Fulton, which he accepted. The agreement, however, had to be approved by the defense directors, who turned it down. Fulton would have to await a change in leadership before broaching his ideas again, but he did not have to wait too long.

In an effort to interest the French government in his system of small canals, he wrote to Citizen General Napoleon on May 1, 1798, promoting canals as an important internal improvement that would stimulate the economy of France. The canals, he argued, would provide transportation to allow a larger number of individuals to produce and sell goods, and pointing out what he had learned from Adam Smith, he added, "Labor is the source of all wealth of all kinds."[38] Napoleon was busy winning a war and was not impressed, but already Fulton was waiting for the appropriate time to resubmit his submarine concept.

There is no evidence that Fulton ever received the second installment of five hundred pounds from John Barker Church for the U.S. rights to his small canals that was due on November 17, 1797. It is most probable that the speculator had taken Fulton's treatise back to America, where canal engineers had found nothing in his approach to warrant the amount of money promised. Also, by this time, Chapman's rebuttal had been available for almost one year, which certainly would have not helped matters as far as Fulton was concerned. To make things even worse, the first payment had been in the form of three hundred pounds cash and a two-hundred-pound note. When Fulton attempted to draw on the promise to pay, it was not honored. This produced for Fulton a desperate need to find other sources of income, and his brilliant mind was turned to moneymaking ideas.

At that time the French navy and merchant fleet provided a steady demand for rope, which was laboriously hand-woven in long, narrow buildings. Thinking back to Cartwright's weaving devices in England, Fulton understood that a mechanical rope-making device, a so-called cordelier, would be of great benefit to France. He interested an American, Nathaniel Cutting, in financing its development and patent costs, with Cutting becoming the owner of the patent.

An agreement was signed, and after some haggling with the French authorities, a patent was issued, and Fulton set about constructing the device. His attention, however, and some of the money supplied by Cutting were diverted to another project, the panorama. Because of poor progress on the cordelier, bitterness grew between the two individuals, and when it became apparent a successful machine was not within reach, an embittered Cutting withdrew, claiming fraud. Fulton would later regret this bad blood between the two men, because Cutting would join the opposition forces attacking Fulton's steamboat monopoly on the Hudson River.

Fulton's panorama, however, was a tremendous success in novelty-loving Paris. Based on a similar display patented by Richard Barker in England that he had seen while in London, Fulton applied for and was awarded a ten-year French patent on April 26, 1799. In his application, he stated in a straightforward manner, "It is already established in London and is admired as a work of genius and utility"[39]—one of the rare times he does not claim an idea for himself, but credits someone else with its development.

The panorama was in a circular building, forty-six feet in diameter, located on the south side of the Boulevard Montmartre. Around the inside circumference hung a number of canvases depicting a 360-degree view of Paris as seen from the Tuileries Palace.[40] It was illuminated by a central ceiling window and was viewed from an elevated platform accessed by a spiral stairwell. Success was immediate, and in December 1799, Fulton sold his patent to James W. Thayer and his wife but retained a percentage of the receipts, which helped finance his experiments.

Eustice Bruix, a younger man who was an excellent strategist, replaced the marine minister Pleville-le-Pelley, who had failed to convince the defense directors of the merits of Fulton's initial submarine proposal. As a consequence of a coup on May 11, 1798, some men came into power that Fulton believed were more favorably disposed to his project. So he commenced what was later described as a "beautiful model of the *Nautilus* five feet long complete in all its parts" (see fig. 11.14).[41] On June 23, 1798, he resubmitted his proposal to Bruix, who appointed a commission of

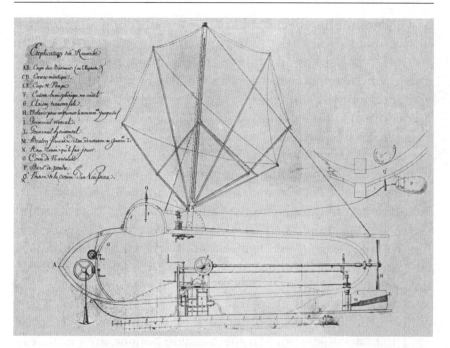

Figure 11.14. Fulton's submarine, the *Nautilus*, with the sail structure in the vertical position. It folds down when the vessel submerges. The propeller in the stern, driven by a hand crank, accomplishes horizontal movement. A second propeller provides for vertical motion. (From Dickinson, *Robert Fulton*, between pp. 82 and 83.)

experts in a number of areas to examine the model and evaluate Fulton's plans for a full-size submarine.

The minister's handpicked group met in early August at Fulton's residence in Barlow's home and carefully went over the model, suggesting several improvements. "Most of the faults that the commission found with the conception," notes biographer John S. Morgan, "arise where Fulton had departed from Bushnell's ideas."[42] Three of the recommended changes brought the design much closer to the *American Turtle*: the use of a second propeller, instead of horizontal rudders, for vertical movement; the improvement of the method of renewing the air; and the addition of a barometer to determine depth. Bushnell had previously employed all of these concepts. The commission also expressed concerns about countermeasure netting and questioned the effectiveness of the weapon, but it was nevertheless fascinated with the weapon's potential.

These suggestions and improvements no doubt gave the committee a modicum of ownership, and though recognizing the design was "undoubtedly imperfect," they recommended that Fulton be funded for his proposal to construct a full-size submarine, as the weapon was the "first conception of a man of genius."[43]

Despite this encouraging report from the committee and letters from Fulton to the ministry encouraging French action, several months lapsed without an approval to proceed. In frustration, Fulton sent word through Joshua Gilpin to Lord Stanhope about his secret weapon, hoping to elicit some British interest. None was forthcoming at that time. He also tried to interest Bavaria and Holland, in both cases finding investors, but there was no interest in their governments.

By early November, there was another change in the French government, and P. A. L. Forfait, a member of the commission that had written a favorable report on the *Nautilus*, became the minister of the navy. Forfait must have provided Fulton encouragement, if not financial support, because he then set about constructing the full-size submarine. It was launched on July 24, 1800, and five days later, a successful public demonstration was held on the Seine (see map in fig. 11.15), where Fulton and his two-man crew stayed under water first for a period of eight minutes, then for seventeen minutes.

After additional testing in Le Havre in August 1800 and much haggling with the French over the terms of his agreement for remuneration, he received permission to proceed against the British. He planned to attack two brigs that were blockading the French coast near the harbor of Growan. The following is Fulton's account of his attempts:

> On the 28th (15 Sept.) I put into a little harbour called Growan near Isigny at 3 leagues from the islands of Marcou. On the 29th the equinoctial gales commenced and lasted for 25 days. During the time I tried twice to approach two English brigs which were anchored near one of the islands, but both times, whether by accident or design, they set sail and were quickly at a distance.

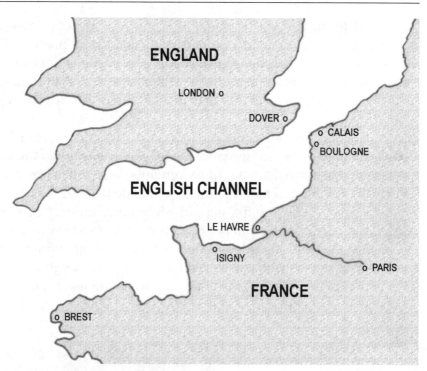

Figure 11.15. Map showing the locations where Fulton operated his submarine and engaged in naval operations. (Rendered by Jack L. Shagena.)

The weather being bad, I remained 35 days at Growan and seeing that no English vessel returned, and that winter approached, besides my Nautilus not being constructed to resist bad weather, I resolved to return to Paris and place under the eyes of Government the results of my experiments.[44]

Despite the fact that he did not destroy a single British ship, "not even a canoe," Fulton did not hesitate to claim success, extolling the virtues of undersea warfare. In the same letter, describing his attempts, he claimed to have conceived additional ideas that would require development, anticipating that the French government would provide the required finding. Not receiving a timely reply, a frustrated Fulton wrote in December, "I retain the most ardent desire to see the English Government beaten," but having worked on the submarine for three years, he would not continue unless the French government treated him "in a more friendly and liberal manner."[45] He went on to threaten the French that unless relations improved, he would seek encouragement in the United States, and in a most disingenuous manner (as he had already been rejected there) included going to Holland.

"Naive and contemptible as this tactic was, it worked,"[46] observes Philip, but most likely because of political considerations, as the approach would have most certainly been offensive and resisted by the bureaucracy. Fulton was provided 10,000 francs to build a larger *Nautilus*, and a new schedule for remuneration was worked out based on the number of guns on a destroyed vessel.

By July 13, 1801, his enlarged boat, which could accommodate four men, was tested with success. Fulton wrote a very lengthy letter to members of the National Institute and commissioners appointed by the first consul, describing in detail the experiments, which are at one point summarized:

> Having thus succeeded
>> To sail like a common boat.
>> To obtain air and light.
>> To plunge and Rise perpendicular.
>> To turn to the right and left at pleasure.
>> To steer by the compass under water.
>> To renew the Common Volume of air with facility.
>
> And to augment the respirable air by a reservoir which may be obtained at all times.
>
> I conceive every experiment of importance to be proved in the most satisfactory manner. Hence I quit the experiments on the Boat to try those of the Bomb Submarine.[47]

He then proposed to construct one or two even larger submarines, capable of carrying an eight-man crew, and to use them covertly to mine the harbors where English ships passed. However, at this point, Fulton did something that at first glance appears totally irrational but was no doubt well thought out. He dismantled his boat, writing, "I am sorry that I had not earyler [*sic*] information of the Consuls desire to See the Plunging boat, when I finished my experiments. She leaked Very much and being but an imperfect engine I did not think her further useful."[48]

The reason, however, becomes clear upon some analysis. To move a submarine under water required the exertion of a great amount of effort by men sometimes in an oxygen-starved environment. It did not move well and, as a result, was not an effective or useful weapons delivery system.

One or more men rowing a surface boat can achieve a speed of two to three miles per hour, and this human generation of power is more efficient than turning a crank, which was the means of propulsion for Fulton's submarine.[49] The resistance encountered to the motion of a boat is proportional to the design (sleek, long, and narrow hulls with shallow draught being preferred) and the amount of wetted surface. For Fulton's submarine, the diameter was roughly dictated by the size of a seated man, and the length by the number of men. Employing several men, these constraints resulted in a shape somewhat like an elongated egg, not like a cigar, making the wetted surface large, hence offering significant resistance. Furthermore, when several men operate a common crank, as opposed to independently rowing on the surface, the power generated does not add algebraically. The draft (pulling power) of horses working in teams is a fraction of that of the same horses working independently, and the situation is analogous. For two horses, the efficiency is 98 percent; for four, 80 percent; and for eight, 49 percent.[50] Hence, using the cranking power of several men, it was impossible to effectively move the boat against any kind of current.

Fulton's first sub carried three men; his second, four; and now he was proposing a crew of eight, but for mining a harbor, not to deliver a mine to a ship. He had concluded that it was not possible to maneuver a sub with only the use of human labor, and some form of mechanical power would be required. It is worthwhile noting that nowhere in his list of accomplishments did he state the *Nautilus* was capable of satisfactory motion in a horizontal direction. He claimed successful testing, and his report was probably accurate as far as it went, but he cleverly omitted a crucial parameter.

Had the submarine remained intact, it is likely that the French would have asked for performance data on speed, and he knew this.

Bushnell had found this out twenty-five years earlier and abandoned sub-based delivery of weapons, and Fulton was to follow a quarter-century later, failing to learn what had previously worked and what had not worked, and more importantly not understanding why. Despite Fulton's hyped achievements, he, to the contrary, was not a very good engineer. As other engineers before him and engineers since, he had to learn by failure; but in his case it could have been easily avoided by studying Bushnell's work. His ego would not let him learn from others. He perceived himself a bit smarter, smoother, and more convincing, which he was, but the would-be engineer was blindsided by his own genius. He depended too much on achieving success through his social skills and did not work hard enough on achieving an understanding of fundamental engineering principles. Eventually, any entrepreneur proposing engineering projects must perform, and when it came to performance, every technically complex undertaking Fulton proposed was a failure. But he was undauntedly immersed in his own self-esteem.

Napoleon, upon learning that the enlarged *Nautilus* had been dismantled, spoke of Fulton "as a charlatan and a swindler, intent only on extorting money" (but see the engraving in fig. 11.16 for a supposed meeting between the two).[51] When word of this reached the inventor, the handwriting was on the wall, and Fulton knew his work for the French had come to an end. The steamboat, however, was still in his future, and at the age of thirty-six perhaps his last chance for fame and fortune.

Morgan notes, "No one ever embraced a project that would bring him fame and apparent fortune more reluctantly than did Fulton take up the steamboat."[52] There were at least two reasons for this. The device was old, not new; hence, becoming recognized as the inventor, regardless of his contribution, did not appear likely. As it had been tried before with significant expenditures of money, but without success, it was also doubtful that the eventual payoff would follow. Combining both of these factors, it is fairly obvious the steamboat did not fit Fulton's vision as the vehicle to ride to fortune and fame.

Figure 11.16. According to tradition, Robert Fulton showed Napoleon a drawing of his submarine. This engraving was done in Philadelphia after the death of Napoleon in 1821. (From Sutcliffe, *Robert Fulton and the* Clermont, p. 162.)

Whereas Fulton had sought to sell his concept of small canals, espousing the economic benefits, he did not, despite his visionary thinking, recognize the full benefits of the steamboat. In reality, steam navigation would come far closer to uniting the world into one trading market than his small-canal system and would eventually prove to be far more important for civilization than manmade waterways. Not seeing this, however, he was fortunate that a man with money, power, and a monopoly was to come into his life and reignite his passion to be successful. It was Chancellor Robert R. Livingston, who, having been appointed minister to France by U.S. president Thomas Jefferson, arrived in Paris late in 1801. Fulton was to meet him in the early part of the following year.

The Chancellor (see fig. 11.17), as he was often called to distinguish him from his father, was influential in New York, having served three terms in the State Provincial Congress, and was a two-term del-

Figure 11.17. Known as the Chancellor, Robert R. Livingston was appointed as the American representative to France by Thomas Jefferson in 1801. In Paris he met Fulton and they signed an agreement to develop a steamboat. (From Sutcliffe, *Robert Fulton and the* Clermont, p. 183.)

egate to the Continental Congress. He also had been a member of the committee that drafted the Declaration of Independence, but pressing state matters prevented him from being present for the signing. He was chancellor of New York from 1777 to 1801, and in this capacity, as the state's highest judicial officer, administered the oath of office to President George Washington on April 30, 1789.

In addition to serving in various political capacities, the chancellor was also interested in science, reading every available book and noting "mechanicks is my hobby horse."[53] He also kept current on the progress of steamboat developments and, as mentioned in chapter 10, personally witnessed a demonstration of Samuel Morey's steamboat in 1796. His wife was the sister of Col. John Stevens of Hoboken, New Jersey, who had followed and improved upon the steamboat efforts of John Fitch and who along with Fitch and Rumsey had received a patent for a steamboat in August 1791 (see next chapter).

The interests of Livingston, Stevens, and Nicholas J. Roosevelt, who owned America's first machine shop, coalesced, and a partnership was formed in April 1798 to build a steamboat. Livingston had obtained from the New York legislature one month earlier a monopoly, previously granted to Fitch, to operate on the waters of

New York State, provided he could demonstrate within one year a boat capable of four miles per hour. The Chancellor, however, was unwilling to leave the design to Roosevelt and his proven mechanics, insisting instead on incorporating an ill-fated propulsion idea of a horizontal wheel hung below the keel. The engine worked well but the wheel did not, and a speed of only three miles per hour was achieved, not fast enough to meet the requirements of the New York State law. Had he followed Roosevelt's idea to use a pair of wooden wheels over the sides, each with eight paddles, it is likely the boat would have been successful, and Fulton would have missed his chance for fame.

Livingston and Fulton had a lot in common. They were both entrepreneurs, each somewhat hardened by failures and cautious. Each had their own egotistical ideas for constructing a successful steamboat, and they shared a common vision for fame and fortune and a tenacity to achieve it. Fulton, however, was far more analytical, understanding more thoroughly the laws of science and what can and cannot be done. Livingston, on the other hand, believed that the idea was the essence of success and, as Stevens and Roosevelt had learned, woefully failed to understand the principle of development, with its attendant consumption of time and money.

At the end of April 1802, Fulton escorted Ruth Barlow to Plombières, a spa noted for it curative powers, where she hoped to obtain relief from tumors growing on the lower part of her body. Joel did not accompany them and through his letters was apparently delighted that the two of them enjoyed the vacation together. It was not all play for Fulton, though, as he undertook an analysis of a boat's hull resistance to moving through water. Colonel Mark Beaufoy had published the original work, and Fulton adapted it to his planned steamboat. Like Rumsey and Fitch before him, he believed that the resistance to a hull through the water varied as the square of the speed and with the hull design.

Before Fulton left Paris, he had ordered a three-foot model boat, eight inches wide, powered by two strong clock springs. Barlow pressed the model maker to complete the small boat and soon had it shipped to the spa. Creating a small basin, Fulton proceeded to make measurements regarding the best means to propel the model through the water. He initially concluded chains with paddle boards to be the most efficient, but upon learning from Barlow in Paris that a Frenchman had displayed a model there proposing the same thing, Fulton switched to paddle wheels. Barlow informed Livingston of the planned means of propulsion, and writing Fulton on July 18, 1802, he stated, "Toot, I had a great talk with Livingston. He says he is perfectly satisfied with your experiments and calculations, but is always suspicious that the engine beating up and down will break the boat to pieces. . . . You have converted him as to the preference of the [paddle] wheel above all other modes, but says they cannot be patented in America because a man (I forget his name) has proposed the same thing there."[54] Combining the factors of resistance versus speed with the effect of the paddle wheel, he was able to calculate the horsepower of the required steam engine and predict the speed of the boat. At various times, Fulton estimated the speed to be eight, twelve, or even sixteen miles per hour, with the last prediction provoking Barlow to suggest he had taken leave of his senses.

Fulton and Ruth remained at the French spa until September, then returned to their Paris home at 50 rue de Vaugirad. By October 10, 1802, Fulton and Livingston signed an agreement to jointly develop a steamboat to run between New York and Albany at the rate of eight miles per hour. First, an experimental boat would be constructed, and upon successful testing, they were to proceed to the United States to build the final model, for which a patent would be taken out in Fulton's name. The patent would be divided into one hundred shares, with each owning half, and they would divide the profits equally. If successful, Livingston would provide all the funding; however, should the experiment fail, Fulton was to repay, within two years, one-half of the initial five hundred pounds advanced by Livingston.

Trial 1. The agreement called for the experiment to take place in England, but it was decided to conduct the trials in Paris on the Seine (see fig. 11.18). Fulton contracted to have the boat and paddle wheels built and arranged to borrow a steam engine of about eight horsepower. The steamboat was apparently ready by the spring of 1803, but the hull, having not been built strong enough, broke in half one night, sinking the boat with the equipment into the river. Temporarily set back but determined, Fulton set about recovering the undamaged equipment and installed it on a new and stronger craft.

Trial 2. By July, the steamboat was nearing completion. On July 24, Fulton wrote his friend Skipwith a congratulatory letter on the birth of his first child, empathizing about the anxiety of seeing the child grow up. In a playful analogy Fulton referred to his own steamboat, saying, "My boy, who is all bones and corners just like his daddy, and whose birth has given me much uneasiness, or rather anxiety,—is just learning to walk and I hope in good time he will be an active runner."[55] He invited Skipwith to the trial that took place on August 9 (see fig. 11.18), which was reported in the *Journal des Debats* of 23 Thermidor (August 11):

> On the 21st Thermidor [August 9, 1803] a trial was made of new invention, of which the complete and brilliant success should have important consequences for the commerce and internal navigation of France. During the past two or three months there has been seen at the end of the quay Chaillot a boat of curious appearance, equipped with two large wheels mounted on an axle like a cart, while behind these wheels was a large stove with a pipe, as if there

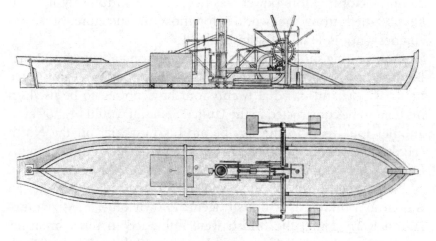

Figure 11.18. Fulton's drawing of his Paris steamboat. It successfully ran on the Seine at about three miles per hour. (From Dickinson, *Robert Fulton*, p. 153.)

were some kind of a small fire engine intended to operate the wheel of the boat.

At six o'clock in the evening, assisted by three persons only, he put the boat in motion with two other boats in tow behind it, and for an hour and a half he afforded the curious spectacle of a boat moved by wheels like a cart, these wheels being provided with paddles or flat plates and being moved by a fire engine.

In following it along the quay the speed against the current of the Seine appeared to us about that of a rapid pedestrian, that is about 2400 toises [i.e., 2.9 miles] per hour; while going down stream it was more rapid: it ascended and descended four times from Les Bons-Hommes as far as the engine of Chaillot; it was manúuvred with facility, turned to the right and left, came to anchor, started again, and passed by the swimming school.

. . . The author of this brilliant invention is M. Fulton, an American and a celebrated mechanician.[56]

This description doth praise the inventor much too much (as have artists done, such as in fig. 11.19). The trial did prove some principles of Fulton's design, but the boat did not achieve the optimistic speed he had predicted. He and Livingston knew that a speed of at least four miles per hour was critical to success in the United States, and his calculations had failed to take into account some fundamental factors.[57] More power was needed, but fortunately it could be obtained from the world's foremost manufacture of steam engines—not in France, but in England.

Anticipating success prior to the trial, Fulton had already written Boulton and Watt on August 6, 1803, requesting an engine three times as powerful, rated at twenty-four horsepower, to be used on his boat to be constructed in the United States. It would be, however, another four years to the day before his next trial on the North (Hudson) River in New York. The British, who feigned interest in his first-loved project, the submarine, would divert his attention.

Sometime in 1803, Fulton had been contacted in Paris by an American working as a British agent, known only as Mr. Smith. Fulton learned that the British were interested in his submarine, whereupon he provided a proposal to be carried back to the British government. A subsequent meeting was arranged in Amsterdam but

Figure 11.19. A romanticized illustration of Fulton's first successful steamboat on the river Seine, with the rear of Notre Dame cathedral shown in the background. (From Bachman, *Great Inventors and Their Inventions*, p. 39.)

may not have taken place, as Smith again came to Paris in March 1804 with an offer alluding to a favorable agreement but vague in details. Fulton nevertheless accepted.

Fulton was skeptical that the British would actually develop and use his submarine, since with the most powerful navy in the world, it would be to their detriment to have such technology in the arsenal of warfare. Even so, he reasoned that he would likely be paid to keep the device secret, and this would suffice as achieving the most important part of his struggle for fortune and fame. Some would argue that offering for sale a weapon of destruction to the enemies of both France and England was treasonous, but Fulton rationalized his penchant for riches by believing the submarine was so terrible it would end wars, completely ignoring the ingenuity of naval officers to develop countermeasures. He also had another reason to cooperate: the Boulton and Watt engine for his U.S. steamboat would require an export license, and this would allow him to be physically closer to the British government. Furthermore, he could more easily establish a working relationship with Boulton and Watt in Soho for the construction of the engine, thereby achieving a more superior engine configuration for his boat.

Leaving Paris on April 29, 1804, and traveling under the assumed name of Mr. Francis, Fulton reached London by a circuitous route on May 19. He drafted and presented a proposal to the British for his submarine and also for torpedoes launched by a surface vessel. Fulton became frustrated with what he perceived as a bureaucratic delay in the evaluation of his plans, but he was finally invited to breakfast with Prime Minister William Pitt. The evaluation committee had concluded the submarine to be impractical (which indeed it was at the time); however, Pitt offered Fulton a salary of two thousand pounds per month plus expenses to develop a surface delivery system for mines or so-called torpedoes. Should the system be deemed to be effective and the government decide to suppress it, he would be paid forty thousand pounds. Furthermore, he would receive a share of the value of destroyed ships—a rather generous offer.

Fulton was disappointed that his first-loved project, the submarine, was not being developed but recognized that the salary and the potential payoffs were significant and agreed to the terms. He set about constructing the boats and bombs to attack the French flotilla in the shallow waters of the coastline not accessible to the English fleet.

Called the "Catamaran Expedition," the first attempt, made on the night of October 2, 1804, was unsuccessful, but Fulton convinced Pitt that the commanding officer failed to follow his instructions. Another attempt, on December 8, led by a Fulton sympathizer, also failed. The government had to take some time and study the problem.

While waiting for the government to authorize additional trials, Fulton had time to acquire some art from his old friend Benjamin West and also met a rich English widow, Clarissa, with whom he established a relationship. For much of 1805, Fulton was preoccupied with artistic and social pursuits but, recalling his mission for financial success, wrote Pitt on August 9, 1805. The letter was designed to prompt the British to proceed with, or withdraw from, the torpedo plan. He did not really care which way they went, for in either case he anticipated to be rewarded financially:

If the invention is insignificant, I do not expect anything for it. If it is an invention which is capable of working a total revolution in marine war *and which I believe*, I of course must have a high idea of its value to myself and country. But of this his Majesty's Ministers will judge.

These considerations lead me to the following conclusions: Will Ministers form it into a system, as before mentioned, so as to give it full effect? If not will they agree with me to let it lie dormant? If not, I am willing to retire. I have so equally balances each of those cases in my own mind that either of them will be equally agreeable to me.[58]

Napoleon had increased the size of his army and the number of barges, prompting British concerns that a French invasion was mounting. So Pitt decided once again to launch an attack using Fulton's system. On the night of September 20, 1805, the British quietly rowed into the Boulogne harbor, attached torpedoes to a brig, activated the timers, and retreated. The devices exploded but did no damage to the ship.

In addition to the problems with the weapon delivery system, now questions were raised about the effectiveness of Fulton's torpedoes in destroying a ship. The inventor had to regain credibility, so Fulton convinced the government to let him try to sink a captured Danish brig anchored at Downes. After several trial runs timed to the coincide with the appropriate tide, on October 16 the brig was completely destroyed—success at last.

The skeptical British admiralty was much surprised at the devastation, but it was against a hapless, undefended ship, and then only successful after practiced dry runs—hardly a real-world scenario. Buoyed by the success, though, they ordered another attempt on the French on October 27. "The torpedo 'exploded and made a similar crash as the Brig lately blown up in the Downes,' without, however, the same destructive effect."[59]

A short time thereafter, the news of Nelson's October 21 victory off Trafalgar against the French and Spanish reached England. This great British triumph mitigated the French threat, torpedoing Fulton's hopes for further development of his engine of warfare. On

June 16, 1806, he again wrote to Pitt, this time an arrogant, greedy, and threatening note: "I did not come here so much with a view to do you any material good as to show that I have the power, and might in the exercise of my plan to acquire fortune, do you an infinite injury, which Ministers, if they think proper, may prevent by an arrangement with me."[60]

In other words, "Give me the money, else I will take my naval weapon of destruction elsewhere." For someone who achieved only limited and qualified success, this threat was purely a bluff. Difficult negotiations ensued, where Fulton requested an annuity of twenty thousand pounds a year to buy his silence, but in the end the British arbitrators ruled the invention was impractical and awarded him the one-time sum of fifteen thousand pounds. Although he publicly raged, he seemed satisfied in a private letter to Barlow. By September 1806, the ordeal was over, and Fulton started making plans to return to the United States, sell his torpedoes to the U.S. government, and also—almost incidentally, perhaps—to complete his contract with Livingston to build a steamboat.

During this period in England, he visited and corresponded with Boulton and Watt about the construction of his steam engine, which had been completed to his specifications. The permission to export it had been obtained, so he sailed for home at the end of October 1806.

After a nine-week voyage in the *Windsor Castle*, including a layover in Halifax, Nova Scotia, Fulton arrived in New York City on December 13, 1806. His planned art study of several years had stretched into nineteen, and he returned home not as an artist but an engineer. One of his first activities was to "toot his own horn" by providing and promoting the newspaper *Aurora* to tout his American arrival. The paper wrote, "[Fulton's] return will be an important acquisition to our country in the various branches of public improvement, of which it is so susceptible; and we cannot but hope that his system of *submarine navigation* may be advantageously united with that of our gun boats to form the cheapest and surest defense of our harbours and coasts."[61] He still envisioned the submarine providing him fortune and fame, cleverly tying in harbor defense, which was so important to the continued commercial success of the thriving New York port—hence the newspaper's readership.

After dashing off a letter to Livingston suggesting they launch steamboat service on the Mississippi, not the Hudson, he dashed off for Washington to be blissfully reunited with Joel and Ruth Barlow, who had returned to America several years earlier. In the new and fast-growing capital, he met numerous influential individuals, including William Thornton, the former steamboat partner of John Fitch, now commissioner of the U.S. Patent Office. Through this connection, he no doubt was able to read the steamboat patents of Rumsey, Fitch, Stevens, and nine others that had been issued in his absence. Afterward he reported to Livingston, "Not one of which approach to practical."

With Barlow's help in assembling a group of politically influential people, including Secretary of State James Madison and Secretary of the Navy Robert Smith, Fulton was able to provide a demonstration of his torpedo system, resulting in a funded agreement for a full-scale trial in New York harbor in mid-1807.

Returning to Philadelphia on January 24, Fulton found a letter from Livingston. His partner was miffed by his inattention to the steamboat over the last 3½ years and was further disturbed by Fulton's suggestion that the Mississippi, not the agreed-upon Hudson, become the location for his next activity. Apparently neither of the partners could find a copy of their 1802 Paris agreement, so, probably to get Fulton's attention, Livingston decided to challenge his partner on some of the details of their understanding. As the issues could not be resolved through correspondence, Fulton traveled to Clermont in the early part of March and hammered out a new contract, where each partner funded one-half of the cost of the new steamboat and shared one-half in the profits. As Fulton has already expended 24,384 livres (equivalent to $5,111) on the Paris boat, the estimated $5,000 additional cost to build the Hudson River boat fell entirely to Livingston. There is no mention of the Mississippi; apparently Fulton finally realized the significant value of the New York State monopoly and Livingston's power and prestige in the state. In April, Livingston had the twenty-year monopoly renewed a second time, and one year later, on April 11, 1808, had the act amended to include five more years for each additional steamboat put into operation, up to a maximum period of thirty years.

Fulton proceeded to New York, where he engaged Charles Browne to construct the hull in eight weeks at a cost of $1,666. On March 16, 1807, he wrote Livingston, "The boat is now building," made plans for the other equipment, and retrieved the Boulton and Watt steam engine from customs, where it had languished since November 1806. Although his activities pacified Livingston, he still had not fully grasped the importance of his work, as he wrote Barlow, "I will not admit that [the steamboat] is half so important as the torpedo system of defense and attack, for out of this will grow the liberty of the seas."[62] Though incorrectly assessing the relative importance of his two ongoing activities, he nevertheless finally committed himself to doing what all good engineers must do, that is, *pay attention to the details.* Understanding the complexity of the steamboat system and his integration task, he listened to the concerns expressed by the workmen and in a gentlemanly manner addressed and resolved each issue. He fully realized that a malfunction in any of dozens of places could cause the entire system to fail, and even if steamboat success would not achieve for him the same level of wealth and recognition as his first-loved torpedo, he desperately coveted success to sustain his credibility as a competent engineer.

In late July, he tested his torpedo system against a two-hundred-ton brig, the *Dorothea*, anchored in New York harbor. After two unsuccessful trials, on the third try the unmanned and defenseless ship was destroyed (see fig. 11.20). Fulton claimed complete success, but the newspapers were not so kind and ridiculed the experiment. One noted that "all that's necessary is that the ships must come to anchor in a convenient place; watch must be asleep, or so complacent as not to disturb any boats paddling about them—fair wind and tide—*no moonlight* . . . bang's the word, and the vessel's blown up in a moment."[63] President Jefferson was more succinct, noting the weapon delivery system as impractical. At this point, Fulton knew his steamboat needed to be more successful, else his detractors, calling it "Fulton's Folly," would have a field day at his expense.

Figure 11.20. Robert Fulton blew up the brig *Dorothea*, anchored in New York harbor, proving the destructive power of his torpedo, but the staged event did not represent a real-world scenario and was ridiculed. (From Knox, *The Life of Robert Fulton*, p. 51.)

Trial 3. By Sunday, August 9, 1807, exactly four years to the day of his second and successful demonstration on the Seine, Fulton was ready to conduct a quiet test run on the waters of New York State. The paddle wheels were only partially complete, but three miles per hour was achieved, allowing the elated inventor to report to Livingston, "she will, when in complete order, run up to my full calculations. I beat all the sloops that were endeavoring to stem the tide with the slight breeze which they had; had I hoisted my sails, I consequently should have had all their means added to my own. Whatever may be the fate of steam-boats on the Hudson, everything is completely proved for the Mississippi, and the object is immense."[64] Fulton returned to the East River port and made some adjustments to the machinery. On Sunday, August 16, with a few dignitaries aboard, he moved the boat to a dock on the Hudson River near the state prison. Storing provisions aboard, he was ready for his historic trial to Albany the following day.

It has been reported in various accounts that a large group of about forty guests, mostly family and a few friends, came on board for the voyage. The story has been repeated from Sutcliffe's account, published in 1909, and is charming but improbable. Why would Fulton, who had proceeded so cautiously thus far, take the risk? The steamboat had never been tested over such a long distance of three hundred miles. Furthermore, it was absolutely critical that the running time not exceed seventy-five hours to garner the New York State monopoly, and the less weight, the better.

Forty guests and a crew of ten, along with luggage, food, etc., would have added about five tons to the boat's one-hundred-ton weight, or 5 percent. Though seemingly insignificant, Fulton wanted every margin available; this is why he carried more expensive coal rather than wood, thereby reducing the weight of fuel by one-half.[65]

Trial 4. On Monday August 17, 1807, the steamboat cast off at about one o'clock with a skeleton crew aboard, consisting of "Fulton, his captain, David Hunt, and an English engineer named George Johnson."[66] Twenty-four hours later, the steamboat docked at Livingston's Clermont estate, departing the next morning for Albany, arriving at 5 PM (see fig. 11.21). Miles traveled were 150, with a running time of thirty-two hours—slightly less than five miles per hour. The return trip would be even faster—thirty hours—meaning exactly five miles per hour, and the Hudson River monopoly was at last secured.

This account, though accurate, does not produce the drama reported by Sutcliffe. So that the reader can appreciate some of the myth surrounding Fulton, Trial 4 has been rewritten according to popular legend, which is the way it has been presented to the public.

THE TRIAL ACCORDING TO TRADITION

Trial 4. On Monday, August 17, 1807, the historic voyage was set. A group of forty guests, mostly family and a few friends, boarded the vessel, simply called the *Steam Boat,* for the epic voyage to the state capital. Many had trepidations in taking part in a ridiculous experiment, but at 1 PM the steamboat, described as "an ungainly craft looking precisely like a backwoods' sawmill mounted on a scow and set on fire,"[67] left her berth and headed up the Hudson River. Fulton tells the story:

Figure 11.21. One of the many illustrations that purports to be Fulton's first trial of his *North River Steam Boat* going up the Hudson River to Albany. The steamboat in these old accounts is always identified as the *Clermont*. (From "Steam Navigation," *Harper's Encyclopaedia of the United States History from 458 A.D. to 1912*, vol. 8 [New York: Harper & Brothers, 1912].)

The moment arrived in which the word was to be given for the boat to move. My friends were in groups on the deck. There was anxiety mixed with fear among them. They were silent, and weary. I read in their looks nothing but disaster, and almost repented of my efforts. The signal was given and the boat moved on a short distance and then stopped. To the silence of the preceding moment, now succeeded murmurs of discontent, and agitation, and whispers and shrugs. I could hear distinctly repeated—"I told you it was so; it is a foolish scheme: I wish we were well out of it."

I elevated myself upon a platform and addressed the assembly. I stated that I knew not what was the matter, but if they would be quiet and indulge me for half an hour, I would either go or abandon the voyage for the time. This short respite was conceded without objection.

Fulton's outward demeanor in front of his patron's influential friends belied his gut reaction. There was a knot in his belly; his vocal chords tightened, his speech confidant yet strained. He desperately did not want to fail again, this time before the already skeptical crowd, as surely the steamboat would become known as Fulton's Folly. He had invested every penny available on the venture, as the budget had been much exceeded, and Livingston had refused to provide more money. Fulton had been forced to borrow, even beg money from his friends, who advanced small amounts very reluctantly. His reputation as an engineer was at stake, and failure was not an option. To complicate matters, Livingston, just prior to departing, had announced Fulton's engagement to Harriet Livingston, one of the Chancellor's nieces, a lady nineteen years his junior. If he did not succeed, the marriage would probably never take place, because the family would oppose Harriet's marrying a failure. Socially, professionally, financially, and procreatively, with a famous and influential family, his future (or so it must have seemed) was squarely on the line. His actions over the next few precious minutes would significantly impact the rest of his life.

He was not sure of the source of the problem, but, extremely confident, he knew every inch of the steamboat's construction, and unless there had been a catastrophic failure of a critical component, most likely he could remedy the problem in a few minutes. His chief mechanic was immediately at his side, and they started their inspection. The vessel's hull had been made as light as possible, and this was the first time that forty passengers, crew, and baggage, adding about five tons of weight, had been on board. Most likely this caused a slight warping of the boat's structure, creating some minor interference in the gearing. Fulton continues:

> I went below and examined the machinery, and discovered that the cause was a slight maladjustment of some of the work. In a short time it was obviated. The boat continued to move on. All were still incredulous. None seemed willing to trust the evidence of their own senses. We left the fair city of New York [see fig. 11.22]; we passed through the romantic and ever-varying scenery of the Highlands; we descried the clustering houses of Albany; we reached its

Figure 11.22. Departure of Fulton's *North River Steam Boat* from New York City, with many passengers visible on the deck, on its first voyage up the Hudson River to Albany. (From E. Benjamin Andrews, *History of the United States* [New York: Charles Scribner's Sons, 1915], p. 287.)

shores,—and then, even then all seemed achieved, I was the victim of disappointment.

Imagination superseded the influence of fact. It was then doubted if it could be done again, or if done, it was doubted if it could be made of any great value.[68]

It was a historical triumph. The thirty-two-hour voyage, not counting a stopover at Clermont, had covered a distance of 150 miles, or slightly less than five miles per hour. The return trip would require only thirty hours, averaging exactly one mile per hour faster than the requirement of the New York law.[69] The monopoly had thus been secured, and now ahead was the challenge of finding paying passengers to recover the investment and, hopefully, accrue a profit.

END OF THE TRIAL, ACCORDING TO TRADITION

Operations continued for the rest of the season, until ice on the Hudson River finally prevented safe navigation. Fulton moved the steamboat to port and then had time to reflect on the first season's operation. In a letter to Livingston on November 20, he reported a 5 percent profit on the total investment of twenty thousand dollars ($8,500 from each and $3,000 borrowed) and projected an eight- to tenfold increase for the next season. There were, however, substantial problems with the vessel's structure, equipment, and accommodations, and in a letter to Barlow, Fulton stated that the boat was "cranky" and needed to be three to four feet wider for stability. Therefore, Fulton recommended completely rebuilding the steamboat, which took a bit of convincing on the part of Livingston.

On January 7, 1808, Robert Fulton, 43, like his father before him, married a much younger woman. Although the couple was engaged, the nuptials caught most of the Livingston family by surprise. Harriet Livingston, 24, was not pretty (see fig. 11.23) but was well educated, played the harp, and painted. Philip observes, "If Fulton's miniature portrait of her is reliable, she was endowed with a *belle poitrine* and chose gowns that it would be displayed to good advantage."[70] As Fulton planned to remove to New York in the

Figure 11.23. Harriet Livingston, wife of Robert Fulton, and two of the four Fulton children. Fulton plowed so much of the steamboat profits into new vessels that his wife complained she had no pocket money. (From Alice Crary Sutcliffe, *Robert Fulton* [1915; reprint, New York: Macmillan, 1925], between pp. 184 and 185.)

spring, they temporarily moved into her parents' home, Teviotdale, eight miles northeast of Clermont. They also spent some time at Livingston's home, from which Fulton was able to more easily supervise the reconstruction of the steamboat, which had been brought to nearby Red Hook.

The rebuilt boat was sixteen feet longer and five feet wider than the original. The depth and draught remained unchanged, at seven feet and two feet, respectively. With its overall length of 149 feet and breadth of 17.9 feet, it was substantially heavier, at 183 tons, versus the first boat, which was 100 tons.[71] Whereas the original had been registered on September 3, 1807, as simply the *Steam Boat* in Fulton's name only, Livingston had now provided one-half of the cost, so the renamed *North River of Clermont* was registered on May 14, 1808, showing the ownership of both partners. The route and times of departure were advertised (see fig. 11.24).

In addition, the schedule and rates for travel to various other points along the way were provided, as well as cabin accommodations and regulations. Success was immediate, as the timing of the steamboat's introduction coincided with a pent-up demand. Although some considered steamboat travel risky because of a possible boiler explosion, the vessel, with its low-pressure engine, proved to be safe, and the number of passengers continued to grow throughout the sailing season, providing much-needed profits to construct new boats.

⤙⤛✿⤚⤚

Over the next seven years, Fulton would father four children and more than one dozen steamboats. Following the rebuilt *North River of Clermont*, which would later simply become known as the *Clermont*, he designed or built boats that operated on the Hudson, Rariton, Potomac, and Mississippi rivers, firmly establishing steam navigation in America. Much of his time, however, would be spent in defending the New York State monopoly. In his writings and professed economics policies, he was strongly in favor of fair trade and an open market, but when it benefited his own personal finances, he

THE STEAM-BOAT.

For the Information of the Public.

THE STEAM-BOAT will leave NEW-YORK for ALBANY every Saturday afternoon, exactly at 5 o'clock—and will pass

West Point about	3 o'clock on Sunday
Newburgh,	6 do. [morning,
Poughkeepsie,	10 do.
Esopus,	1 in the afternoon,
Redhook,	3 do.
Catskill,	6 do.
Hudson,	8 in the evening.

She will leave ALBANY for NEW-YORK every Wednesday morning, exactly at 8 o'clock, and pass Hudson about 3 in the afternoon,

Esopus,	8 in the evening,
Poughkeepsie,	12 at night,
Newburgh,	4 Thursday morning,
West Point,	7 do.

on board; dinner will be served up exactly at 2 o'clock; tea, with meats, which is also supper, at 8 in the evening; and breakfast at 9 in the morning; no one has a claim on the steward for victuals at any other hour.

REGULATIONS,
FOR THE NORTH RIVER STEAM-BOAT.

The rules which are made for order and neatness in the boat, are not to be abused   Judgment shall be according to the letter of the law.   Gentlemen wishing well to so public and useful an establishment, will see the propriety of strict justice, and the impropriety of the least imposition on the purse or feelings of any individual.

Figure 11.24. Although registered as the *Steam Boat*, Fulton's first United States vessel was advertised as the *North River Steam Boat* to identify its route. In this advertisement, only a portion of the information presented to the public has been reproduced. (From Sutcliffe, *Robert Fulton and the* Clermont, between pp. 268 and 269.)

could convincingly defend the monopoly. This law provided Livingston and Fulton with cash flow, and Fulton spent money as fast as he earned it, causing Harriet to complain to Livingston that she wanted for pocket money.

After an unfortunate exposure to cold weather followed by exhausting work, Fulton died on February 23, 1815, in his New York home, which overlooked the harbor. Nevertheless, steamboat transportation was by then firmly in place, and others were already offering competition. This was to be his legacy. Cadwallader D. Colden wrote a eulogy that was read before the Literary and Philosophical Society of New York, and his address, titled *The Life of Robert Fulton by His Friend* (and, it might be added, his attorney), was printed in 1817 as the first of many Fulton biographies. This glowing account—emphasizing, as one might suspect, only Fulton's positive attributes—was to remain his official story for many years. This, combined with the influence of Livingston, who published his own portrayal of the first commercially successful steamboat,[72] and the public's view at that time that successful inventors were heroes, all account for the reasons the title "inventor of the steamboat" fell to Fulton. It is, however, not accurate.

The former sign painter, jeweler, portrait artist, canal engineer, submarine and torpedo proponent, and finally steamboat builder was laid to rest in the Livingston family vault in Trinity Churchyard in New York. The cold, gray, and snowy day was February 24, 1815. Many officials and prominent individuals attended his funeral, and offices and shops were closed as a sign of respect. Fulton was forty-nine and had become the United States' best-known entrepreneur. He was imaginative, successful, persistent, and eminently convincing, and he hustled all his life for, and finally did achieve, fame and fortune. He was not, however, terribly creative, preferring to use proven ideas by others, adding improvements from his active and fertile mind. With his passing, the United States lost a very significant and important contributor to steam navigation.

Notes

1. Several references were consulted for general information for this chapter. Among them were John S. Morgan, *Robert Fulton* (New York: Mason/Charter, 1977); Cynthia Owen Philip, *Robert Fulton, a Biography* (New York: Franklin Watts, 1985); William Graham Sumner and Robert H. Thurston, *Makers of American History: Alexander Hamilton and Robert Fulton* (New York: University Society, 1904); Alice Crary Sutcliffe, *Robert Fulton and The Clermont* (New York: Century, 1909); H. W. Dickinson, *Robert Fulton, Engineer and Artist: His Life and Works* (London: John Lane, 1913); and James T. Flexner, *Steamboats Come True: American Inventors in Action* (1944; reprint, Boston: Little, Brown, 1978).

2. Thomas W. Knox, *Life of Robert Fulton and a History of Steam Navigation* (New York: G. P. Putnam's Sons, 1886), p. 5.

3. Mathew S. Henry, *Life of William Henry* (Philadelphia: American Philosophical Society, 1860), p. 59.

4. Morgan, *Robert Fulton*, p. 11.

5. Flexner, *Steamboats Come True*, p. 120.

6. A guinea was an English gold coin last minted in 1813 and worth twenty-one shillings (twenty shillings equaled one pound).

7. Quoted in Morgan, *Robert Fulton*, p. 27.

8. Quoted in Philip, *Robert Fulton*, p. 18.

9. Quoted in Morgan, *Robert Fulton*, p. 31.

10. Philip, *Robert Fulton*, p. 23.

11. Quoted in Sutcliffe, *Robert Fulton and the* Clermont, p. 330.

12. Dickinson, *Robert Fulton, Engineer and Artist*, p. 128.

13. Quoted in Philip, *Robert Fulton*, p. 16.

14. Morgan, *Robert Fulton*, p. 37.

15. Flexner, *Steamboats Come True*, p. 220.

16. Quoted in J. Franklin Reigart, *Life of Robert Fulton* (Philadelphia: C. G. Henderson, 1856), pp. 156–57.

17. Morgan, *Robert Fulton*, p. 39.

18. In general, *force* can be substituted for *purchase*. Fulton later used the word to describe the effectiveness of water propulsion techniques, such as the "purchase" of the paddle wheel.

19. Sutcliffe, *Robert Fulton and the* Clermont, pp. 303–304.

20. Dickinson, *Robert Fulton, Engineer and Artist*, p. 38.

21. Fulton, *Treatise on the Improvement of Canal Navigation*, p. vii.

22. Ibid., pp. ix–x.

23. Quoted in Dickinson, *Robert Fulton, Engineer and Artist*, p. 51.

24. Quoted in Philip, *Robert Fulton*, p. 53.

25. William Chapman, *Observations on the Various Systems of Canal Navigation, with Inferences Practical and Mathematical; in Which Mr. Fulton's Plan of Wheel-Boats, and the Utility of Subterraneous and of Small Canals, Are Particularly Investigated* (London: I. and J. Taylor, 1797), p. 2.

26. When the Chesapeake and Delaware Canal was constructed in 1824–29, the slope of 1.5 to 1 was also used. It was found to be too steep, with an excessive amount of sloughing off into the channel, and over the years has been increased to 2 to 1.

27. Chapman, *Observations on the Various Systems of Canal Navigation*, p. 34.

28. Dickinson, *Robert Fulton, Engineer and Artist*, p. 29.

29. Ibid., p. 37.

30. Ibid., p. 22.

31. Several biographers have suggested that the American purchaser was Joshua Gilpin, who was instrumental in promoting the Chesapeake and Delaware Canal. In her thoroughly researched 1985 biography, *Robert Fulton*, Philip identifies Church on p. 61 but does not provide a reference.

32. Philip, *Robert Fulton*, p. 61.

33. Cadwallader D. Colden, *Life of Robert Fulton* (New York: Kirk & Mercein, 1817), pp. 26–27.

34. Dickinson, *Robert Fulton, Engineer and Artist*, p. 68.

35. Morgan, *Robert Fulton*, p. 59.

36. Ibid., p. 21.

37. Philip, *Robert Fulton*, p. 75.

38. Quoted in Dickinson, *Robert Fulton, Engineer and Artist*, p. 69.

39. Quoted in Philip, *Robert Fulton*, p. 89.

40. Several Fulton biographers have reported that the scene was the burning of Moscow, but the information supplied by Philip, *Robert Fulton*, p. 90, for his first scene is used here.

41. Quoted in Philip, *Robert Fulton*, p. 82.

42. Morgan, *Robert Fulton*, p. 62.

43. Ibid.

44. Quoted in Dickinson, *Robert Fulton, Engineer and Artist*, p. 108.

45. Quoted in Morgan, *Robert Fulton*, p. 71.

46. Philip, *Robert Fulton*, p. 95.

47. Quoted in Sutcliffe, *Robert Fulton and the* Clermont, pp. 93–94.

48. Morgan, *Robert Fulton*, p. 78.

49. *Mechanical Engineers' Handbook*, ed. Theodore Baumeister (New York: McGraw-Hill, 1958), p. 9-228.

50. Ibid.

51. Morgan, *Robert Fulton*, p. 79.

52. Ibid., p. 117.

53. Quoted in George Dangerfield, *Chancellor Robert R. Livingston of New York, 1746–1813* (New York: Harcourt, Brace, 1960), p. 287.

54. Quoted in Dickinson, *Robert Fulton, Engineer and Artist*, p. 146. In a footnote, the individual is identified as Capt. Samuel Morey. "Toot" was Barlow's familiar name for Fulton.

55. Quoted in ibid., p. 156.

56. Quoted in ibid., pp. 157–58.

57. Fulton and others had understood that the force to move a boat through the water varied as the square of its speed. Intuitively, this can be understood by considering the water to be comprised of a large number of very small particles, each of which offers resistance to movement. As the number of particles encountered increases directly with speed, and furthermore, the resistance each offers varies directly with speed, compounding these two effects results in the opposing force being proportional to the speed squared. What Fulton, as well as others, did not recognize was that horsepower is work performed in a specific amount of time (one horsepower is 550 foot-pounds of work per second); hence, a third factor, time, is added to the equation, resulting in the required horsepower varying as the cube of speed. Fulton would modify his calculations to account for this discrepancy but would erroneously draw the conclusion that a speed of more than six miles per hour was impossible.

58. Quoted in Dickinson, *Robert Fulton, Engineer and Artist*, p. 190.

59. Ibid., p. 195.

60. Quoted in ibid., p. 196.

61. Quoted in Philip, *Robert Fulton*, pp. 183–84.

62. Morgan, *Robert Fulton*, p. 136.

63. Quoted in Philip, *Robert Fulton*, p. 196.

64. Quoted in Morgan, *Robert Fulton*, p. 139.

65. The energy of coal, measured in British Thermal Units (BTUs) per pound, is about twice that of wood.

66. Cynthia Owen Philip, "Paddles & Profits: Robert Fulton Brings Steam to the Hudson," *Seaport* 19, no. 3 (fall 1985): 23.

67. Dickinson, *Robert Fulton, Engineer and Artist*, p. 217.

68. Quoted in Sutcliffe, *Robert Fulton and the* Clermont, pp. 202–203.

69. The apparently lost contract of 1802 between Livingston and Fulton called for eight miles per hour, but only four miles per hour was required to meet the requirements of the New York monopoly law.

70. Philip, *Robert Fulton*, pp. 216–17.

71. Dickinson, *Robert Fulton, Engineer and Artist*, pp. 326–27. Depending on the source, there is some confusion about the size of Fulton's original steamboat. For the rebuilt boat, the dimensions provided are from New York Custom House records obtained by Dickinson.

72. Robert R. Livingston, "The Invention of the Steamboat," *American Medical and Philosophical Register* (January 1812). In the article, Livingston mentions other steamboat developers, but he credits himself and Fulton with the invention. A copy can be found in "Steamboats, Hudson River,"

Harper's Encyclopaedia of United States History from 458 A. D. to 1912, vol. 8 (New York: Harper & Brothers, 1912).

Figure 12.1. Col. John Stevens. (Rendered by William T. Sisson from Robert H. Thurston, *History of the Growth of the Steam-Engine* [New York: D. Appleton, 1901], p. 178.)

12.

John Stevens
1749–1838

Biographies are too often eulogies. Yet it seems impossible to trace the career of John Stevens and his sons (Robert Livingston and Edwin Augustus) without, apparently at least, joining the ranks of the hero worshipers.

—J. Elfreth Watkins, *John Stevens and His Sons* (1892)

"The Colonel," as John Stevens was commonly known for most of his life, died at the age of eighty-nine and, over a long career of securing his personal fortune, pursued a vision to enhance America's transportation infrastructure. He firmly believed, "The wealth and prosperity of a nation may be said to depend, almost entirely, upon the facility and cheapness with which transportation is effected internally."[1] He was America's first great civil engineer and mechanical-engineering visionary, and he also made contributions to the nation's public health and defense. Opposing the Livingston-Fulton Hudson River monopoly, he designed, constructed, and operated steamboats on the eastern waters of the fledgling United States but generally opposed canals in favor of what he correctly envisioned to be a better transportation system—railroads. In 1825, he built America's first experimental railroad around his Hoboken estate, understanding that skeptics would eventually be convinced by "ocular demonstrations." About thirty years after his death, and in accordance with his wishes, his son Edwin provided funding for the Stevens Institute of Technology as "an academy for teaching fundamental subjects and the elements of science."[2]

In 1699, at the age of seventeen the first John Stevens, progenitor of the family and grandfather of Col. John Stevens, arrived in America. This lad was indentured for seven years to Barna Cosans, who was secretary to the governor and clerk of the New York royal provincial council.[3] Young John Stevens quickly became skilled in the law and the language, and after completing his servitude in 1708, he was one of thirteen individuals who received a grant of land in New York. Hearing stories of veins of copper in an area called the "Devil's Feather Bed," he ventured to New Jersey, but after arriving, decided to buy land rather than mine it.

About 1706, perhaps later, he married Ann Campbell, daughter of a wealthy real estate dealer, and immediately assumed the status of Mr. J. Stevens.[4] He and his father-in-law both bought and sold land, while in the process holding onto the best parcels for themselves. Noting the deplorable conditions of roads and river crossings, J. Stevens wrote, "If this Province is to have its chance to develop, we must give it better roads and ferries."[5] This perceptive and visionary understanding prompted him to become part owner in Redford's ferry to Amboy. As the family's first venture into intercolonial transportation infrastructure, this small beginning would later become the primary focus of Col. John Stevens's entire life.

At the time Mr. J. Stevens died, at the age of fifty-five in 1737, he had acquired a fair amount of wealth, leaving it to his sons: Campbell, John, Lewis, and Richard. Of these offspring, Stevens's biographer Archibald Turnbull describes the second-born, John, at the age of twenty-one as "already dominating the family. His was the controlling spirit under which his brother Campbell began the operation of a small fleet of merchantmen."[6] John soon acquired a reputation of being willing to sail to any part of the world and transport any cargo and supported these activities by establishing warehouses at Perth Amboy (see fig. 12.2). He assumed management of the family's affairs and became a prolific writer of letters finding the pen an easier way to express his ideas rather than through conversation.

This second John Stevens, later known as "Honorable John," became acquainted with James Alexander, a leading lawyer, surveyor general of New Jersey and New York, mathematician, and landowner of note. It is most probable that this relationship stimulated John's interest in public service as well as matrimony, leading to his marriage of James's daughter Elizabeth in 1748. Honorable John and Elizabeth spent time in both Perth Amboy and in New York City, where she preferred to live. It was in New York during the midsummer of 1749 that son John (called Johnny, John Stevens Jr., and later simply the Colonel) was born. The union also produced a daughter, Mary, who was called Polly by some of her family. Polly was to grow up and marry Robert R. Livingston, heir to the Clermont estate on the Hudson River and the future steamboat partner of Robert Fulton.

Figure 12.2. The father of Col. John Stevens shipped cargo to many parts of the world, storing the goods to be exported and those imported in his warehouses at Perth Amboy. (From William McDowell, *The Shape of Ships* [New York: Roy Publishers, n.d., ca. 1948], p. 115.)

Continuing his successful shipping business, public service drew the Honorable John into politics. Encouraged by his friends in 1751, he ran for the assembly in New Jersey and, upon being elected, decided to become active. In 1756 he became the paymaster of the Jersey Blues and in 1765 was appointed to a commission to negotiate a treaty with the Delaware, Susquehanna, and other Native American tribes.

His wife, not content with country living, encouraged him to buy a house in New York City, and when No. 7 Broadway became available in 1760, this was done (see fig. 12.3). This precipitated many journeys between New York and Perth Amboy providing Stevens firsthand knowledge of the poor condition of intercolony transportation infrastructure. As a member of a commission to recommend how to improve New Jersey roads, a report was issued, encouraging the General Assembly to establish "a Lottery to defray the Expense of the same," a fund-raising solution typical of the time.[7] In 1763, Honorable John was appointed to the council of Governor Bernard, a function apparently focused on organizing social functions.

Figure 12.3. Broadway in New York City, with the home acquired by the Colonel's father at No. 7 Broadway being located down the street where the coach and horses are shown. (From Archibald Douglas Turnbull, *John Stevens, An American Record* [New York: Century, 1928], between pp. 240–41.)

In 1768, Johnny graduated from King College with a BA, and in the same class, receiving an MA, was Robert R. Livingston. Some time later, Robert and Johnny's sister Polly were to become engaged and marry.

The seeds of rebellion were germinating in America, and Johnny volunteered his services to the Continental Army. He received a commission but resigned some time later to serve the state of New Jersey. On July 15, 1776, at the age of twenty-seven, Captain Stevens assumed the duties of New Jersey state treasurer, moving from place to place to keep documents and funds safe from enemy hands.[8] He was promoted to major and later to colonel, achieving the rank that would identify him for the rest of his life. This progression would suggest that he performed the duties of treasurer rather well; however, it would be years after the war before his accounting of the funds could be sorted out.[9] As he intermingled his personal business with funds borrowed from the state treasury, his record-keeping was jumbled, and he found it necessary to borrow from his father to finally settle with the state.

The young officer had fallen in love with Rachel Cox, whose father, Col. John R. Cox, eventually became the deputy to Nathaniel Greene, quartermaster general to George Washington. While the war was still officially being fought, John Stevens and Rachel were married on October 17, 1782. After the close of hostilities, they moved to New York into the family home at No. 7 Broadway, which was in need of much repair, having been occupied during the war by the British. Relations with close neighbors were not the same, though, as war loyalty had divided friends. For a social outlet, the Colonel joined the St. Andrew's Society but did not find New York City to his liking.

Casting about for another location, he soon found a tract of land across the Hudson River, called Hobuck, that was owned by William Bayard, who had remained loyal to the crown. The property was confiscated by New Jersey, and through his political influence, Stevens was able to acquire 689 acres for 18,340 pounds on March 16, 1784. There he built a house, kitchen, and barn, which were completed in 1787, the same year John Fitch and James Rumsey were running their steamboats on the Delaware and Potomac rivers.

With the little funds he had left, he also purchased another 150 acres of land near Tappan.

While awaiting the completion of their New Jersey home in New York, Rachel had delivered the first of their eleven children, a boy named John Cox in honor of both grandfathers. By the time the house was completed and overlooked what was planned to be a beautiful English park (see fig. 12.4), Rachel, still in New York, had another boy, Robert Livingston Stevens, who was named for the Colonel's sister's husband.

With frequent trips across the Hudson, John Stevens was frustrated that the service had little improved since the earliest days of the ferry that was authorized in 1654. Turnbull notes, "To await any other man's pleasure instead of crossing the river at his own will was galling to the Colonel,"[10] so the answer seemed obvious—buy the ferry. This was done for the amount of 1,250 pounds, and the acquisition included an additional 100 acres, bringing his New Jersey–side total to 789 acres.

During this period, the U.S. Constitution had been framed in Philadelphia and submitted to the states for ratification. Honorable

Figure 12.4. The Villa Stevens in Hoboken, across the Hudson River from their Broadway Street home in New York City, was completed in 1787. The Colonel purchased the ferry he used in his travels back and forth making his first entry into interstate transportation. (From Turnbull, *John Stevens*, between pp. 160 and 161.)

John believed that it should be ratified at once by New Jersey and sought the opinion of Robert R. Livingston, who wrote him on December 8, 1787, "I am very glad to hear the choice your county had made of members of the convention, & hope from the general complection of your state that you will have the honor of being the first in acceding to the new constitution. In saying this, I answer your question and let you know that it meets with my sincere concurrence, & indeed I sh'd censure a constitution which I had no small agency in framing, if I were not to approve it." Livingston had served on the drafting committee for the Declaration of Independence (see fig. 12.5) and was a member of the group that wrote the constitution for the state of New York. He continues:

Figure 12.5. The drafting committee of the U.S. Declaration of Independence. Primary architect Thomas Jefferson of Virginia is shown reading a draft to Benjamin Franklin (Pennsylvania); Robert Livingston (New York) is seated, John Adams (Massachusetts) is standing, and Roger Sherman (Connecticut) is standing on the far right. (Print, titled *Drafting the Declaration of Independence*, from a painting by Alonzo Chappel [New York: Johnson, Fry, 1870].)

It is expressly formed upon the model of our state government. My vanity is not a little flattered to find that the only *new idea* in government which has been started in America, where so many have thought on the subject, owes its birth to me, & has been adopted by such respectable bodies as Massachusetts, New York, and the general convention. I mean the council of revisions, tho' the alteration that have made, in vesting this power of revision in the executive magistrate *alone*, rather than, as with us, in the Executive and Judicial, the latter of whom are independent, is a material defect, since the legislature have always been solicitous to encroach on both.[11]

The letter goes on for two more paragraphs to speak of "the wise & the weak, the ignorant & the experienced, will divide the influence, & each must be gratified; their favorite child, like the son of the patriarch, will wear a coat of many colours." About this Turnbull writes, "Since he claimed so large a share in its making, it has been astonishing to discover how the chancellor, as lawyer and as business man, ultimately interpreted it. Twenty years later he was inclined to blend the 'coat of many colours' into a Livingston tartan. However, it is a fair presumption that he was always an opportunist; for the moment he was backing Hamilton in the fight for ratification by New York."[12]

After reading a publication written by John Adams, the Colonel took up the cause to have the Constitution ratified, writing a relatively short pamphlet that appeared anonymously with the lengthy (though not uncommon for the time) title of, *Observations on Government, Including Some Animadversions on Mr. Adams' Defense of the Constitution of Government of the United States, and Mr. DeLome's Constitution in England.* It was printed by W. Ross, Broad Street, New York, and signed, "A Farmer of New Jersey." He sent copies of it to his family and friends, soliciting their opinions. In the pamphlet, he declared, "the security of the liberties of a people or state depended wholly on a proper delegation of power. The several component parts of government should be so distributed that no one man, or body of men, should possess a larger share thereof than what is absolutely necessary to the administration of government."[13] The separation of powers to safeguard the freedoms of the people seems obvious today, but at the time, it was a new experiment in government.

The Colonel referred to a mechanized meat-roasting spit that would be used over an open fire as an analogy to explain his ideas. The weight, hence the prime moving source, was the people; the mechanical parts of the spit, the machinery of government; and the flywheel and regulator, the checks and balances built into the operation of government: enough flywheel momentum to keep the machinery running if bureaucrats fail to act, but also enough checks to keep bureaucrats from acting outside the power granted to them by the people. This delicate balance was elegantly explained in language that could be understood by a large segment of the populace.

Honorable John presided over the New Jersey convention to decide whether to ratify the proposed Constitution, and it is not known if the Colonel's pamphlet had any effect. The delegation was favorably disposed at the outset, and on Feb 11, 1788, Honorable John reported New Jersey's ratification "a fitting climax" to many years of dedicated service to the state. When it was decided to have Congress meet in New York, the Colonel briefly lobbied to be a candidate; however, upon learning that his father-in-law, Col. John Cox, also wanted a position, he withdrew his name.

In January 1788, James Rumsey had published a small quantity of pamphlets in Berkeley County, Virginia, claiming priority to the invention of the steamboat. Upon hearing that a competing steamboat developer, John Fitch, was planning a rebuttal, Rumsey decided to have his original pamphlet, with minor changes, reprinted in Philadelphia. Both pamphlets appeared in May 1788. A few months later Rumsey's partner, Joseph Barnes, replied to Fitch with another pamphlet, and the two documents describing Rumsey's work found their way into the hands of the Colonel, who maintained an extensive library.[14]

On September 5, 1788, the Colonel wrote Rumsey with some suggestions:

> I shall not trouble you with an apology for thus addressing you, although to you an utter stranger. If the following hints prove of any service towards perfecting your project you are heartily welcome to them. Though they may not suggest anything new yet, as the writer thereby manifests his good wishes towards the promotion of useful discoveries, you cannot but be pleased with them.

I have not been able to procure any other information respecting your plan of propelling a Boat by means of Steam than what I could collect from your pamphlet published at Philadelphia and from another published at the same place, entitled "Remarks on Mr John Fitch," etc. Your invention of generating steam by means of a worm [see fig. 12.6] is certainly of the utmost importance, but most particularly so when applied to the purpose of navigation. An insuperable objection to the application of steam in any other mode hitherto made use of is the very great proportion of room which the boiler and other apparatus must necessarily occupy on board a vessel.

Recognizing the importance of an efficient and lightweight steam generator, the Colonel continued providing some suggestions for

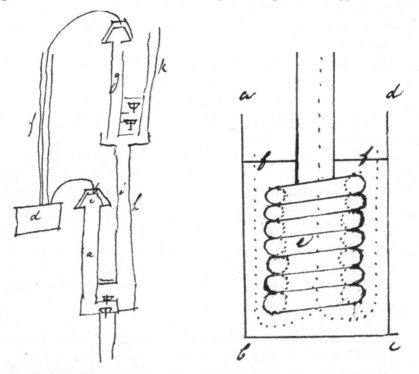

Figure 12.6. The reference to a "worm" for generating steam in Stevens's letter was the approach Rumsey used in his steamboat. The Colonel also sketched such a device, as well as a two-stage water injection pump, shown on the left. (From Turnbull, *John Stevens*, between pp. 128 and 129.)

improvements to Rumsey's boiler. For improved propulsion of the boat he added the last paragraph: "With very little additional machinery, the piston, which is raised by means of the steam and depressed again by the pressure of the atmosphere, might be made to force water, two trunks alternately; through one at the rising, and one at the falling, of the piston; by which the force of the engine would be doubled."[15] This observation was partially correct, and being fairly obvious, it must have occurred to Rumsey himself. Rumsey no doubt chose the descending stroke to power his water jet pump, because an atmospheric force approaching fifteen pounds per square inch was consistently much greater than the five to ten pounds per square inch used to raise the piston. Hence, the Colonel's suggestion would have provided some improvement in performance but would not double it.

For someone always concerned with speed, more speed, and then greater speed, the Colonel did recognize that the use of a water jet would not be practical to achieving "any great degree of velocity" when applied to vessels. At that time and with available technology, he was indeed correct; however, the idea did intrigue him. He later pursued water jets, and when he applied to the New York legislature for patent protection on a steamboat, he proposed this technique for propulsion. Engineering details were not his strong suit, but as a visionary of the future needs of society, he was at his finest. Turnbull correctly notes, "No one more clearly foresaw the future's demand for rapid transit and delivery; no one, in the face of every discouragement, worked harder to meet that demand."[16]

Prior to writing James Rumsey, it appears that observing John Fitch's steamboat being operated on the Delaware River had already stimulated the Colonel's interest in steam navigation. Dr. Charles King, who at the time was president of Columbia College, gave an address titled, "Progress of the City of New York," in which he said,

> Mr. Stevens' attention was first turned, or rather the bent of his genius was developed and directed, towards mechanics and mechanical philosophy, by the accident of seeing in 1787 the early and as now may be said imperfect steamboat of John Fitch, navigating the Delaware river. He was driving his phaeton on the banks

of the river, when the mysterious craft, without sails or oars, passed by. Mr Stevens' interest was excited—he followed the boat to its landing—familiarized himself with the design and the details of this new and curious combination, and from that hour became a thoroughly excited and unwearied experimenter in the application of steam locomotion on the water, and subsequently on the land.[17]

It is doubtful that Stevens ever saw Rumsey's steamboat, as his work had been carried out on the far away Potomac River near Shepherdstown, although Fitch and Rumsey did have some knowledge of each other's activities. Since the publication of the pamphlets, the Colonel knew he would have to act quickly to secure any legal advantage against either inventor. To this end, he wrote John Watt in the New York legislature,

> I see by the proceedings of the Legislature in Child's paper, that Mr Rumsey petitioned for an "exclusive privilege" of building Steam Boats, and Mr Fitch has petitioned "that the prayer of the petition of the Rumseyan Society not be granted." As neither of these gentlemen, from what has yet appeared, have brought their Schemes to that degree of perfection as to answer an valuable purpose in practice, I conceive that neither of them are entitled to such an exclusive privilege as may secure to them all the benefits arising from the improvement invented by others. . . . I have been induced to draw up the enclosed petition, which I request the favor of you to present.[18]

The Colonel's petition, along with those of Rumsey and Fitch, was presented to the legislature on February 9, 1789, and referred to a committee. As the members of the committee were not knowledgeable in the science and technology required for a comprehensive evaluation, they chose to give priority to the petition first filed, which was Rumsey's. His New York steamboat patent, pressed by the Rumseian Society, was granted on February 25, 1789.

Not succeeding at Albany, the Colonel decided to obtain Federal protection. Although the Constitution provided for such coverage, no legislation had yet been passed to enable the patent provision to be brought into law. It has been reported that through the Colonel's efforts, the patent law was passed,[19] but there does not appear in the records information to support such a claim.[20] In any case, the enabling legislation was passed on April 10, 1790, but a delay ensued. Stevens learned in a letter dated January 25, 1791, from the commission's secretary, Henry Remsen, that "The Commission named in the act for the promotion of the Useful Arts, judging it most expedient not to proceed further . . . until a bill supplemental to the said [1790] Act, and which is now before Congress, passes."[21] The new legislation failed, and on May 4, 1791, Remsen informed the Colonel about his patent's status: "The investigation of the claims of Rumsey & Fitch to a priority in the application of Steam to navigation, which was to have been made by persons mutually appointed by them, was attempted but has not completed; and the Commissioners as a late meeting agreed to grant patents to them— and to all claimants of steam patents—according to their respective specifications."[22]

The mutually conflicting patents issued to Rumsey, Fitch, Stevens, Nathan Read, and Englehart Cruse were all granted on August 26, 1791—the most unfortunate turn of events for the speedy development of the steamboat in the United States. Without unambiguous federal patent protection, it could not be argued by any one of the applicants that a particular steamboat configuration would be protected. In particular, this was devastating to John Fitch, as his financial backers were not assured of a return on their investment, and he was abandoned. Thus the Constitution's promise to provide a short-term monopoly for any useful art that could be beneficial to the public was stalled, and it would be another sixteen years before the first commercially successful steamboat would be introduced, and then initially without any federal patent protection.

Meanwhile, the Colonel had not been neglecting his wife; Rachel had been adding to the nursery, with James Alexander Stevens born on January 29, 1790. Barely two years later came Richard, and not too long thereafter Francis Bowes. These were happy events, but not all was joy, as Honorable John Stevens passed away in May 1792. A sixth son, Edwin, was added to the family some years later, in July 1795.

On April 11, 1795, John Fitch wrote the Colonel with a proposal, offering 40 percent of an interest in the patent of a horse boat for the very modest sum of twenty dollars. It was not pursued, as Stevens was only interested in steam, and it is surprising that Fitch did not approach the Colonel in that arena, as they both held patents for steamboats. Most likely, if the subject had been broached, the Colonel had his own ideas, as during the summer of 1796, he and Robert R. Livingston, along with several others, received a demonstration of a five-mile-per-hour steamboat built by Samuel Morey, the subject of chapter 10. As a result of the Colonel's discussions with Nicholas J. Roosevelt, who had an interest in the Schuyler foundry called Soho, at Belleville on the Passaic River, a steamboat plan emerged.

As John Fitch had obtained in 1787 an exclusive right to navigate on the waters of New York for fourteen years, and James Rumsey had obtained a patent for his steamboat in New York in 1789, Stevens and Roosevelt would need political help and money. Hence, they turned immediately to include Livingston in the partnership. This arrangement was satisfactory to the Chancellor, a gentleman inventor—provided, however, the boat and engine were designed in accordance with his plan.

Immediately his ideas were called into question, as the foreman chosen for the task protested that the Chancellor's water jet approach would not work. Although the Colonel had also supported the water jet several years earlier, in 1788, he now had reservations. It was not too long, however, before the Chancellor had a better idea, described in a January 22, 1798, letter: horizontally mounted wheels 4½ feet in diameter should be employed under the keel in a well. A vertical axle on each wheel would be powered by the steam engine, causing water to be thrust to the rear, propelling the boat forward.

Livingston was also at work on the political front, having a bill introduced into the New York Assembly at Albany that transferred the Fitch fourteen-year monopoly to himself. A condition attached to his March 1798 legislation required that within one year, a boat of at least twenty tons must be demonstrated on the Hudson River at a speed of not less than four miles per hour.

Writing to Roosevelt, whose contribution to the steamboat effort was labor, the Chancellor allowed, "You have 12/100 in the patent rights and also the rights under the [state] law; with the exception I mentioned to you, of all ferries to the New Jersey shore; to prevent any interference with my ferry or Mr Stevens', the *exclusive* use of boats to which we claim."[23] It would appear that the 88/100 remaining share of the patent and New York monopoly was divided between Livingston and Stevens.

Trial 1. Sometime toward the end of the second half of 1798, a trial was conducted, which according to Roosevelt achieved a speed of three miles per hour, not fast enough to secure the New York monopoly.[24] An after-the-fact description has been provided by New York mayor Abram S. Hewitt, quoting the oft told story by his father, who worked at Soho, and a passenger on the trial run:

> There [to Soho] John Stevens came and built the first low-pressure engine. He himself ordered and paid for the first non-condensing double-acting engine that was built on the American continent. That engine was put in a boat in which I traversed the route from Belleville to New York and back again; John Stevens being the owner, builder, and captain of the boat; Mr Smallman, Mr Rhodes, and myself being the passengers. And we came to New York in that boat nine years before Fulton put the *Clermont* on the Hudson.[25]

Roosevelt had run tests and determined to his satisfaction that the engine was too small for the wheels specified by Livingston and had suggested in a letter to the Chancellor on September 6, 1798, "I would therefore recommend that we throw two wheels of wood over the sides fastened to the axes of the flys with 8 arms or paddles, that part which enters the water of sheet iron to shift according to the power they require either deeper in the water or otherwise."[26] By this time, Morey had successfully demonstrated his paddle wheels on the Delaware River, but Livingston wrote back, expressing confidence in the approach he had proposed.

After the New York monopoly lapsed in March 1799, the Chancellor had it renewed, this time incorporating the names of all three parties: Robert R. Livingston, John Stevens, and Nicholas Roosevelt. In 1800, the parties entered into a formal written agreement for a period of twenty years, each having a one-third share and sharing equally in the expenses. Should a partner decide to dispose of his interest, it should first be offered to the others.

It was agreed to build a new boat and engine employing a sixteen-inch, double-acting cylinder, with a three-foot stroke. Propulsion was to be by means of a stern paddle, as used by Morey in his 1796 demonstration and now embraced by Stevens. An amount of eight hundred dollars for the total estimated cost of twelve hundred dollars was to be supplied to Roosevelt in equal, weekly installments, with him providing the labor and materials. A completion date of May 1801 was anticipated.

When the date passed, an irritated Livingston wrote and complained to the Colonel about what he perceived to be Roosevelt's lethargic performance: "I think you should bring R. to some conclusion. If he will do nothing for us, it is not right that he should share in our inventions."[27] Far more accurate would have been an argument by Roosevelt that the Chancellor was sharing his money but not his inventions, as the difficult development of the engine and boat constituted far more of the invention than the initial idea. The impatient Livingston, however, had no appreciation of design, development, and systems integration; but, of course, his capital and that of the Colonel was an absolutely necessary ingredient to the process.

Roosevelt provided a written answer to Stevens, who was the principal contact, advising he had been diligent: "Every part of the work is now finished except the cylinder (which will require 10 night's work, the only time I can devote to it) and the paddles. . . . In two weeks, I expect to be moving."[28]

Trial 2. The next trial took place in the summer of 1801. Unfortunately, the pulsations caused by the double-acting cylinder caused excessive vibration of the boat, no doubt jeopardizing the integrity of the structure. After several trials, this difficulty led the Colonel to exclaim, "Scrap her and build another"; however, the Chancellor had a different view, now wanting to adopt the Barker wheel (a turbine) as the most promising way to propel the boat. But because they couldn't agree, an attempt was made to sell the engine, leaving the partners dejected.

In 1800, Livingston had made a losing run for the New York governorship against John Jay, and President Thomas Jefferson had offered him the position of secretary of the navy. Livingston declined, but Jefferson persuaded him to accept the post of minister to France in 1801. For the next few years, this appointment would prevent his dabbling in U.S. steamboat efforts, but in Paris, he would meet Robert Fulton, and his interest would be rekindled.

When the Colonel's effort to build a turnpike road from the town of Bergen to his Hoboken ferry succeeded in 1802, the owners of the nearby Paulus Hook ferry were quick to organize to improve the road leading to their service. Stevens correctly reasoned that the way to garner the majority of the ferry business was to provide better and faster service, so again his interest turned to steam power. His wife's sister, Sarah, was married to Dr. John Coxe, who lived in Philadelphia, where high-pressure steam pioneer Oliver Evans was working (see chapter 8). Gregg reports, "On February 18, 1803, Stevens wrote Dr. John Redman Coxe for detailed information about Evans' public claim that he had invented a successful high-pressure steam engine."[29]

Coxe visited the steam facilities of Evans, explained his mission, and, with a cooperative Evans, obtained answers to the Colonel's mechanical questions: the pistons were packed with hemp, the valves worked by gears from the flywheel, etc. Using this information, Stevens later published, in the same issue of the *Medical Repository* that carried Evans's pamphlet on "Operation of the Steam Engine," a letter attacking Evans and claiming he, not Evans, had priority for some of the steam engine work. Evans fired back, and an angry series of letters ensued, with claims and counterclaims, but there seemed little doubt that at that time, the brilliant Evans had a knowledge of steam engines, in particular a vision of high-pressure steam, far ahead of Stevens.

Later, Evans was to offer his engine for a steamboat and would "warrant [it] to work for four years without repairs."[30] He further provided steamboat performance beyond what could be obtained from a Boulton and Watt engine with an engine cost of twenty-five hundred dollars per mile per hour, noting the Colonel's vision, which he shared, "because speed is what counts." It was one of Evans's creative performance guarantees that was years ahead of its time, but it appears that the egos of the two entrepreneurs kept them from achieving a business arrangement. Had this been overcome, it seems likely that the Colonel's penchant for better ferry service and that of Evans to build lighter and more powerful steam engines would have coalesced, and the United States would have had steamboats plying its waters many years earlier.

Meanwhile, on April 11, 1803, the Colonel received a patent on a tube boiler whereby water is injected by a forcing pump (see fig. 12.7). This action brought him in contact with Dr. William Thornton, commissioner of the U.S. Patent Office, and he would correspond with him later in regard to Fulton's steamboat patent.

By this time, the family had exploded with daughters: Elizabeth, Juliana, Mary, Harriet, and Ester Cox. The large family and turnpike activities kept the Colonel poor in cash, although he owned much land. He decided to lease fifty lots at Hoboken but enforced restrictions so severe that few tenants subscribed.

In a letter from Paris dated April 28, 1804, the Chancellor made the first mention of Robert Fulton to Stevens, apparently anticipating the letter to be hand carried by Fulton to the Colonel but being mailed instead: "The principal object I have in view, is to

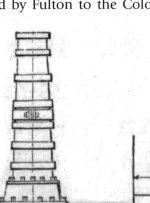

Figure 12.7. Water tube boiler invented by John Stevens about 1803, which would be used later to provide high-pressure steam to a small steamboat called *Little Juliana*. (From Thurston, *History of the Growth of the Steam-Engine*, p. 264.)

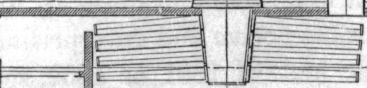

introduce to your acquaintance Mr Fulton, with whom you will be very much pleased. . . . He was my partner in an experiment made here on the Steam Boat, and his object is to build one in the United States."[31] This must have come as a surprise and shock to Stevens, for the Chancellor had a new steamboat partner, and together they planned to introduce steamboat service in the United States. The twenty-year agreement constructed in 1798 and formalized in 1800 between the Colonel, Livingston, and Roosevelt was not mentioned. This was to strain the relationship with his brother-in-law for the rest of their lives. Fulton, however, would not arrive until late in 1806; meanwhile, Stevens and his boys had their own steam navigation plans.

Trial 3. The following month, May 1804, a group of three students at Kings College were strolling Battery Park and witnessed a truly remarkable sight. A boat was moving ahead without any visible means of propulsion. One individual in the group, James Renwick, later described the event:

> As we entered the gate from Broadway, we saw what we, in those days, considered a crowd, running towards the river. On inquiring the cause, we were informed that "Jack" Stevens was going over to Hoboken in a queer sort of boat. On reaching the bulkhead by which the Battery was then bounded, we saw lying against it a vessel about the size of a Whitehall rowboat, in which there was a small engine no *but no visible means of propulsion*. The vessel was speedily underway, my late much-valued friend, Commodore Stevens, acting as the coxswain, and I presume the smutty-looking personage who fulfilled the duties of engineer, fireman, and crew, was the more practical brother, Robert L. Stevens.[32]

What Renwick had presumed was correct. John Cox (Jack) was steering the small craft, and his brother Robert was handling the task of keeping the unusual boat filled with equipment moving at a speed of four miles per hour. Named for their eldest sister, the *Little Juliana* (see fig. 12.8), propelled by "wheels" in the stern, was the forerunner of modern-day ships with counterrotating screw propellers.

The idea of a propeller, however, was not new, as it had been

Figure 12.8. The twin-screw *Little Juliana*, driven by a high-pressure steam engine, achieved four miles per hour but was not successful because of boiler technology limitations. (From Thurston, *History of the Growth of the Steam-Engine*, p. 269.)

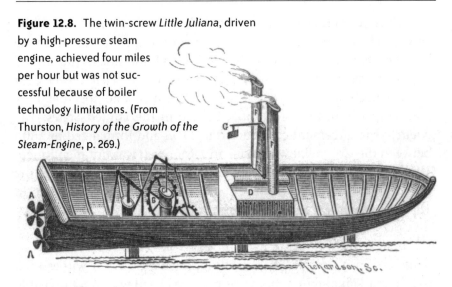

proposed by Bernoulli in 1753, and in 1776, American David Bushnell employed two hand-operated propellers to move his submarine both horizontally and vertically. Rumsey had proposed using a screw propeller in his 1792 British patent, and Fulton, copying Bushnell, used what he called "flips" on his *Nautilus* submarine. But this demonstration was the first time that counterrotating twin screws had been put to a practical use. Here is the Colonel's description of his historic equipment, which had begun in 1802 with a single screw propeller (see fig. 12.9):

> For simplicity, lightness, and compactness, the engine far exceeded any I have yet seen. A cylinder of brass, about eight inches diameter and four inches long, was placed horizontally in the boat; and, by alternating pressure of the steam on two sliding wings, the axis passing thro' its center was made to revolve. On one end of this axis, which passed through the stern of the boat, wings like those on the arms of a windmill [propellers] were fixed, [and] adjusted to the most advantageous angle for operation on the water. This constituted the whole of the machinery.
>
> Working with the elasticity of steam merely, no condenser, no air pump was necessary, And, as there were no valves, no apparatus was required for opening and closing them. This simple little steam engine was, in the summer of 1802, placed on board a flat-

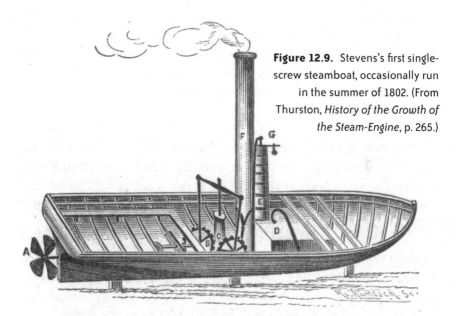

Figure 12.9. Stevens's first single-screw steamboat, occasionally run in the summer of 1802. (From Thurston, *History of the Growth of the Steam-Engine*, p. 265.)

bottomed boat I had built for the purpose. She was occasionally kept going until cold weather stopt us. When the engine was in the boat, her velocity was about four miles per hour.[33]

One important piece of equipment, the steam generator, was omitted in this description. The high-pressure boiler on the *Little Juliana* eventually "gave way so that it was incapable of operation," and the experiment came to an end.

This was not a practical vessel, as the equipment occupied most of the hull, making it incapable of commercial success. Furthermore, the technology of the time barely supported the portable steam generator. It did, however, portend the future, as Stevens was a visionary thinker.[34]

Reading about high-pressure steam activities in England in February 1805, the Colonel sent his oldest son, John Cox, to investigate.[35] John Cox had three missions: first, to find out about what was being accomplished in England with high-pressure steam; second, to obtain a British patent on his steam boiler; and third, to interest Boulton and Watt in the future of high-pressure steam, hence his boiler patent. John Cox succeeded on the first two, but meeting with

Matthew Boulton's son, who had assumed responsibility for the business, he could not engender any enthusiasm for steam pressures higher than those currently being used. Low-pressure steam had proven very successful for the firm, and development of high-pressure steam in England would have to wait, along with the possible licensing of his boiler.

A slight digression is worthwhile to understand the Colonel's interest in many other areas. He had become involved with the Manhattan Company in working to provide water through holes in wooden pipes to New York City. The water was not only to be used for drinking, bathing, and fire protection but also to wash the filthy streets, improving the health of the residents.

Noting a recent fire causing a damage of one million dollars, he turned his attention to another subject, writing Mayor DeWitt Clinton on March 4, 1805, proposing "a steam engine which will be connected with a fire engine and forcing pump; the whole to be placed on board a vessel, under cover and well secured from the weather. The vessel to be capable of being easily transported from place to place on the river, as occasion may require," would pump water into an elevated pond for washing streets and also for firefighting.

The Colonel also suggested establishing "The U.S. Bridge Company" to erect a floating bridge, with draws, across all rivers served with ferries in the New York City area. As was predictable, ferry operators objected and prevailed. It was also argued that the floating bridge would interfere with shipping.

Undeterred, and with a vision for improved transportation infrastructure, he proposed a fixed bridge of such a height as not to hinder the passing of ships. This also met with opposition, so he reverted in 1806 to the use of a vehicular tunnel for both the East and Hudson rivers. With this forward-thinking idea, he was unfortunately nearly a century ahead of his time.

Believing that a war with Great Britain was surely coming, the Colonel wrote on August 12, 1807, to Selah Strong, chairman of the Corporation of New York, where he stated his view that "war with England is inevitable."[36] To prepare, he proposed a system to defend the New York harbor, for which one hundred thousand dollars was appropriated by Congress. In December 1808, when he learned that the federal government was contemplating building a number of gunboats, he proposed using his shipyard at Hoboken, as it was strategically located and well equipped. Nothing came of his proposals, so the Colonel reverted to his commercial pursuits.

According to an April 2, 1813, letter from Fulton to his attorney, Cadwallader Colden (see chapter 12), Fulton stated that his partner Livingston, in the winter of 1806–1807, proposed to the Colonel an equal, three-way partnership with the steamboat venture, provided that Stevens would pay one-third of the expenses up to that time, amounting to $1,666; however, Stevens declined.[37] In turning down the proposal, Stevens pointed out that the agreement signed in 1800 already provided him, as well as Roosevelt, with a one-third interest in the exclusive right of a monopoly for operating steamboats on the waters of New York: "It was not, therefore, from motives of generosity that this offer was made to me, but from a sense of equity and justice."[38]

Although this was indeed true from a legal standpoint, it was equally true that the time provision to perform for the New York monopoly had expired. Livingston later had renewed it in his and Fulton's name, so the Colonel was not in a good negotiating position. Proud, though, he was and believing that Fulton would fail, he confidently declined the offer. As events that followed created a significant amount of consternation among Livingston, Fulton, and Stevens, Turnbull correctly observes, "[Stevens] was about to be thrust out into the cold unless he accepted Livingston's offer; he would have saved much worry and a great sum of money if he had then and there put his pride in his pocket and taken his wallet out of it."[39]

There were at least two reasons for the Colonel to act as he did. First, he had seen Fulton's design and was convinced, at this time, that high-pressure steam was the way of the future, and the low-pressure Boulton and Watt engine, connected to paddle wheels, would not succeed with any reasonable speed. Secondly, he had already spent a large sum of money pursing steamboat design and believed he would eventually secure the upper hand, putting Livingston in a position to *deal* with him. His ego got in the way, and his visionary high-pressure steam plans were not to be quickly supported by technology.

Had the Colonel, when he sent John Cox to England in 1805 to convince Boulton and Watt of the merits of high-pressure steam, instructed his son to first understand what the world's foremost steam engine producers were doing and why, he might have acted differently. They certainly instinctively knew that high pressure yielded more horsepower per pound of engine, a measure of mobility efficiency, but they also knew the difficulties of development that high-pressure boilers posed. At this time, the steam generator was the weak link in the strong steam equation, and even after introduction, it would present serious safety problems for another quarter-century. Unless vision is supported by technology, it remains prophecy, not product—a very important point that Stevens missed.

Still another reason for the Colonel's refusal to accept Livingston's new partnership proposal was grounded, as reflected in a series of letters with the Chancellor, in his interpretation of the Constitution. It was Stevens's view that the monopoly was unconstitutional, hence of little value. Livingston, on the other hand, argued that the states alone had the right to regulate traffic on their waterways. He cited that New York, which claimed the entire ownership of the Hudson River, provided for licensing for crossing ferries. He went on to argue that there could be no difference between a boat crossing the river and traveling along it. The Colonel, however, did make such a distinction; and when Justice John Marshall handed down the 1824 Supreme Court decision, the monopoly was declared unconstitutional. The federal government had reserved the right to regulate interstate commerce, and no distinction was made to crossing or to traveling along a river.

After the August 1807 successful run of the *North River Steam Boat*, the Colonel reevaluated his high-pressure steam position and decided to construct a steamboat along the lines of Fulton's. He informed Livingston of his plans, and on January 7, 1808, contracted for a 120-foot vessel to be delivered by April 1. This alarmed Fulton and Livingston, because if Stevens could build a boat, they most likely could keep him off New York waters, but not elsewhere. To forestall this possibility, they drafted a lengthy letter, dated January 13, 1808, which the Colonel received about one week later. In it they argue at length their monopoly rights in the state of New York and ended with much patent protection bluster. About this, Roy Louis DuBois writes,

> At the close of the long reply they warn Stevens that if he proceeded with his present plan of construction he would interfere with their right under a patent from the United States. That statement was more than misleading; it was blatantly dishonest. In fact it was no more than an outright falsehood. They had no patent whatsoever in January of 1808, nor had they even applied for one or expressed their intention of applying for one. It was not until January 19, 1809, a year later, the Fulton presented his specifications and petition for a United States patent.[40]

Fulton and Livingston had plans not only to control steamboats in the waters of New York but in all of the United States and would use intimidation, threats, and even lies to do so. At this point, however, Stevens was not aware that a patent had not been obtained, but he would find out later, when he wrote William Thornton at the U.S. Patent Office.

The very last sentence in Fulton's letter read, "You will give a written acknowledgment that your boat is worked by our permission under our patent,"[41] stabbing the Colonel in his considerable ego. He responded by taking a wait-and-see approach, noting in the reply that his lack of rebuttal was not because he could not, but because he found it "unnecessary."

His beautiful steamboat, the *Phoenix*, was launched Saturday, April 6, 1808, at four o'clock in the afternoon and was highly touted: "For model and workmanship, she is allowed by judges to be the superior of any in this country" (see fig. 12.10).[42] By late July, the machinery had been installed, and the boat was about ready. The Colonel wrote to Livingston, testing the water (so to speak), and informing him,

> As my vessel and machinery are now nearly completed, I wish to know whether she will be permitted to navigate the Hudson between the Cities of New York and Albany, provided she accomplishes the journey in a shorter time than you can possibly do with

Figure 12.10. The *Phoenix*, Stevens's low-pressure steamboat, and the first steamboat ever to operate on the ocean. (From an old black-and-white postcard in the possession of the author. The image is from a painting attributed to Charles B. Lawrence. Reprinted by permission of the Mariners' Museum, Newport News, Virginia.)

your boat and machinery. Should you signify your permission, I shall have no objection to navigate under a license granted by you. Whereas, in the case this permission is not obtained, she will, for the present at least, be cleared out of Perth Amboy, and passengers to and from the different landing between this place and Amboy will be taken in and landed at Hoboken.[43]

The canny, confident, and perhaps a bit cocky Stevens was laying down a challenge and attempting to secure the route on the lower Hudson by suggesting that the monopoly would only be practically enforced on the profitable New York–Albany route. No doubt he expected a refusal for this run, because if his boat should prove faster, he would garner the lion's share of the business, but in any case negatively impact their profits. In the temporary absence of Fulton, Livingston replied on July 26, 1808, suggesting that it would have been better for the Colonel to go elsewhere, "As the waters of this State have been preoccupied, first by Fitch and then myself for years before you thought of a steam boat," implying that Stevens had no right conferred by their twenty-year agreement signed in 1800. Fulton, upon seeing the Colonel's letter, wrote Livingston with a give-no-quarter approach: "I will never admit that he shall navigate the waters of the State of New York during the terms of years which the law of that State had secured to us."[44] So much for compromise.

Fulton's letter was sent to Stevens, who meanwhile had tested the *Phoenix* beginning on September 27, 1808.

Trial 4. Running several hundred miles with his new steam engine, boiler, and "wheels at the sides," the appropriately named *Phoenix* (as she had risen from the ashes of the failed agreement with Livingston) lived up to the Colonel's expectations. Traveling between Hoboken and Perth Amboy, a distance of thirty miles, the running time had been five hours, twenty-six minutes, just over 5½ miles per hour. There were a number of problems; however, Stevens believed that the steamboat, one hundred feet long with a sixteen-foot beam, could be improved and achieve an even greater speed, but for the moment he was gratified.[45]

He now replied to Fulton's letter, in particular addressing the vertical wheels, and challenged him to a lawsuit. He believed that if he lost he would resort to another propulsion technique, but if he won, Fulton's alleged patent would be badly damaged, and this would "prove fatal to [him]." Regarding the monopoly on the Hudson, he suggested, "surely a sister state [New Jersey] would have an equal right to grant her citizens similar privileges to the exclusion of all others. The consequence is obvious."[46] Stevens certainly had a way to "politely pick scabs," and now the lines in the sand had been drawn—but financially, he was in no position to sustain a drawn-out litigation.

This affront to the Fulton-Livingston monopoly brought the expected threat of lawsuit, but more pressing was the danger of attachment of the Colonel's steamboat. So he was faced with the choice of fight or flight. Being cash-poor, and with his lots in

Hoboken not selling well enough to produce a sustainable income, he started to evaluate the economic side of the situation and considered moving his boat to the Delaware River, out of reach of the monopoly.[47] In the interim he kept his boat operating, awaiting Fulton's next move.

About this time, he began to wonder about the specific claims of Fulton's patent, so he wrote William Thornton of the patent office to request a copy. On November 2, 1808, Thornton wrote back, "You write for a copy of the Specification of Mr. Fulton's Patent for Steam Boats. Not being his Invention, he has never applied for a Patent."[48] Thornton, of course, considered his former steamboat partner, John Fitch, to be the real inventor.[49] The surprised Stevens assumed that this answer must have been a mistake, as abundant correspondence from Fulton and Livingston had made statements about the patent and its coverage, so he again wrote Thornton, who replied and confirmed his previous response. Now Stevens was not only surprised but also "stunned." Turnbull observes, "The boot was decidedly upon the other foot and their standing was not to be compared with his own."[50]

It was not beyond Fulton to "stretch the truth," as he had done in his submarine proposals, or to become an opportunist, seizing the monopoly although it was in opposition to his avowed free market views; but now the charge of subterfuge could be squarely leveled. This approach was the direct result of a strategy to make the steamboat operations profitable. Of the monopoly, Gregg observes, "The chief techniques were first to secure the monopoly vantage point of a state grant and/or patent right through political pressure, and then to defend this position by threats, intimidation, coercion, cutthroat competition, amalgamation, seizures, and reprisal. As Stevens bitterly remarked, this was a competition 'not of merit but of the purse.'"[51]

A patent was subsequently issued to Fulton on February 11, 1809, and Stevens requested a copy to ascertain what, if anything, he may be infringing. The patent office's policy at that time required the inventor to approve dissemination of his patents, and Fulton refused to allow Stevens a copy. The Colonel hired a respected Philadelphia attorney, Horace Binney, to act as his Washington representative, and Binney reported that the secretary of state had decreed that anyone wanting a copy of a patent (see fig. 12.11) was entitled to one. Continuing, he added,

> Mr. Fulton's patent appears to be held in no very great estimation by Dr. Thornton. He told Fulton that it was not worth sixpence, and that he thought the monopoly of the North River to be in point of law of as little value; so that as far as the intimacy of this gentleman with the patent law of the U.S. and with the opinion of the Secretary of State upon this subject, may be considered as the ground of his sentiments, the circumstance is in your favor. I have little doubt that they are the correct sentiments upon this matter, for which I hope to be able to give you my reasons upon my return.[52]

Regarding the patent, he was undoubtedly right, as it dealt with discoveries, not even his own, on the proportions of a vessel and its resistance to water. Furthermore, paddle wheels were not new, and certainly not Fulton's invention. Nevertheless, a patent was issued, which contained the further surprising statement that a steamboat with a Boulton and Watt engine could not exceed six miles per hour.

Meanwhile, Fulton had completed his second steamboat, the *Raritan*, but turned its operation over to John R. Livingston, the Chancellor's brother. John Livingston served notice to the Colonel

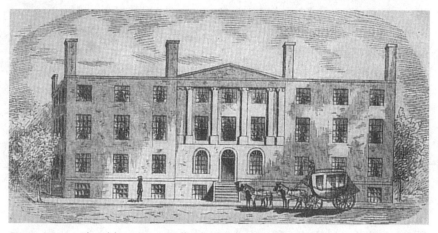

Figure 12.11. The old U.S. Patent Office building in Washington. (Pen drawing from *Real America in Romance, Valor and Victory: The Age of Vindication, 1783–1824*, ed. Edwin Markham, 13 vols. [New York: William Wise, 1912], 10:379.)

that he planned a competitive New York–New Brunswick run on the same schedule and from the same location as the *Phoenix*, but he would charge passenger fares "only one-third as high" as those of Stevens. This deliberate predatory pricing scheme caused Stevens to lament, "had Mr Livingston [the Chancellor] confined himself to threats of prosecution merely, I should have probably have remained and waited the event of a suit. But when he had recourse to a species of warfare which it was presumed my limited finances would not admit me engaging in, I retired."[53]

The fight on the lower Hudson was over. It was time for flight, and the Colonel anxiously watched the weather, waiting for an opportunity for Robert Stevens to take the *Phoenix* out to sea, down the coast, and up the Delaware River. The day came on June 10, 1809, and, accompanied by a small schooner, the steamboat safely arrived in Philadelphia after a historic thirteen-day voyage.

About July 10, steamboat service between Philadelphia and Trenton was inaugurated, and the forty-mile trip one way was made in eight hours. This provided the Colonel with a source of income, bolstering his confidence. So when he received a joint letter from Livingston and Fulton offering the use of their machinery and experience and "privately" rather than publicly acknowledging working under their patent, he replied that his boat in no way infringed on their patent but acknowledged, "I am probably doomed to be pursued and persecuted, on the Delaware, in a manner similar to that employed on the New Brunswick run." In addition to some financial prosperity, he had obtained legal opinions that the Hudson monopoly was unconstitutional and somewhat belligerently replied that should he be attacked, "all prospects of accommodation would end."[54] He was ready to fight!

Probably recognizing the truth in the Colonel's response and detecting a tone that would not allow further intimidation, Fulton arranged a meeting, which was held on November 27. An agreement was reached to split up the United States, which Fulton and Livingston had believed was theirs. They, the monopoly, would have New York State, Lake Champlain, and the Brunswick, Ohio, and Mississippi rivers; Stevens would have the Delaware, Chesapeake,

Santee, Savannah, and Connecticut rivers for seven years. However, if no operations were established in that time, the waterways would revert to the monopoly. At last there was an agreement with the monopoly that portended, for the moment at least, peace.

The Colonel turned his attention to improvements in the *Phoenix* and obtained a patent for his steamboat. Issued on June 3, 1810, it covers the use of a crankshaft, double-stroke air pump (air and water injection), steam cutoff valves, and a fire tube boiler. With a vision of intercoastal water transportation, he surveyed the Chesapeake Bay and advertised a plan for a steamboat run from Baltimore to Elkton, then a stage to Wilmington, and another boat to Philadelphia.[55] Looking at the northern end, he obtained a fourteen-year lease for the Hoboken ferry and in the fall of 1811 christened the *Juliana*, advertising the world's first steam ferryboat service.

About this time, a group of investors in Albany launched the steamboat *Hope* and introduced Hudson River service in direct competition with the monopoly. Fulton asked the Colonel to assign the monopoly his steamboat patents to aid in his defense, reassigning them when the court action was over. Using only his own knowledge of the law, Stevens concluded such an action to be subterfuge and totally inadmissible; hence, he declined to become involved.

The Colonel proposed ferry service across the Delaware River at Philadelphia, but another builder was selected, who built a steamboat that appeared to have infringed on Stevens's patent for the *Phoenix*. Rather than instigate legal action, the Colonel, with his partner, Maj. Theodosius Fowler, decided to invest their energies and money in another, elegantly appointed steamboat, with a keel of 136½ feet and a twenty-one-foot beam. Completed at the end of 1812 and named the *Philadelphia*, she soon was affectionately known as "Old Sal." The fastest steamboat afloat, she achieved eight miles per hour, exceeding Fulton's stated 1809 patent maximum by two miles per hour.

In October, the Colonel received a rather impudent and arrogant letter from Fulton, accusing him of trying to lure away some of his workmen and threatening legal action. Stevens replied that he found the letter offensive and knew nothing of the allegation, closing with, "be assured, Sir, that the very indecent threat contained in your letter has not, nor never will have, any influence whatsoever on my conduct."[56]

The Colonel had "thrown a rock through a hornet's nest," and Fulton retaliated on a front where he believed he had a strong legal position:

> Your letter appears to imply that you have not encroached on my rights. It is time, however, that point should be settled by our counsel. You are running a steam boat [*Juliana*] without a license from my partner or me. When the boat was talked of, and long after the Paulus Hook was formed, the most which was contemplated was to carry foot-passengers. But horses, gigs, carriages, and cattle are being carried, to the injury of the Paulus Hook Company, and in contempt of my United Sates Patent and Livingston and Fulton's State Grant. These, Sir, I believe to be encroachments which cannot be maintained in law.

In the two next paragraphs, it becomes clear that what is "sticking in Fulton's craw" is the Colonel taking credit for the success of the steamboat, an affront to his inflated ego:

> My letter, although conditional, seems to have given you some offence. Your letters, last winter, to Mr Boyd, making yourself the inventor of steamboats and I mere cypher, had something in them really offensive and extemely injurious, by raising a cabal against extending patents to twenty-one years; by which you extremely injured me, and yourself through me, for your only protection is in the protection of my rights. This you will discern when suits arise on the Delaware or Chesapeake.
>
> You had no claims on me, either as relation of friend, yet I came forward and granted you great privileges which I hoped you would at least see the policy of acknowledging and maintaining. But the arduous desire to be thought an inventor has kept you at constant war with Livingston and Fulton, and particularly with me, my interest, and your own through me. This you have evinced in your writings, conversations, and acts. But, Sir, here ends writing on this subject, on my part. When justice fails, the law must make up the deficiency.[57]

Fulton was upset and distressed that the Colonel was taking credit for what he believed was rightfully his invention, something he, in his quest for lasting recognition, desperately wanted for himself. As Dorothy Gregg correctively observes, "Each one passionately desired to go down in the annals of posterity as *the* inventor of the steamboat and neither could ever forgive the other his 'pretensions.'"[58]

The Fulton letter ended portending legal action, but the Colonel hoped to avoid this and, rather than further aggravate the situation in the north, looked south. By December 24, 1812, a bill was passed by North Carolina, providing him a twenty-year monopoly in the state, only to have it challenged by Fulton's agent John Devereau DeLacy as having been obtained under false pretenses.

In a drafted, unsigned document in Fulton's hand, dated February 1813, the monopolist detailed his strategy on how to block the Colonel's North Carolina activities and also attacked Oliver Evans, saying he had "contrived to patent almost the whole machinery of mills and . . . written a book of dreams on steam-carriages."[59] He went on in the third person tooting his own horn, adding, "the first useful steamboat . . . required the fortunate circumstance of adequate genius and capability in the same person or persons and [Fulton and Livingston] persevered to success."

Although throwing verbal rocks at Stevens—who was building steamboats with all American parts and "stealing his thunder"—was perhaps to be expected by someone with Fulton's inflated ego, it does, however, seem odd that he would also attack Evans, someone he barely knew. Evans was truly an original thinker, years ahead of Fulton in ideas and concepts, and was most deserving of patent protection. Fulton, on the other hand, had no real patent position, as he had not invented anything, only arranged the development of others into a configuration that was commercially successful. This contribution, though, was significant, as it demonstrated to everyone the viability of steam navigation, but as such was not new or novel. Furthermore, he sought to keep ahead of his competition by using the Fulton-Livingston monopoly to threaten legal action against more innovative developers.

The Colonel considered ways to satisfy Fulton, outlining some ideas that he wanted to discuss in a letter to his son John Cox but, noting the Chancellor's ill health, had not written him. Livingston died one week later, and though his rights were passed to his heirs (represented by Robert L. Livingston), practically speaking, the partnership power rested with Fulton. The Colonel decided to kill two birds with one stone and offered the younger Livingston (1775–1843) and Fulton each one-third of his North Carolina monopoly and to relinquish rights to all waters of the United States, provided he would be given the right to navigate the Hudson River. All parties could use patents held by each, but Stevens refused to acknowledge the superiority of Fulton's patent. Copies of the agreement were sent to each party; Livingston rejected it, but Fulton made a counterproposal.

Fulton wanted the Colonel to provide him "recognition" and in addition place more severe conditions on his ferry operation. Turnbull notes, "Peace, except at this own price, was the last thing Fulton wanted," and when the Colonel turned down the offer, "his next step was swearing of injunctions" leading to seizure of the steam ferry *Juliana*.[60] Again the Colonel was faced with the choice of fight or flight and, wanting to avoid litigation that was "so pregnant with expenses," sent the *Juliana* to safe waters on the Connecticut River, barely escaping six barges filled with men intent on confiscation.

To continue his ferry, he cleverly instituted horse boat service. Then Fulton, through DeLacy, pursued the Colonel in North Carolina, petitioning the legislature to repeal the monopoly. He argued that the privilege had been obtained, "surreptitiously and by false suggestions," also belittling the Colonel's steamboat contributions in the process.[61] Stevens waged a defense before a committee appointed to examine the matter, and after due deliberation, the charges brought by Fulton were determined to be "unfounded and unbecoming." This gave the Colonel a much-needed and deserved victory, but unless he bowed to Fulton's ego, there could be no peace with the monopoly.

Stevens often would refer to his sons as the "boys." Robert had developed into a fine engineer, and Edwin into a capable businessman, upon whom could he could rely to keep his steamboat operations going. He then turned to another frontier, one that promised faster and faster speeds, the steam carriage, which was to become known as the railroad. Instead of the state of New York digging a much-discussed canal between Erie and Albany, he wrote Governor DeWitt Clinton on February 24, 1812, proposing a railroad. He provided a detailed proposal showing a cost of only one-quarter of that of the canal, with a speed of four miles per hour; however, it was very politely turned down. Turnbull notes the explanation given: "The commission, having given the proposals that consideration due to a 'gentleman whose scientific researches and knowledge of mechanical powers entitles his opinion to great respect,' found it self unable to concur with him."[62] Being rejected, Stevens prepared a rebuttal asserting that the commission's objections appeared "so void of real foundation," but again the Colonel was too forward-thinking in his ideas for the time. He could, however, have taken some solace from fellow visionary Oliver Evans: "One step in a generation is all that we can hope for. If the present shall adopt canals, the next may try the rail-way with horses, and the third generation use the steam carriage."[63]

Stevens instinctively knew he was on the right track, so to promote his new idea, the Colonel wrote and published in 1812 *Documents Tending to Prove the Superior Advantage of Rail-Ways and Steam-Carriages over Canal Navigation* (see fig. 12.12). In this visionary pamphlet, he states,

So many and so important are the advantages which these States would derive from the general adoption of the proposed rail-ways, that they ought, in my humble opinion, to become an object of primary attention to the national government.

I can see nothing to hinder a steam-carriage from moving on these with a velocity of one hundred miles per hour. This astonishing velocity is considered to be merely possible. It is probable that it may not in practice be convenient to exceed twenty or thirty miles per hour. Actual experiments, however, can alone determine this matter, and I should not be surprised at seeing steam-carriages propelled at the rate of forty or fifty miles per hour.[64]

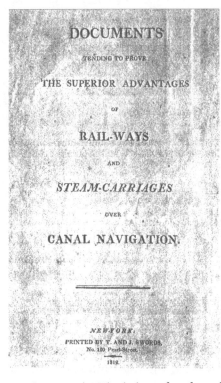

Figure 12.12. John Stevens correctly foresaw the coming of the railway that would overshadow canals in this remarkable 1812 document later called the birth certificate of U.S. railroads. (From John Stevens, *Documents Tending to Prove the Superior Advantages of Rail-Ways* [New York: T. & J. Swords, 1812], title page.)

Many years later his sons reprinted this publication. An official of the Pennsylvania system, upon reading it for the first time, declared the document "the birth-certificate of all rail roads in the United States."[65]

The Colonel traveled to wherever a railroad might provide an economic benefit, proposing one in Virginia to haul coal; one across the Delmarva peninsula; and still another from Philadelphia to Harrisburg. Not able to stimulate the same enthusiasm as he had previously for steamboats, he was labeled a "visionary projector." The time was not right in the United States, and Turnbull correctly notes, "The chance to be the first in the field with railroads slipped out of America's hands and was drowned in the canals."[66]

At the urging of Stevens, a railroad charter was passed on February 6, 1815, by the New Jersey legislature to establish a line from the Delaware River near Trenton to the Raritan at or near New Brunswick. Less than one week after the birth of the first U.S. railroad act, the United States' first commercially successful steamboat designer, Robert Fulton, died. The two events marked battles still yet to be won: to give life to railroad service in the United States, and to kill the Livingston-Fulton monopoly in New York. On both fronts, the Colonel was active.

John DeLacy, who had since severed his agent ties with Fulton, wrote Roosevelt that Fulton's "brazen-faced effrontery, as if he was entitled to violate the law as well as private rights," were for naught, as he was "down never to rise again." Turnbull observes that "[Stevens] had no love for Fulton and no cause to regret his passing."[67] Acknowledged by everyone who knew him, Fulton was brilliant, a genius, persuasive and charming. He was a driven man, willing to use any device and technique to succeed. He had many influential friends, and his attorney, Cadwallader Colden, would eulogize him in an 1817 biography, which would produce positive and lasting impressions on future generations.

Stevens maintained a reciprocal agreement with the Livingston heirs to use Fulton's patents, but when the Colonel considered instituting a suit in Philadelphia for infringement, he consulted his attorney, Horace Binney. After some investigation and consultations with other attorneys, Binney wrote Stevens on April 13, 1819. In about two pages, the attorney tore the heart out of Fulton's patents, noting that the one issued in 1809 was for proportions, not a machine, as Fulton did not claim inventing the machine, and proportions are not a patentable discovery. Furthermore, the proportions are not accurate, as Fulton concluded a maximum velocity of six miles per hour, which on examination of the boats on the Delaware is false. With regard to his 1811 patent, Binney correctly asserted that this was not the invention of a machine with numerous parts, but for a miscellany of individual parts, again not patentable. He pointed out that many of the parts, such as the sails, were not patentable at all and concluded the infringement suit would be without merit.

This was the patent position always shared by Stevens and Thornton, but it was never so elegantly and concisely stated with such impact. Thus two legs of the three-legged monopoly stool were now broken. The first leg was lost upon the deaths of the persuasive and capable Livingston and Fulton, who both had important political connections. The second leg fell when Binney shredded Fulton's patents; now all that remained was the legal monopoly itself, which was by this point under vigorous attack. The milk stool was about to topple.

Five years later, in February 1824—nine years after Fulton's death—the case of *Gibbons v. Ogden*, which was to decide the legality of the New York steamboat monopoly, finally reached the U.S. Supreme Court. With Chief Justice John Marshall presiding, Gibbons was seeking the right to navigate the waters of New York State and had Daniel Webster as his attorney (see fig. 12.13). When the decision was handed down, the monopoly was broken as unconstitutional. The Colonel had long held this position, and now it was official: he could run his steamboats on the Hudson, and the business of doing so was firmly in the hands of his "boys."

As mentioned earlier, in promoting the railroad, Stevens wrote, "The wealth and prosperity of a nation may be said to depend, almost entirely, upon the facility and cheapness with which transportation is effected internally."[68] The railroad, he believed, was a far better way to effect speedy transportation of goods and people. Pursuing this dream, he finally succeeded in having the state of Pennsylvania pass legislation for establishing a railroad from Philadelphia to Columbia on the Susquehanna on March 21, 1823. Subsequently, an amended bill was enacted on April 27, 1825, and two

Figure 12.13. Daniel Webster, left, successfully argued the case in which Chief Justice John Marshall, right, declared the Fulton-Livingston New York steamboat monopoly unconstitutional. (Portrait of Webster from D. H. Montgomery, *Leading Facts of American History* [Boston: Ginn, 1891], p. 243; portrait of Marshall from Allen C. Thomas, *History of the United States* [Boston: D. C. Heath, 1901], p. 197.)

years later the large sum of $2 million was appropriated for construction. When the line began operations, it was to become the first commercial railroad in America, finally reaching Pittsburgh and known as the Pennsylvania Railroad, a name suggested by the Colonel. Considering the section of track into New Jersey, which had been previously authorized, it can even be argued that the railroad was chartered eight years earlier, on February 6, 1815. Other railroads were soon authorized: the Mohawk & Hudson, April 17, 1826; the Baltimore & Ohio, February 28, 1827; the South Carolina Canal & Railroad, December 27, 1827. But Stevens was the prime mover in the success of the first railroad, growing out of his experimental circular track at his Hoboken estate, providing skeptics with their first "ocular demonstration" (see fig. 12.14).[69]

The Colonel and his boys continued to improve steamboat service, introducing the *Burlington*, the *Fairy Queen*, and the *New Philadelphia*, running at twelve miles per hour. The *North America* was placed into service and, with a pair of steam engines of the overhead beam type, achieved a speed of fifteen miles per hour. "Speed was his hobby; he looked for it in man, beast, or machine and kept records of the performances of all three."[70] In addition, Stevens's

Figure 12.14. Seeing is believing, according to the old saw, so John Stevens constructed a circular railroad track on his Hoboken estate to provide demonstrations to those unconvinced of the power of steam on rails. (From Turnbull, *John Stevens*, between pp. 448 and 449.)

stagecoach lines connected the various steamboat routes, providing the traveler or shipper with long-distance carriage.

Needing a better source of power to pull the trains, Stevens took notice of the Rocket locomotive, introduced by Stephenson in England in 1829. Two years later, the Colonel sent his son Robert (see fig. 12.15) to investigate, and he brought back two significant railroad improvements. The first was the T-rail, which he had designed on the voyage over. He had it fabricated at an iron foundry in Dowles, Wales, being made into sixteen-foot lengths weighing at first thirty-six pounds per yard, then later a heaver version, at forty-two pounds. He shipped back to America a quantity of 550 sections, each sixteen feet long. The second improvement was a new locomotive designed by Robert Stephenson called the Planet. When the younger Stevens saw trials of it in December 1830, he ordered one, which was later assembled in the United States. By November 12, 1831, it was renamed the John Bull and ran on the Camden and Amboy Railroad.

Because of a scarcity of stone blocks upon which the T-rail was laid,[71] Robert used cross-timbers, also called sleepers, secured by stone. He also developed the six-inch spike with a hook for attachment—the same construction method used today. Turnbull concisely sums up the father-son contribution to railroading: "Just as the labyrinth of threads on a railroad map of the United States leads inevitably to John Stevens, so the miles of actual steel are Robert's imperishable monument."[72]

The Colonel believed

Figure 12.15. The Colonel's son, Robert L. Stevens, became a fine engineer and carried on the vision of his father in steamboats and with the railroad. (From Thurston, *History of the Growth of the Steam-Engine*, p. 270.)

that "Good Morals and good Government in a Republic are only attainable and maintainable by knowledge and information pervading the whole mass of Society."[73] Stevens was a staunch supporter of education and even at age eighty considered running for office on an education program. When he passed away on March 6, 1838, "it was the Colonel's hope that some of his estate might be devoted to founding and sustaining an 'academy' for teaching fundamental subjects, and the elements of science."[74]

Edwin inherited much of the Colonel's fortune and in his will provided that a block of land, a building fund of $150,000, and an endowment of $500,000 be used to establish an academy. In September 1871, the Stevens Institute of Technology accepted its first students and has contributed as a school of higher learning to the present time—a living tribute to America's first great civil engineer, mechanical engineering visionary, godfather of the U.S. transportation infrastructure, and perhaps "the Leonardo da Vinci of colonial America."[75]

First presented in chapter 4, the following table of steamboat pioneers and their contributions has been updated with the efforts of Robert Fulton and John Stevens. The table will be augmented with three other contributors in chapter 13, where the inventor of the steamboat is identified.

Table 12.1. Summary of the efforts of steamboat pioneers through Stevens

Steamboat pioneer	Concept date	Steamboat trial date(s)	Notes
Papin	1690	1707	Human, not steam power
Hulls	1736	None	Concept only for steam tug
D'Auxiron	1770	None	In 1774 boat accidentally sunk
De Jouffroy	ca. 1771	1778 Unsuccessful 1783 *Pyroscaphe*	Some success on 1783 trial, but work interrupted by the French Revolution
Miller	ca. 1786	1787 *Edinburgh* 1788 Steamboat	Human, not steam power Engine by Wm. Symington
Dundas	ca. 1798	1802 Tugboat *Charlotte Dundas*	Hull by Alexander Hart with engine by Wm. Symington; towed 140 tons on a canal
Henry	1779	None	Concept only
Rumsey	1783	1786, 1787, 1790, 1792, 1793	Limited success with water jet propulsion
Fitch	1785	1787, 1788, 1789, 1790	Introduced steamboat service on the Delaware in 1790, which was not financially viable
Evans	1785–86	1805 Amphibious 1812 Middlesex 1816 *Aetna* 1819 *Philadelphia*	About 1787 proposed boat with paddle wheels to Fitch and yielded priority to him
Read	1788	1789	Human, not steam power
Morey	1790	1793, 1794, 1795, 1796, 1797	Demonstration for Livingston in 1796 at 5 mph; side paddle wheels in 1797
Fulton	1793	1803 Paris 1807 New York	Structure failed in first 1803 attempt; second achieved 2.9 mph; in New York in 1807 achieved 5 mph, securing monopoly
Stevens	1787	1798, 1801, 1804, 1808, 1812	1798, 3 mph 1808 *Phoenix*, 5.5 mph 1812 *Philadelphia*, 8 mph

Notes

1. Quoted in Archibald Douglas Turnbull, *John Stevens, an American Record* (New York: Century, 1928), p. 465. Turnbull viewed the Colonel in a very favorable light, as does the author of this account. Much of the information in this chapter is taken from his book, with the chronology following the order he presented. Other principal sources of information have been Roy Louis DuBois, "John Stevens: Transportation Pioneer" (PhD diss., New York University, 1973), a thoroughly annotated work; and Dorothy

Gregg, "Exploitation of the Steamboat: The Case of John Stevens" (PhD diss., Columbia University, 1951).

2. Quoted in Turnbull, *John Stevens*, p. 527.

3. DuBois, "John Stevens," pp. 1–2.

4. Ibid., p. 3n.

5. Quoted in Turnbull, *John Stevens*, p. 14.

6. Ibid., p. 23.

7. Ibid., p. 36.

8. Gregg, "Exploitation of the Steamboat," p. 37, states he served as state treasurer from 1777 to 1782.

9. DuBois, "John Stevens," p. 64.

10. Turnbull, *John Stevens*, p. 87.

11. Quoted in ibid., pp. 88–89.

12. Ibid., p. 89.

13. Quoted in ibid., p. 91.

14. Gregg, "Exploitation of the Steamboat," p. 97.

15. Quoted in Turnbull, *John Stevens*, pp. 101–102.

16. Ibid., p. 103.

17. Quoted in R. H. Thurston, "Messrs. Stevens, of Hoboken, as Engineers, Naval Architects and Philanthropists," *Memoirs and Professional Reports*, reprinted from the *Journal of the Franklin Institute*, October 1874, p. 2.

18. Turnbull, *John Stevens*, p. 103.

19. Ibid., p. 108.

20. Gregg, "Exploitation of the Steamboat," p. 112n.

21. Quoted in Turnbull, *John Stevens*, p. 108.

22. Quoted in ibid., pp. 109–10.

23. Quoted in ibid., p. 133.

24. J. H. B. Latrobe, *Lost Chapter in the History of the Steamboat* (Baltimore: John Murphy, 1871), p. 40.

25. Quoted in Turnbull, *John Stevens*, p. 135.

26. Quoted in Latrobe, *Lost Chapter in the History of the Steamboat*, p. 35.

27. Quoted in Turnbull, *John Stevens*, p. 161.

28. Quoted in ibid., p. 162.

29. Gregg, "Exploitation of the Steamboat," p. 78.

30. Turnbull, *John Stevens*, p. 176.

31. Quoted in ibid., p. 183.

32. James Renwick, letter to Frederick De Puyster, quoted in ibid., p. 185.

33. Quoted in Turnbull, *John Stevens*, p. 189.

34. In 1844, an independent committee of the American Institute of New York reconstructed the 1804 *Little Juliana* for its annual fair, held at Niblo's Garden. The refurbished original engine and boiler, along with reconstructed propellers, were placed in a boat and achieved a speed of eight miles per hour. One member of that committee was James Renwick, to whom we are indebted for the description of the 1804 trial.

35. Gregg, "Exploitation of the Steamboat," p. 115.

36. Quoted in Turnbull, *John Stevens*, p. 225.

37. Cadwallader D. Colden, *Life of Robert Fulton* (New York: Kirk and Mercein, 1817), p. 167, states that prior to the completion of Fulton's steamboat, offers were made to "several gentlemen" to share in one-third of the cost. Turnbull, *John Stevens*, p. 235, cites Fulton's April 2, 1813, letter to Colden, where Livingston made the offer to Stevens of $1,666 for one-third of the total cost. DuBois, "John Stevens," p. 177, states that Stevens was made the one-third offer for $5,000 and an acknowledgment that Fulton was the inventor of the steamboat. In any case, Stevens did not accept.

38. Quoted in Turnbull, *John Stevens*, p. 235.

39. Ibid., p. 236.

40. DuBois, "John Stevens," pp. 234–35.

41. Quoted in Turnbull, *John Stevens*, p. 254.

42. Ibid., p. 257.

43. Quoted in ibid., p. 258.

44. Quoted in ibid., p. 259.

45. DuBois, "John Stevens," pp. 184–85.

46. Quoted in Turnbull, *John Stevens*, p. 260.

47. See DuBois, "John Stevens," pp. 186–93, for a discussion of the Colonel's financial woes during the period of the steamboat's construction.

48. Gregg, "Exploitation of the Steamboat," p. 128.

49. W. Thornton, *Short Account of the Origin of Steamboats* (Washington, DC: Rapine and Elliot, 1814), p. 1.

50. Turnbull, *John Stevens*, p. 267.

51. Gregg, "Exploitation of the Steamboat," p. 139.

52. Quoted in ibid., p. 133.

53. Quoted in Turnbull, *John Stevens*, p. 271.

54. Quoted in ibid., p. 283.

55. Later, in June 1812, he would advocate the establishment of the Union Line to construct a turnpike between Elkton and Wilmington.

56. Quoted in Turnbull, *John Stevens*, p. 331.

57. Quoted in ibid., p. 332.

58. Gregg, "Exploitation of the Steamboat," p. 94.

59. Turnbull, *John Stevens*, p. 335.

60. Ibid., p. 340.

61. Gregg, "Exploitation of the Steamboat," p. 109.

62. Turnbull, *John Stevens*, p. 371.

63. Oliver Evans, "Memoir 'On the Origin of Steam Boats and Steam Waggons,'" ed. Arlan K. Gilbert, offprint of *Delaware History* 7, no. 2 (September 1956): 164.

64. John Stevens, *Documents Tending to Prove the Superior Advantages of Rail-Ways and Steam-Carriages over Canal Navigation* (New York: T. and J. Swords, 1812), pp. 3, 6, 6n.

65. Quoted in Turnbull, *John Stevens*, p. 373.

66. Ibid., p. 381.

67. Ibid., pp. 434–35.

68. Quoted in ibid., p. 465.

69. Quoted in ibid., p. 478.

70. Ibid., p. 487.

71. In a discussion of the Frenchtown and New Castle Railroad, George Johnston, *History of Cecil County, Maryland* (1881; reprint, Baltimore: Genealogical Publishing, 1989), p. 426, provides a description of the stone blocks. They were about ten to twelve inches square and had two holes drilled through, into which wooden plugs were inserted. The blocks were set upon the ground the distance between the rails, and the rails were attached to the stone blocks with spikes driven into the wooden pegs.

72. Turnbull, *John Stevens*, p. 509.

73. Quoted in ibid., p. 526.

74. Ibid., pp. 526–27.

75. J. K. Finch, *Early Columbia Engineers: An Appreciation of John Stevens, James Renwick, Horatio Allen, and Alfred W. Craven* (New York: Columbia University Press, 1929), p. 6.

Part 3.

Conclusion and Afterthoughts

Figure 13.1. *The Thinker*, an 1880 sculpture by Auguste Rodin (1840–1917). (Rendered by William T. Sisson from an engraving by André Leveillé [ca. 1860–1920].)

13.

The Steamboat Inventor

Every man of genius is considerably helped by being dead.

—Robert S. Lynd

I n assessing claims as to which candidate is most deserving of the title "Inventor of the Steamboat," it should be noted that between the times of William Henry's concept and Fulton's conquest (with the first commercially successful steamboat), there was a continuum of effort in America to bring steam navigation into existence. Following Fulton, there were many new entries, and by 1816, the time of his death, seven steamboats were being built each year, going to fifteen per year by 1824. In another decade, the yearly number had jumped to fifty and continued to grow, as the United States took the world steamboat lead. The time, the technology, and the opportunities in the first part of the nineteenth century in the United States were finally right.

But before launching into a discussion of the inventor of the steamboat, the contributions of several other individuals should be mentioned. Their efforts were not sustained over a long period of time, however, so they can be covered in a few pages.

Other Contributors

Elijah Ormsbee

Elijah Ormsbee, a Connecticut carpenter by trade and inventor by avocation, teamed up with David Wilkinson of Pawtucket, Rhode Island, to, as they used to say in that part of the country, "get up" a steamboat.[1] To do this, Ormsbee borrowed a long boat belonging to the ship *Abigail* and obtained the loan of a 150-gallon copper kettle from a distiller. Wilkinson set about making a steam cylinder and piston as well as producing some of the other iron components.

When the parts were ready, Ormsbee retreated to a place called Windsor Cove, or Cave, near Providence, Rhode Island. There he fabricated the remaining necessary components, installed the engine and boiler, mounted the goose-foot paddles, and connected every-thing together. Loading on a supply of wood and making last adjustments to his machinery, he was ready for a trial.

On an afternoon or early evening in the autumn of 1792, or perhaps as late as 1794, he fired up his boiler, engaged the steam engine, and glided out into the river. The goose-foot propulsion spurted the little boat along to a long wharf in Providence. Enthralled with the steamboat's success, the next day Ormsbee proceeded up the Seekonk River to Pawtucket to demonstrate the steamboat to his partner Wilkinson (see fig. 13.2).

For several weeks he ran the boat up and down the river, to the astonishment of every observer. The piston in the steam cylinder was pushed upward by several pounds of steam; water was then injected and the steam condensed; and the atmospheric pressure forced the piston back down. The goosefoot paddles folded together when they were driven toward the front of the boat, then opened to eighteen to twenty-four inches in width for the aft stroke, which imparted forward motion. A pulsating speed of three to four miles per hour was reported.

Not finding anyone to finance a larger and more practical model, Ormsbee disassembled the steamboat, returning the long boat and copper still to their owners. Here the fun adventure ends.

Figure 13.2. View of Pawtucket, Rhode Island, as it appeared in an early sketch. (From Henry Howe, *Memoirs of the Most Eminent American Mechanics* [New York: J. C. Derby, 1856], p. 93.)

William Longstreet

William Longstreet was born October 6, 1759, near Allentown, New Jersey, and grew up in Monmouth County.[2] About 1784, he married Hanna Randolph, who upon the death of her father inherited a considerable fortune. With their new-found wealth, Hanna and William decided to start again in Augusta, Georgia, moving with their new son in the fall of 1785. They joined some friends already there and found the land less expensive. Of this union was also born Augustus Baldwin Longstreet in 1790, who throughout his life would pursue many vocations, becoming well known before he died.

William was a practical and capable man, often turning his attention to invention of mechanical devices such as the cotton gin, "something prophetic of the sewing machine,"[3] and the steam engine. Teaming up with Isaac Briggs, who provided some financial support (and who would later apply for a federal patent), a steam engine was developed, and the General Assembly of Georgia passed an act on February 1, 1788, securing to "Isaac Briggs and William Longstreet, for the term of fourteen years, the sole and exclusive privilege of using a newly constructed steam engine invented by them."[4]

On September 26, 1790, Longstreet wrote a letter to the governor of Georgia, Edward Tefair, in Savannah:

> Sir:—I make no doubt but that you have often heard of my steamboat, and as often heard it laughed at. But in this I only shared the fate of all the projectors, for it has uniformly been the custom of every country to ridicule even the greatest inventions until use had proven their utility.
>
> In not reducing my scheme to practice has been unfortunate for me, I confess (and perhaps the people in general), but until very lately I did not think that either artist or material could be had in this place sufficient.
>
> However, necessity—that grand source of invention—has furnished me with an idea of perfecting my plan, almost entirely with wooden materials, and by such workmen as may be gotten here; and, from a through confidence of its success, I propose to ask for your assistance and patronage.[5]

Nothing came of the request for governmental assistance, but many years later, by the middle of the first decade of the nineteenth century, Longstreet had acquired enough funds to begin. Activity was initiated and was greeted by skeptics, which he anticipated and placated, but he was not very happy when he attended the theater one evening and found himself the butt of a song:

> Billy boy, Billy boy, can you steer the ship to land?
> Billy boy, Billy boy, can you steer the ship to land?
> Yes, I can steer the ship to land,
> Without a rudder in my hand.
>
> Billy boy, Billy boy, can you row the boat ashore?
> Billy boy, Billy boy, can you row the boat ashore?
> Yes, I can row that boat ashore
> Without a paddle or an oar.

When the amused audience started "shaking with laughter" and turned to see Longstreet's reaction, the chagrined inventor rose from his seat and "strode majestically out."[6]

The incident no doubt made him even more determined to be

successful, and in 1807, a few days after Fulton's *North River Steam Boat* demonstration, Longstreet's "clumsy wooden-machinery boat," propelled by long wooden poles that pushed against the bottom, moved up the Savannah River at five miles per hour.[7]

Compared to other steamboats being built at the time, the vessel was not practical, and after the trial, no further mention of it has been found. Longstreet died on September 1, 1814, and his tombstone in the churchyard of St. Paul's in Augusta attests to his blithe spirit: "All the day of the afflicted are evil, but he that is of a merry heart hath a continual feast."[8] Such was the life of William Longstreet, a man of many interests and talents.

Green and Backus

In 1789 after working as the master of the Potowomut anchor works at Coventry, Rhode Island, for some time, Griffin Green moved to Belpré, Ohio.[9] He was to become involved in building and operating a floating gristmill on the Ohio River. About the same time, Elijah Backus, formerly of Nantucket, Connecticut, purchased an island in the river opposite the town of Belpré.

Over the next several years, these two individuals became known to each other and decided to produce what was probably a model steam engine. Most likely this effort was started early in 1796, as by May 18, Green signed a contract with John Patterson of Philadelphia for "what certainly must have been a full-sized steamboat."[10] On June 2, 1796, a formal agreement between the two investors was drawn, with Backus having five-twelfths and Green seven-twelfths of the rights to the steam engine. It seems that Patterson was an experienced engine builder, having built the steam engines at London Bridge.

By June, however, the project ran into financial difficulties, so more partners were brought into the agreement, providing new sources of money. Despite this effort, it appears the partnership was forced into bankruptcy, and the steam engine and ill-fated boat were never completed.

Inventor of the Steamboat

An easy way to conclude this chapter would be to assume the position taken by many historians, namely, that it was not an invention at all, but rather a series of developments leading to eventual success. The book's title belies such a conclusion, though, and the facts uncovered herein also indicate otherwise, so this conclusion is excluded. Another way to wrap up the discussion without too much controversy would be to declare the school textbook favorite, Robert Fulton, as the inventor, but again the information presented points in another direction.

Therefore it is required to briefly reexamine the concept of invention and compare this against the work of all the steamboat pioneers. The first patent board was faced with a similar decision, so first let us revisit what they did, perhaps to gain some insight.

The U.S. Patent Board in 1791—consisting of Secretary of State Thomas Jefferson, Secretary of War Henry Knox, and Attorney General Edmund Randolph[11]—did not choose to unravel the conflicting claims of Rumsey, Fitch, Read, and Stevens. We can, however, over two hundred years later revisit their responsibility and determine what, if any, conclusions could have been drawn. The patent legislation of 1790 authorized the three members of the board, or any two of them, to grant patents for "any useful art, manufacture, engine, machine, or device, or any improvement therein, not before known and used . . . [if the patent board] . . . shall deem the invention or discovery sufficiently useful and important."[12] To qualify for a patent, an applicant needed to describe, with specifications and drawings, (1) something tangible such as a machine or device, (2) something new that was not before known or used, and (3) something useful and important. Formulas and principles were excluded.

As we are dealing with a steamboat, this invention clearly qualifies as a machine or device, and furthermore, the patent board and nearly every man of science agreed that the steamboat was useful and important. We, therefore, are only left to resolve the question of newness. For the patent board to determine which applicant had priority, it would have been necessary to ascertain who first conceived of a steamboat *that could be brought to a practical reality.* For a part-time board with other pressing duties, and in the face of claims and counterclaims, this task most likely would have appeared daunting, although an unsuccessful attempt was made.[13] In a letter from patent board secretary Remsen to John Stevens on May 4, 1791, it was stated, "The investigation of the claims of Rumsey & Fitch to a priority in the application of Steam to navigation, which was to have been made by persons mutually appointed by them, was attempted but not completed; and the Commissioners at a late meeting agreed to grant patents to them—and to all claimants of steam patents—according to their respective specifications."[14] Indeed, two hundred years later, it still poses some difficulties.

As steamboats had been prophesied for many years, any inventor claiming to be first could always face the possibility that another individual could argue that he had the idea first, leading to another claimant, etc. However, recalling the discussion in chapter 2, a new idea, even though it may be novel, does not qualify as an invention unless the two pillars of science and technology can support it. Although this is understood today, two hundred years ago, the concept had not been evolved to the point it could be used to discriminate. Therefore, using current knowledge of the nature of invention along with the methodology of engineering product development, a researcher today can proceed with a clearer mental image of the critical issues that must be examined and resolved.

To produce a visual depiction of an overview of the development of the steamboat and the contributions of each inventor, the reader is invited to examine the graph in figure 13.3. On the horizontal axis is the time period over which the invention of the steamboat took place, running from 1770 to 1820. On the vertical axis are three labels:

- *Concept.* This is the date the idea of a steamboat occurred to the inventor and is indicated by a dot on the graph generally in the center of the band, but offset in some cases to allow for a less cluttered presentation.
- *Trials.* In order for a trial to take place, the inventor underwent a process of design, development, and integration that was

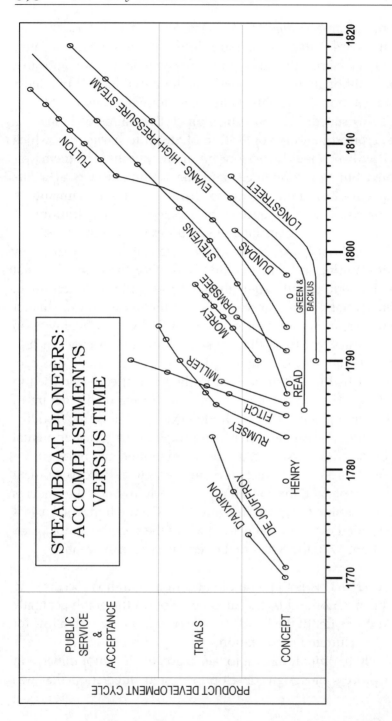

Figure 13.3. Graph showing accomplishments of the steamboat pioneers. (Graph by Jack L. Shagena.)

followed by testing. Again, dots on the graph represent approximate trial dates by the successful inventors.

- *Public Service & Acceptance.* If the inventor was able to offer steamboat service to the public, this is shown by a dot in the lower part of this band. If the service was accepted by the public, resulting in a financially viable operation, a line is extended upward toward the top of the band.

These three bands embody the three phases of the engineering development cycle, as discussed in chapter 3: concept, design and development, and verification.

Table 13.1 is an updated form of the table previously presented, showing the contributions of steamboat pioneers including the efforts of Ormsbee, Longstreet, and Green and Backus. One minor change incorporated into the table is that now the contributions have been sorted in order of concept dates, making it easier to follow the individuals as presented on the graph.

The Contributors Reviewed

- *Denis Papin.* Denis Papin is not a viable inventor candidate, because the vessel he ran was powered by human exertion, not steam. Therefore he is not shown on the graph. The difficulties that Papin faced will be better understood by examining appendix B.
- *Jonathan Hulls.* Although Jonathan Hulls was given credit by early researchers for having constructed a steam-powered vessel, there is no evidence he ever built a steamboat. As in the case of Papin, the steam engines at the time of his proposal were too large and heavy to have supported steam navigation. Therefore Hulls is not shown on the graph; again, the reader is referred to appendix B.
- *Comte d'Auxiron.* Comte J. B. d'Auxiron conceived of a steamboat about 1770 and formed a society for the vessel's construction in 1772, with one of his partners being Claude de

Table 13.1. Final summary of the efforts of steamboat pioneers

Steamboat pioneer	Concept date	Steamboat trial date(s)	Notes
Papin	1690	1707	Human, not steam power
Hulls	1736	None	Concept only for steam tug
D'Auxiron	1770	None	In 1774 boat accidentally sunk
De Jouffroy	ca. 1771	1778 Unsuccessful 1783 *Pyroscaphe*	Some success on 1783 trial, but work interrupted by the French Revolution
Henry	1779	None	Concept only
Rumsey	1783	1786, 1787, 1790, 1792, 1793	Limited success with water jet propulsion
Fitch	1785	1787, 1788, 1789, 1790	Introduced steamboat service on the Delaware in 1790, which was not financially viable
Evans	1785–86	1805 Amphibious 1812 Middlesex 1816 *Aetna* 1819 *Philadelphia*	About 1787 proposed boat with paddle wheels to Fitch and yielded priority to him
Miller	ca. 1786	1787 *Edinburgh* 1788 Steamboat	Human, not steam power Engine by Wm. Symington
Stevens	1787	1798, 1801, 1804, 1808, 1812	1798, 3 mph 1808 *Phoenix*, 5.5 mph 1812 *Philadelphia*, 8 mph
Read	1788	1789	Human, not steam power
Morey	1790	1793, 1794, 1795, 1796, 1797	Demonstration for Livingston in 1796 at 5 mph; side paddle wheels in 1797
Longstreet	1790	1807	Savannah River, 5 mph
Ormsbee	ca. 1791	1792–94	Seekonk River with goosefoot propulsion, 3–4 mph
Fulton	1793	1803 Paris 1807 New York	Structure failed in first 1803 attempt; second achieved 2.9 mph; in New York in 1807 achieved 5 mph, securing monopoly
Green & Backus	1796	None	Effort failed due to lack of funds
Dundas	ca. 1798	1802 Tugboat *Charlotte Dundas*	Hull by Alexander Hart with engine by Wm. Symington; towed 140 tons on a canal

Jouffroy d'Abbans. Construction was started on the Seine late in 1772, but a counterweight fell through the hull in 1774, and his efforts were doomed. A trial was never conducted.[15]

- *Claude de Jouffroy.* As a member of the company organized by d'Auxiron, Marquis Claude de Jouffrey d'Abbans decided to construct his own steamboat. In 1778, he ran a steam-powered vessel on the river Doubs, but the trial was not successful.

Five years later, in 1783, he and a partner ran the *Pyroscaphe* on the river Saône near Lyons against the current for about fifteen minutes. Some accounts claim the vessel self-destructed at the end of the trial. In any case, support for further testing did not materialize, and the French Revolution interrupted his work. Much later, in 1816, he built the steamboat *Charles-Philippe*, which ran on the Seine,[16] but not being a candidate for inventor, this later effort is not shown on the graph.

- *William Henry.* According to the account by Thurston, a German writer named Schoepff visited Henry during the period of 1783–1784 and was shown a drawing of a steamboat. The sketch was probably created in 1779, and this is the date shown for Henry's steamboat conception, although he discussed his idea with friends several years earlier. There is no evidence, however, that his work went beyond the conceptual drawing.

- *James Rumsey.* The earliest date cited by Rumsey for divulging his steamboat plan is the middle of 1783, when he confidentially told John Wilson of Philadelphia, who was visiting Bath, Virginia. In 1784, he informed his former business partner, Michael Bedinger, and told George Washington in the fall of that same year. His first steamboat trial occurred on March 14, 1786, followed by trials in December 1786, September 1787, and two public trials in December 1787, achieving speeds of three and four miles per hour. The trial of his British steamboat, the *Columbian Maid*, occurred in February 1793, two months after his untimely death.

- *John Fitch.* In his autobiography, Fitch says that the idea for a steamboat came in the spring of 1785. His first steamboat trial was in May 1787, followed by trials in August 1787, July 1788, and September 1789, and finally, in April 1790, the jubilant and self-described "poor Johney Fitch" achieved seven miles per hour on the Delaware River. That summer, steamboat service was introduced, with the boat traveling two to three thousand miles but was not able to carry enough passengers to be financially viable.

- *Oliver Evans.* As early as 1773, Evans had conceived of the idea to use steam to propel a carriage. In 1785 or 1786, he discussed steamboats using paddle wheels with Mr. Samuel Jackson of Redstone (now Brownsville), Pennsylvania.[17] About 1787, he proposed the same concept to John Fitch, who opposed the idea. As he learned that Fitch had conceived of the steamboat in 1785, he yielded priority to him.[18] Evans's first steamboat engine was initiated in 1802 for the ill-fated Mississippi River venture of McEver and Valcourt. His first steamboat was born in the spring of 1805, when he proposed to the city of Philadelphia a steam dredge he named the *Orukter Amphibolos*. It was tested later that year. In 1812, one of his engines powered a steamboat on the Middlesex Canal, and four years later, in 1816, the steamboats *Oliver Evans* and *Aetna* were launched with high-pressure steam engines. Over the next several years, Evans provided similar engines for several boats, inaugurating American high-pressure steam.

- *Patrick Miller.* Miller of Dalswinton, Scotland, experimented with the triple-hulled *Edinburgh*, launched at Leith in October 1786. The vessel, however, was human-powered, and efforts to propel it were exhausting on the crew. The tutor of Patrick Miller's son, James Taylor, suggested the use of steam power, and William Symington, who had patented a steam engine, was engaged to produce one for a twin-hulled vessel. On October 14, 1788, the steamboat, with the poet Robert Burns aboard, achieved the speed of five miles per hour on Dalswinton Lake near Dumfries.[19]

- *John Stevens.* After seeing Fitch's boat on the Delaware River and reading about the Rumsey-Fitch controversy, Stevens became interested in steam navigation. He teamed up with Livingston and Roosevelt and in the latter part of 1798, their steamboat was tested at three miles per hour—not fast enough to secure the New York State monopoly. Another steamboat trial with his partners in the summer of 1801 was also unsuccessful. Working alone, Stevens turned to high-pressure steam and in May 1804, ran the *Little Juliana*, achieving four miles per hour, but boiler technology would not support the steam pressure necessary for this approach. In 1808, Stevens ran a new low-pressure steamboat, the *Phoenix*, at over five miles per hour and successfully introduced public steamboat service. He and his sons were to follow with a number of successful steamboat ventures.

- *Nathan Read.* Around 1788, Read conceived of a portable boiler and an improved steam cylinder. He proposed that these be used to propel a steam carriage and for steam navigation. However, nothing came of his proposals.

- *Samuel Morey.* The concept of a steamboat came to Morey about 1790, and he conducted a trial on the Connecticut River in the spring of 1793. He constructed a larger boat, testing it in New York, during the summers of 1794 and 1795. In 1796, he demonstrated his steamboat with a stern paddle wheel to Livingston, Stevens, and others. The following summer, he operated a steamboat on the Delaware River, using side paddle wheels and achieving success. An investor planned to develop a larger passenger version but because of some financial misfortunes did not have the money to carry out the venture.

- *William Longstreet.* Longstreet, along with Isaac Briggs, developed a steam engine in 1788, and the concept of a steamboat occurred to Longstreet around that time, certainly by 1790. He was unable to find funding, so the idea lay fallow for a decade and a half. By 1807, with his own resources, he finally constructed a pole steamboat, which was tested on the Savannah River at about five miles per hour.

- *Elijah Ormsbee.* Ormsbee tested a steamboat, using many borrowed parts, on the Seekonk River in Rhode Island sometime around 1792–1794. (The latter date is shown on the graph.) The exact date of the concept is not known but was most likely several years earlier and for the purpose of the graph is assumed to be 1791.

- *Robert Fulton.* In a letter to Lord Stanhope in 1793, Fulton said that the concept of a steamboat had occurred to him that summer. He ran his first steamboat on the Seine in Paris in

August 1803 and the *North River Steam Boat* on the Hudson River in August 1807. After that, he constructed or supervised more than a dozen steamboats before his death in 1815, firmly establishing steam navigation in the United States.

- *Green & Backus.* In 1796, Griffin Green and Elijah Backus signed an agreement to construct a steamboat but immediately ran into financial problems, and very quickly thereafter, the effort died.

- *Lord Dundas.* Lord Dundas of Kerse wanted to examine the merits of towing barges on a canal with a steamboat rather than horses. The hull of his steamboat was built at the Grangemouth Dockyard by Alexander Hart and was named *Charlotte Dundas* after Dundas's daughter. The engine was built and installed by William Symington, who had previously provided an engine for Miller's twin-hulled craft. In 1802, the steamboat successfully pulled two vessels, seventy tons each, a distance of 19½ miles in six hours, or slightly over three miles per hour. Concern over damage to canal banks led to abandoning the effort. Writing in *The Birth of the Steamboat*, H. Philip Spratt claims this vessel to be the world's first practical steamboat.[20]

OK, So Who *Really* Invented the Steamboat?

When we left the steamboat inventor discussion, the only remaining open question had to do with who "first" conceived a practical steamboat. Papin, Hulls, and D'Auxiron did not run steamboats, so they cannot be considered. De Jouffroy achieved a very limited fifteen minutes of success but did not accomplish a practical result. In America, William Henry was first in concept, but we know nothing of his design, as a vessel was never built and the drawing he prepared has not been found. Without knowledge of the hardware, it is likely that even if the drawing were available, no conclusions could be drawn. In fact, Henry himself, according to Fitch, made no claims as to being the inventor of the steamboat.

The next potential inventor was James Rumsey, followed closely by John Fitch. Rumsey was two years ahead of Fitch in concept, but his water jet propulsion was not as effective as side paddles or paddle wheels, so some would say that he never achieved a practical design. Much later, technology would bring water jets to the forefront with high-speed boats, but in 1787, the approach he selected was not effective. Fitch took his steamboat design much further than Rumsey, proving technical feasibility at seven miles per hour.

Beyond Rumsey and Fitch are a number of others shown on the graph. Although one or more of them most likely independently conceived of a steamboat, the individual cannot be considered as the inventor unless we, at the onset, dismiss both Rumsey and Fitch. As an inventor has to be "first" with a practical idea, and both Rumsey and Fitch did achieve some degree of success, those who followed can, at this point, be appropriately dismissed. "But what about Fulton?" you ask. Although Fulton was a brilliant entrepreneur—or projector, as they were then called—he did not invent anything, only assembling the best ideas from others and effectively integrating them into a working steamboat. With the New York monopoly, a commercially successful operation ensued; this was not an insignificant accomplishment, but it was *not* invention. Others agree.

Robert Thurston, professor of mechanical engineering at the Stevens Institute of Technology, writing an 1891 biography on Fulton, observes in a section called "Old Legends," "Robert Fulton has often, if not generally, been assumed to have been the inventor of the steamboat, as Watt is generally supposed to be the inventor of the steam-engine, which constitutes its motive apparatus. But this notion is quite incorrect." Thurston was of the opinion that the steam engine and steamboat are both the results of the contributions of many individuals. He goes on to state, "Fulton did nothing to modify the engine, or to improve the steamboat even. He simply took the products of genius of other mechanics, and set them at work, in combination."[21]

In his 1977 Fulton biography, John S. Morgan writes, "Robert Fulton came late and reluctantly on the steamboat scene. He did not invent the steamboat, but he built the first commercially practical vessel."[22] In a biography published in 1985, Cynthia Owen Philip observes in the epilogue that Fulton was "A courageous idealist . . . [and] also an opportunist. Fulton did not invent the steamboat." But, failing to understand the concept of invention, she goes on to assert, "nor did Fitch, Rumsey, Roosevelt, or Thornton."[23] Historian Robert C. Post, who has studied and written about invention, simply states, "Robert Fulton didn't invent the steamboat. Mythology has often been taken too literally."[24]

We are therefore left with two candidates, Rumsey and Fitch. Both were skilled artisans: Rumsey had been a millwright, blacksmith, and tavernkeeper and had constructed and operated a sawmill and bloomery; Fitch had been a brass founder, silversmith, surveyor, mapmaker, engraver, and printer. It is known that Rumsey had some scientific knowledge, owning a copy of Desaguliers's 1744 book *A Course of Experimental Philosophy*, which among other things described a Newcomen atmospheric steam engine. Rumsey appeared to have a contact or confidant at the American Philosophical Society, perhaps through which he obtained some technical information, as he mentions in a letter to Washington on March 10, 1785, his concern for secrecy "as [the steamboat] is easy performed and the method would Come Very natural to a Rittenhouse or an Elieot [Elliot]"[25]—both of whom were members of the Society.

Fitch claimed he had never seen a steam engine at the time he conceived of its power and was later surprised when shown one in a book. His personality and awkward manner did not integrate well into the American Philosophical Society; he met with a number of its members but, unlike Rumsey, was never voted into the body. Rumsey ran both thought experiments and actual experiments to gather scientific data, whereas Fitch was inclined to prefer an intuitive approach, which most often was no more than trial and error. In many cases, he depended entirely on his partner Henry Voight to solve a difficult problem.

Rumsey was secretive but confidentially disclosed to others as early as 1783 his intentions to construct a steamboat, and he told George Washington in the fall of 1784 in Richmond. He later mentioned the meeting in a letter to him on March 10, 1785, where he guardedly elaborated on his ideas. Fitch's conception date is the spring of 1785, generally considered April, and, unlike the cautious and secretive Rumsey, widely broadcast his steamboat ideas, writing to Congress for support and seeking the opinion of members of the American Philosophical Society. George Washington later told Fitch that Rumsey's idea for a steamboat predated his, and this was cause for concern, but Fitch would rely on the fact that he had made an earlier public disclosure for priority.

On April 23, 1781, John Fitch personally presented his steamboat application to the federal patent board and later wrote in the first person, "then requested that the oldest Patent might be granted to me that I might become the defendant in any suit hereafter brought as I could prove the them that Rumsey was twelve months Posterior to me." Fitch noted that, "Mr. Randolph said the oldest applycant must have the oldest Patent," and Fitch, believing this meant Rumsey, whose patent application predated his, would be given an older patent.[26] Fitch responded that he was the first applicant, having communicated his plan to Congress in August 1785, and in March 1786 petitioned them for an exclusive right. As these events predated the patent law, and with no precedents to follow, the part-time patent board was not inclined to spend an indeterminate amount of time sorting out priority; hence, it ruled to issue all patents on the same date, albeit with conflicting claims.

The date of the *first* steamboat trial for Rumsey, though unsuccessful, was March 14, 1786, with Fitch's *first* trial occurring in May 1787, using a twelve-inch cylinder. It has been reported that Fitch made an earlier trial with a three-inch model engine cylinder, but this cannot be verified from Fitch's autobiography or from other documented information. By December 1787, Rumsey had run two public trials on the Potomac River, ending his American efforts. Fitch continued working, and by the summer of 1790, ran two to three thousand miles, firmly establishing the technical feasibility of the steamboat.

It is clear that Fitch took the development to a higher level, leading some to claim that he is the rightful steamboat inventor (see fig. 13.4). We must be careful, however, to distinguish invention from development; otherwise, anyone who makes a significant improvement in a product could become a self-proclaimed inventor. He or she could then be followed by another individual with yet another improvement, hence another a claim to invention, leading to much confusion.

Clearly the Constitution and the 1781 patent legislation meant to reward an individual with a limited-time monopoly for being the "first" to envision a workable and worthwhile new idea. Sometimes the "workability," or functionality, of the invention is immediately obvious upon reading the written description, as it is clear to the reader that all of the science and technology is in place to accomplish it. In other cases, where the feasibility is not obvious, it can

Figure 13.4. Portrait of John Fitch, identified as "Steamboat Inventor," found in the Senate wing, basement floor, of the Capitol. The inventor is shown holding one of the stern paddles of a model of his steamboat. (From a fresco by Constantino Brumidi found in Charles E. Fairman, *Art and Artists of the Capitol of the United States of America* [Washington, DC: Government Printing Office, 1927], p. 205.)

often be convincingly proven through science, and no physical implementation is required. Still, at other times, no one at all will believe the inventor.

Indeed, for the steamboat, the idea of such a device seemed far-fetched, as it required a paradigm shift in conventional wisdom. Therefore, until it was proven to be possible, it would in the minds of most remain impossible. At the time Rumsey demonstrated his imperfect steamboat in December 1887, he proved to those present that it was possible, and their sworn statements attest to this. Fitch, four months earlier in August 1787, had also shown the feasibility of a steamboat. So each inventor, insofar as invention was concerned, had proven his case, though admittedly, "Nothing is invented and perfected at the same time."[27] Then which one deserves the credit?

It was mentioned previously that one advantage of looking at this scenario two hundred years later is that we can better understand what was happening and come to a more informed decision. And in doing this, we have to get back to the basics. The fundamental issue to be resolved is, which inventor had the "first" concept that could be made to work? Thus the 1978 new college edition of *The American Heritage Dictionary of the English Language* points us in the right direction: James Rumsey is the inventor of the steamboat.

Notes

1. George H. Preble, *Chronological History of the Origins and Development of Steam Navigation* (Philadelphia: L. R. Hamersly, 1883), pp. 26–28; see also Thomas W. Knox, *Life of Robert Fulton and a History of Steam Navigation* (New York: G. P. Putnam's Sons, 1886), pp. 82–83.

2. John Donald Wade, *Augustus Baldwin Longstreet: A Study of the Development of Culture in the South* (1924; reprint, Athens, GA: University of Georgia Press, 1969), briefly addresses the life of Augustus's father, William; see also Knox, *Life of Robert Fulton*, pp. 82–83.

3. Wade, *Augustus Baldwin Longstreet*, p. 10.

4. "Longstreet, William." *Dictionary of American Biography*, ed. Dumas Malone, 22 vols. (New York: Charles Scribner's Sons, 1933), 11:393.

5. Quoted in Wade, *Augustus Baldwin Longstreet*, pp. 9–10.

6. Ibid., p. 13.

7. Ibid., p. 12. Other accounts report that the trial occurred at various times between 1806 and 1808.

8. Quoted in ibid., p. 14.

9. Melvin H. Jackson, "The Philadelphia Steamboat of 1796," *American Neptune* 50, no. 3 (summer 1990): 201–205.

10. Ibid., p. 202.

11. David Read, *Nathan Read: His Invention of the Multi-tubular Boiler . . .* (New York: Hurd and Houghton, 1870), p. 116.

12. *Congressional Record*, 1st Cong., 2nd Sess. (1790), p. 110 of 755, http://memory.loc.gov/ (accessed August 2003).

13. Oliver Evans, *The Abortion of the Young Steam Engineer's Guide* (1805; reprint, Wallingford, PA: Oliver Evans Press, 1990), p. 128. The minutes of the patent board, Henry Remsen, Jun. Clerk, for November 23, 1790, contain this entry: "Some of the claims for patents, founded on the supposed discovery of new application of steam to useful purposes, not having been stated so precisely as to be satisfactory to the board, and it being their wish to hear all those claims together, Ordered, That the first Monday of February next be appointed for the hearing of all parties interested; that notice be given to John Fitch, James Rumsey, Nathan Read, Isaac Biggs [i.e., Briggs], and John Stevens of this order; and that each of them be required to transmit in writing to the Board, a precise statement of their several inventions, and of the extent thereof."

14. Quoted in Archibald Douglas Turnbull, *John Stevens, an American Record* (New York: Century, 1928), pp. 109–10.

15. H. Philip Spratt, *Birth of the Steamboat* (London: Charles Griffin, 1958), p. 34.

16. Ibid., p. 38.

17. Oliver Evans, "Memoir 'On the Origin of Steam Boats and Steam Waggons,'" ed. Arlan K. Gilbert, offprint from *Delaware History* 7, no. 2 (September 1956): 155.

18. Evans's concept date for a steamboat with paddle wheels is 1785–86, and since he yielded to Fitch's 1785 date, it seems probable that Evans disclosed his idea in 1786, not a year earlier.

19. Spratt, *Birth of the Steamboat*, p. 49.

20. Ibid., pp. 8, 61.

21. William Graham Sumner and Robert H. Thurston, *Makers of American History: Alexander Hamilton and Robert Fulton* (New York: University Society, 1904), pp. 1–2.

22. John S. Morgan, *Robert Fulton* (New York: Mason/Charter, 1977), pp. 155–56.

23. Cynthia Owen Philip, *Robert Fulton, a Biography* (New York: Franklin Watts, 1985), p. 353.

24. Robert C. Post, *Smithsonian Book of Invention* (New York: W. W. Norton, 1978), p. 23. Mitchell Wilson, *American Science and Invention* (New York: Bonanza Books, 1954), p. 71, makes an argument for the myth that Fulton was the inventor of the steamboat: "As naive as the myth was, it served a purpose. The first two decades of the century had made Americans nakedly aware of all its lacks in the mechanical arts. A new profession was appearing in the American social life; and just as George Washington had been made the inspiration of generations of boys who wanted to become lawyers and politicians, so Robert Fulton was held up as proof that Americans could make anything they turned their hands to, and make it better than anyone else in the world."

25. "Letters of James Rumsey," ed. James A. Padgett, *Maryland Historical Magazine* 32 (1937): 19.

26. Fitch, *The Autobiography of John Fitch*, ed. Frank D. Prager (Philadelphia: American Philosophical Society, 1976), pp. 197–98.

27. Quoted in H. L. Mencken, *New Dictionary of Quotations on Historical Principles* (New York: Alfred A. Knopf, 1991), p. 598. Old English proverb attributed to John Ray.

Figure 14.1. Don Quixote on a quest. (Rendered by William T. Sisson from *Silhouettes*, ed. Carol Belanger Grafton [New York: Dover, 1979], p. 36.)

Epilogue

Giving Everyone His Due

Perhaps some will not agree with the writer's choice of inventor of the steamboat; however, most will not have strong opinions one way or another. Some will be truly surprised that the popular choice, Robert Fulton, as taught in the fourth- or fifth-grade reader, is not the inventor. Unquestionably, he is not. Others, led by descendants of one of the steamboat pioneers or by those with a geographic connection to one of the individuals, will continue to argue that *their* candidate is the one deserving of recognition. If anyone chooses to examine only a portion of the facts available, most any conclusion can be drawn, as has been done in the past.

In unraveling one steamboat priority claim, Ella May Turner perceptively writes, "The Rumsey-Fitch [steamboat] controversy is indeed a labyrinth and he who travels through it should carry with him the lamps of research and fairness, lest he not only lose his own way but mislead others; for 'the evil that men do' by making false and prejudiced statements 'lives after them.'"[1] The author humbly agrees.

It is most important to recognize that many of the pioneers *did* make a very worthwhile contribution to the ultimate success of the steamboat—but not everyone can be the inventor. As these individuals have been closely examined in this account, it is possible to quantify their contribution with a succinct phrase. The author therefore *dares* to provide his list of accolades for the individuals identified in this short history. In case your favorite steamboat "projector" was not the one selected as the inventor, perhaps there can be some solace when you realize he was not completely overlooked for his contribution in the steamboat saga.

Steamboat Pioneers and Their Accomplishments

Comte D' Auxiron: Early pioneer who dared to try

Claude De Jouffroy: Early experimenter who achieved limited success

William Henry: First American conceptualizer of the steamboat[2]

James Rumsey: Inventor of the steamboat

John Fitch: Father of the world's first steamboat line[3]

Oliver Evans: Engineering visionary and high-pressure steam pioneer

Patrick Miller: Validated the use of a paddle wheel as a means of propulsion

John Stevens: Godfather of America's transportation infrastructure

Nathan Read: Transportation conceptualizer

Samuel Morey: Steamboat pioneer and technological visionary

William Longstreet: The steamboat was one element of his life's "continual feast"

Elijah Ormsbee & David Wilkinson: Boys just having fun

Robert Fulton: Father of steam navigation

Green & Backus: A day late and a dollar short

Lord Dundas: Father of England's first practical steamboat, the *Charlotte Dundas*

Enough said.

Notes

1. Ella May Turner, *James Rumsey, Pioneer in Steam Navigation* (Scottdale, PA: Mennonite Publishing House, 1930), p. 137.

2. The suggestion by Herbert H. Beck, "William Henry, Progenitor of the Steamboat, Riflemaker, Patriot," *Papers of the Lancaster County Historical Society* 54, no. 4 (1950): 65, that Henry bear the title "progenitor" cannot be effectively argued, as there is no evidence that his work influenced the other steamboat developers.

3. This is perhaps the most difficult steamboat pioneer to quantify. As a developer, he was not outstanding because of a lack of scientific knowledge. His persistent perseverance was his outstanding trait, which finally eked out some technical success. Unquestionably, however, his steamboat ran the first service ever offered to the public.

Appendix A.

Rumsey's Water Jet Steamboat Explained

On August 30, 1788, a pamphlet, authored by James Rumsey and titled *Explanation of a Steam Engine and the Method of Applying It to Propel a Boat*, was printed in Philadelphia by Joseph James. In the publication, Rumsey provided as Plate 3 the drawing found in figure A.1, which shows an isometric cross-section of a portion of the boat with his equipment installed. The same identifying letters Rumsey used to label components in the drawing have been repeated, for ease of cross-reference, in the schematic in figure A.2, following the reproduced plate. In addition the schematic is augmented with some additional detail, in particular the operation of the walking beam, steam valve, and condenser components, which the secretive Rumsey probably did not want to disclose. These additions have been labeled with Arabic numerals.

Although it is possible to obtain a general understanding of the operation of the engine and boat from Rumsey's description and drawing, the process becomes far easier in schematic form, although some of the spatial aspects are lost. With the combination of the

417

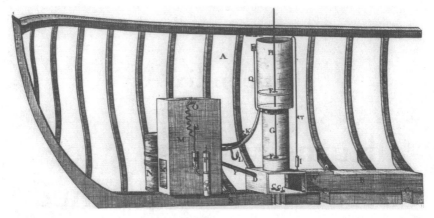

Figure A.1. Rumsey's steamboat in cross-section. (From James Rumsey, *Explanation of a Steam Engine, and the Method of Applying It to Propel a Boat* [Philadelphia: Joseph James, 1788], fig. 5.)

schematic and Rumsey's drawing, a fairly good overall presentation can be formulated.

The configuration depicted in the schematic is essentially that of the half-scale reproduction of Rumsey's steamboat that was completed in 1987 by the Rumseian Society of Shepherdstown, West Virginia, for the two hundredth anniversary of his Potomac River demonstration. The Society named their boat the *Experiment*. Minor differences between the reproduction and Rumsey's incomplete and sometimes confusing description are discussed in the notes.[1]

Steam generator (M). The process starts with what Rumsey called the furnace, which consists of a firebox (2), which is fueled through the opening (X), where the door is omitted. The fire rests on a grate, with the ashes falling into a pit (1). Heat from the burning fuel causes the water in the tube boiler (3) to vaporize. Combustion byproducts are exhausted via the smoke stack (4), fitted with a damper.

Low-pressure steam, between five to ten pounds per square inch, is fed out of the water tube boiler, where an adjustable puppet valve (L) controls the pressure generated.[2] The steam is piped to the switchable two-way valve (K), which, as shown in the figure, directs it into the lower part of the steam cylinder (H).

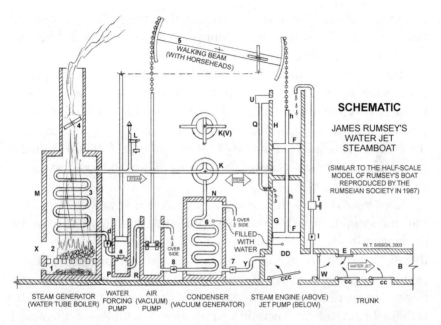

SCHEMATIC

JAMES RUMSEY'S
WATER JET
STEAMBOAT

(SIMILAR TO THE HALF-SCALE
MODEL OF RUMSEY'S BOAT
REPRODUCED BY THE
RUMSEIAN SOCIETY IN 1987)

W. T. SISSON, 2003

STEAM GENERATOR   WATER   AIR
(WATER TUBE BOILER)   FORCING   (VACUUM)   (VACUUM GENERATOR)   STEAM ENGINE (ABOVE)   TRUNK
                        PUMP   PUMP   JET PUMP (BELOW)

CONDENSER

Figure A.2. Schematic of Rumsey's steamboat. (Schematic by William T. Sisson from a draft produced by Jack L. Shagena.)

Steam engine (steam cycle). The pressure of the steam acting on the top piston (F) forces it to the top of the cylinder. The steam piston is directly connected by the rod (h) to the lower water piston (also labeled F), which draws water into the cylinder (G) and chamber (DD) through the large flap valve on the bottom of the boat (ccc), shown partially open. Also, the rod is connected to the walking beam (5) via a chain that communicates with the horsehead on the right-hand end. As the piston rod raises, the weight of the beam on the opposite side of the pivot causes the left-hand end to fall, releasing the tension in the left-hand chain and allowing the air (vacuum) pump piston to descend due to its own weight.

Air (vacuum) pump (R). As the air pump's piston falls, condensated water (which has collected in the bottom of the pump's cylinder) flows through the piston's two check valves (7 and 8), ending up on top of the piston. Check valve 8 keeps the piston from forcing air and water back into the condenser spiral. When the piston ascends on the next half-cycle of the engine, this condensate is supplied to the water forcing pump (P), hence recycled, with any excess being expelled over the side of the boat.

Steam engine, continued (vacuum cycle). When the steam piston (F) reaches the top of the cylinder, the switchable two-way valve is rotated (mechanism not shown) to assume the position in K(V), whereby the steam from the generator is shunted and the steam in the cylinder is connected directly to the condenser (N), also called a vacuum generator. This causes the second half of the steam engine's cycle to commence.

Condenser (N). At this time, the expanding steam finds its way into the spiral tube (6) of the condenser and, upon contact with the cold water in the vessel, collapses or condenses from vapor to water, thereby creating a vacuum. This allows the air pressure (fifteen pounds per square inch) to force the steam piston down. As the steam piston descends, bringing with it the right-hand end of the walking beam, the air (vacuum) pump piston, with the condensated water on top, rises, creating a negative pressure in the condenser's spiral tube, thereby helping to expel the steam from the cylinder (H). This action provides a form of positive feedback, hence speeding up the expulsion of steam from the cylinder and descent of the piston.

Water forcing pump (P). To supply steam for the next cycle of the engine, the water forcing pump is employed. As the steam piston descends, bringing down the right-hand end of the walking beam, it also forces down the piston rod for the water injection pump (note dashed line). This causes water to be injected via the regulator and check valve (d) into the water tube of the steam generator. Check valve a, at the bottom of the pump's cylinder, closes during the forcing cycle and afterward opens to allow the overflow condensate from the air pump to refill the cylinder.

Water jet propulsion pump. As the steam and water pistons (connected by the rod) descend, water is forced into the chamber (DD), causing valve ccc to close and the flap valve (W) to swing open, with the water being rifed or exhausted out the trunk (B).[3] Valve E opens, allowing air to follow the water, the force of which causes the vessel to be propelled forward. At the same time the jet water is expelled, a small amount of cold water is forced through pipe Y and check valve 7 into the condensing vessel (N). As the warmed water has risen to the top of the condenser, it is expelled over the side of the boat.[4]

As the water piston (F) starts to ascend, some amount of water is drawn from the trunk just before the gate valve (W) closes and the bottom flap valve (ccc) opens. The bulk of the water, however, is drawn through the bottom flap valve into chamber DD and the jet pump cylinder (G). Because some water has been removed from the trunk due to the closing delay of valve W, water is allowed to fill the trunk via two flap valves (cc) in the bottom of the trunk. Trapped air in the trunk will prohibit the water from entering; therefore the float valve at the top of the trunk (E) remains unsealed until the air is expelled.[5]

When the steam piston has descended to the bottom, valve K switches to its original position, allowing the steam cylinder to be filled with steam, hence repeating the cycle.

Maintaining the seals. Note the crooked-neck vertical pipe extending from the top of water chamber DD that parallels the water and steam cylinders. When the water piston descends, it forces a small amount of water through check valve I to the top of the steam cylinder. Valve T regulates the volume, and the water keeps the hemp or leather seal between the piston and the cylinder tight. Likewise, the collection box (U) is filled with a small amount of water, which drains via tube Q into the pump cylinder to keep the seal there wet as well. As the water piston ascends, the air in the cylinder (G) is allowed to escape via pipe Q.

Rumsey's implementation of the steamboat was far simpler than Fitch's or Fulton's, and it can be observed that what seemed a simple task, i.e., to propel a boat by steam, is far more complicated than may at first be anticipated.

Notes

1. The author is indebted to the Shepherdstown Rumseian Society, particularly historian Nicholas Blanton and chief engineer Daniel Tokar for their insightful comments on early drafts of this explanation.

2. Rumsey used weights to control the steam pressure.

3. "Rifed" was a descriptive term used by Rumsey.

4. Rumsey actually reversed the process, supplying cold water into the bottom of the condenser and drawing the warmed water from the top via tube Y into chamber DD during the upward stroke of the piston. As some part of the water used for thrusting the boat forward is lost in the method shown, using Rumsey's approach might provide a modicum of performance improvement.

5. For the Rumseian reproduction, the builders chose not to include the flap valve E and achieved satisfactory performance.

Appendix B.

Minimum Horsepower to Weight Examined

D esigning a steamboat or steamship is a very complicated and involved process. There are many factors that affect the vessel's performance, including the boat's length, width, hull shape, wetted surface, and so forth. In addition, there are other tradeoffs that also affect a ship's operation, such as design safety factors, mission, operating environment, available materials, cost of operations, and finally overall construction cost. To examine these, even in a cursory manner, is well beyond the scope of this steamboat history and beyond the knowledge of the author.

Nevertheless, it is desirable to explain why a practical steamboat was not possible until the technology of the motive power (steam engine) advanced to a point that the horsepower-to-weight ratio could support steam navigation. To explore this important factor without embedding a steam engine in a vessel and doing an analysis, it is possible to examine the combination of the two parametrically. The author hypothesized that there existed a relationship between the motive power of a vessel as measured in horsepower divided by the displacement or weight of the vessel, and the speed it could achieve. After a cursory examination, the hypothesis appeared plausible, so an in-depth evaluation was undertaken.

A further consideration related to time, technology, and science, as it was suspected that marine engineers today are able to achieve better performance for a given set of parameters than our ancestors. In this area, however, it is surprising how well some of the pioneers did, but we must factor in today's much more stringent safety requirements, which portend additional weight and other constraints, such as compartmentalization for survivability.

To provide a range of horsepower-to-displacement ratios, the data in Table B.1 were selected on the basis of their availability and because they represent about 150 years of progress. Fourteen steamboats are considered, starting with Fulton's first commercially practical *North River Steam Boat* in 1807 and running through the *John Sergeant*, built in 1956. The distribution between paddle wheel boats and screw propeller boats is approximately equal.

For each steamboat, the horsepower was divided by the ship's displacement for each ten tons, and this value, a figure of merit for speed, is shown in the second column from the right-hand side. In the adjacent column to the left is the speed the steamboat achieved during normal operations.

In figure B.1, the paired values are plotted on a semilogarithmic graph, with a date placed adjacent to each datum point. This way, the date can be looked up in the table to ascertain the associated steamboat and its parameters. As might be expected, some of the data is scattered, more so in the earlier years of construction, but it is possible to comfortably draw a straight line though them that approximates the power, displacement, and speed relationship.[1]

There are a few surprises. Note the 1807 datum point for Fulton's first U.S. effort, the *North River Steam Boat*. The point is well "off the curve"; however, when Fulton rebuilt what he described as a "cranky boat" during the winter of 1807–1808 the new and much heavier design of 1808, with the same identical engine, is much closer to the predicted graphed line.[2] By 1816, with the *Chancellor Livingston*, Fulton had achieved performance that was almost right

Table B.1. Sampling of power, weight, and speed of data for particular steamboats, 1807–1956

Year built	Ship	Horsepower	Propulsion	Displacement (in tons)	Speed (in knots)	HP per 10 tons displacement	Source
1807	North River Steam Boat	20	Paddle	100	4	2.0	Dickinson
1808	North River of Clermont	20	Paddle	182.5	4	1.1	Dickinson
1816	Chancellor Livingston	60	Paddle	526	5.65	1.14	Dickinson
1816a	Aetna	45	Paddle	360	6.75	1.25	Marestier
1840	Acadia	425	Paddle	2050	8.3	2.07	Preble
1852	Arabia	938	Paddle	3950	12.5	2.37	Preble
1864	Cuba	520	Screw	4100	12.8	1.27	Preble
1932	Mariposa	22,000	Screw	26,140	20.5	8.42	Baumeister
1937	Queen Mary	158,000	Screw	77,400	29	20.4	Baumeister
1940	America	34,000	Screw	35,440	22	9.6	Baumeister
1946	Antrim	1,700	Screw	7,435	11	2.29	Baumeister
1951	Constitution	37,000	Screw	30,090	22.5	12.3	Baumeister
1954	World Glory	15,000	Screw	58,265	17.1	2.57	Baumeister
1956	John Sergeant	6,000	Screw	13,570	15.4	4.42	Baumeister

Sources: H. W. Dickinson, Robert Fulton, Engineer and Artist: His Life and Work (London: John Lane, 1913), appendix G, pp. 326–27; Jean Baptiste Marestier, Memoir on Steamboats of the United States of America (1824; reprint, Mystic, CT: Marine Historical Association, 1957), pp. 37, 59; George H. Preble, Chronological History of the Origins and Development of Steam Navigation (Philadelphia: L. R. Hamersly, 1883), appendix, Table XV, pp. 434–35; and Mechanical Engineers' Handbook, ed. Theodore Baumeister (New York: McGraw-Hill, 1958), Table 2, pp. 11-48–49.

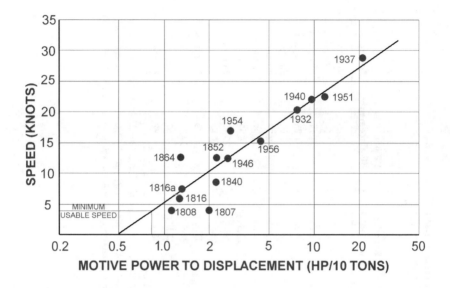

Figure B.1. Graph showing the power propelling a steamboat divided by the vessel's displacement, plotted against the speed achieved. Note: "1816" indicates the *Chancellor Livingston*; "1816a" the *Aetna*. (Graph by Jack L. Shagena.)

on. By comparison, another steamboat, the *Aetna*, constructed at the same time, for which Oliver Evans supplied a high-pressure steam engine, is very slightly above the curve and shown as datum point 1816a. The Evans steam engine operated at about 150 pounds per square inch, compared to the British Boulton and Watt engines of Fulton's boats, which had steam pressures of five to ten pounds. Thus, for the engine alone the horsepower-to-weight ratio was much higher, portending better performance.

In 1937, probably the largest passenger steamship built up to that time, the *Queen Mary*, would appear to have been very well designed, as was the 1932 *America*, both shown on the upper end of the curve.

The lower end of the curve poses the most interest. It shows that the motive power produced must be at least 0.8 horsepower to achieve a speed of four knots when the weight of the hull and steam engine combined equals ten tons. This speed is generally accepted to be the usable minimum because of water currents and wind in the operating environment. A further consideration for practical operations is that the weight of the boat and cargo will be the bulk of the displacement, so assuming one-fifth of the total weight for the engine yields 0.8 horsepower for two tons. This simply was not achievable with the steam engines created before Watt's invention of the separate condenser or the later invention of the double-acting steam engine, eventually followed by high-pressure steam engines.

Rumsey achieved a very lightweight steam engine with his equipment, estimated by various contempories at 500 to 800 pounds, whereas Henry Bedinger estimated John Fitch's equipment as much heavier, at "five tons."[3] The power produced by Rumsey's engine had been roughly estimated by the Shepherdstown Rumseian Society at around one horsepower, so for the time, he achieved a design with an excellent performance-to-weight ratio.

Notes

1. The line may be represented by the formula $S = 16.8 \log [HP/D] + 5.2$, where S is speed in knots, HP is the horsepower of the engine, and D is the ship's displacement in units of ten tons.

2. This suggests there were some problems with the 1807 *North River Steam Boat* that have yet to be understood. Possible issues might be the efficiency of the power takeoff, the number and cross-section of the paddle wheels, the depth of the paddle wheels in the water, and the speed at which the wheels rotated. Another possibility is that the displacement recorded for the 1807 and 1808 steamboats could be incorrect, but based on drawing two feet of water, which both the original and rebuilt vessels had, this does not appear to be the discrepancy. This would be a good subject for investigation for an engineering thesis.

3. See James Rumsey, *A Short Treatise on the Application of Steam Whereby Is Clearly Shewn from Actual Experiments, That Steam May Be Applied to Propel Boats or Vessels of any Burthen against Rapid Currents with Great Velocity* (Philadelphia: Joseph James, 1788), certificate nos. 6, 7, 12.

Index